Mazda Hatchback Automotive Repair Manual

by J H Haynes
Member of the Guild of Motoring Writers
and Trevor Hosie

Models covered:

UK: Mazda Hatchback 323, 985 cc
           Mazda Hatchback 323, 1272 cc
USA: Mazda GLC, 77.6 cu in, 86.4 cu in and 90.9 cu in

Covers 4 and 5-speed manual gearbox and automatic transmission

ISBN 1 85010 051 9

Haynes Publishing Group
Sparkford Nr Yeovil
Somerset BA22 7JJ England

Haynes North America, Inc
861 Lawrence Drive
Newbury Park
California 91320 USA

Acknowledgements

Thanks are due to Toyo Kogyo Company Limited for the supply of technical information and certain illustrations, to Castrol Limited who supplied lubrication data, and to the Champion Sparking Plug Company who supplied the illustrations showing the various spark plug conditions. The bodywork repair photographs used in this manual were provided by Lloyds Industries Limited who supply 'Turtle Wax', 'Dupli-color Holts', and other Holts range products.

A special mention is due to Robin Gould of Pilton Garage, Pilton, Shepton Mallet, Somerset, for the loan of original material. Richard Guy of Guy's Automobile Engineers, Marnhull, Dorset, also provided certain necessary technical information.

Lastly, special thanks are due to all those people at Sparkford who helped in the production of this manual. Particularly, Martin Penny and Les Brazier, who carried out the mechanical work and took the photographs, respectively, John Austin who edited the text, and Ian Robson who planned the layout of each page.

About this manual

Its aims

The aim of this Manual is to help you get the best value from your car. It can do so in several ways. It can help you decide what work must be done (even should you choose to get it done by a garage), provide information on routine maintenance and servicing, and give a logical course of action and diagnosis when random faults occur. However, it is hoped that you will use the Manual by tackling the work yourself. On simpler jobs it may even be quicker than booking the car into a garage, and going there twice to leave and collect it. Perhaps most important, a lot of money can be saved by avoiding the costs the garage must charge to cover its labour and overheads.

The Manual has drawings and descriptions to show the function of the various components so that their layout can be understood. Then the tasks are described and photographed in a step-by-step sequence so that even a novice can do the work.

Its arrangement

The manual is divided into twelve Chapters, each covering a logical sub-division of the vehicle. The Chapters are each divided into Sections, numbered with single figures, eg 5; and the Sections into paragraphs (or sub-sections), with decimal numbers following on from the Section they are in, eg 5.1, 5.2, 5.3 etc.

It is freely illustrated, especially in those parts where there is a detailed sequence of operations to be carried out. There are two forms of illustration: figures and photographs. The figures are numbered in sequence with decimal numbers, according to their position in the Chapter: eg Fig. 6.4 is the 4th drawing/illustration in Chapter 6. Photographs are numbered (either individually or in related groups) the same as the Section or sub-section of the text where the operation they show is described.

There is an alphabetical index at the back of the manual as well as a contents list at the front.

References to the 'left' or 'right' of the vehicle are in the sense of a person in the driver's seat facing forwards.

Whilst every care is taken to ensure that the information in this manual is correct no liability can be accepted by the authors or publishers for loss, damage or injury caused by any errors in, or omissions from, the information given.

Introduction to the Mazda GLC

The Mazda GLC was first introduced in 1977. It is a conventional vehicle, having a front mounted engine and rear wheel drive. The GLC comes as a three-door hatchback through 1980 or as a five-door station wagon through 1982.

The 1977 and 1978 models were equipped with a 1272 cc (77.6 cu in) engine. In 1979 and 1980 the vehicles had 1415 cc (86.4 cu in) engines. The displacement was boosted, once again, to 1490 cc (90.9 cu in) for the 1981 thru 1983 models.

There are four and five-speed manual transmissions and a three-speed automatic transmission used in the model range.

All models in the range are well equipped with items normally regarded as extras, which makes this fine Japanese car a serious rival to the other sub-compact type vehicles on the market today.

Contents

1
2
3
4
5
6A
6B
7
8
9
10
11
12
13

4

Mazda 323 Hatchback (UK Specification)

Buying spare parts and vehicle identification numbers

Buying spare parts

Replacement parts are available from many sources, which generally fall into one of two categories – authorized dealer parts departments and independent retail auto parts stores. Our advice concerning these parts is as follows:

Retail auto parts stores: Good auto parts stores will stock frequently needed components which wear out relatively fast, such as clutch components, exhaust systems, brake parts, tune-up parts, etc. These stores often supply new or reconditioned parts on an exchange basis, which can save a considerable amount of money. Discount auto parts stores are often very good places to buy materials and parts needed for general vehicle maintenance such as oil, grease, filters, spark plugs, belts, touch-up paint, bulbs, etc. They also usually sell tools and general accessories, have convenient hours, charge lower prices and can often be found not far from home.

Authorized dealer parts department: This is the best source for parts which are unique to the vehicle and not generally available elsewhere (such as major engine parts, transmission parts, trim pieces, etc.).

Warranty information: If the vehicle is still covered under warranty, be sure that any replacement parts purchased – regardless of the source – do not invalidate the warranty!

To be sure of obtaining the correct parts, have engine and chassis numbers available and, if possible, take the old parts along for positive identification.

Vehicle identification numbers

Modifications are a continuing and unpublicised process in vehicle manufacture. Spare parts manuals and lists are compiled on a numerical basis, the individual numbers being essential to identify correctly the component required.

Chassis number: The chassis number is stamped on the right-hand side of the engine rear bulkhead.

Engine number: The engine number is stamped on a flat surface on the engine block, behind the fuel pump.

Model plate: The vehicle model, etc, are stamped on the model plate which is attached to the right-hand side of the engine rear bulkhead, adjacent to the chassis number.

The chassis number

The engine number

The model plate

Use of English

As this book has been written in England, it uses the appropriate English component names, phrases, and spelling. Some of these differ from those used in America. Normally, these cause no difficulty, but to make sure, a glossary is printed below. In ordering spare parts remember the parts list may use some of these words:

English	American	English	American
Accelerator	Gas pedal	Locks	Latches
Aerial	Antenna	Methylated spirit	Denatured alcohol
Anti-roll bar	Stabiliser or sway bar	Motorway	Freeway, turnpike etc
Big-end bearing	Rod bearing	Number plate	License plate
Bonnet (engine cover)	Hood	Paraffin	Kerosene
Boot (luggage compartment)	Trunk	Petrol	Gasoline (gas)
Bulkhead	Firewall	Petrol tank	Gas tank
Bush	Bushing	'Pinking'	'Pinging'
Cam follower or tappet	Valve lifter or tappet	Prise (force apart)	Pry
Carburettor	Carburetor	Propeller shaft	Driveshaft
Catch	Latch	Quarterlight	Quarter window
Choke/venturi	Barrel	Retread	Recap
Circlip	Snap-ring	Reverse	Back-up
Clearance	Lash	Rocker cover	Valve cover
Crownwheel	Ring gear (of differential)	Saloon	Sedan
Damper	Shock absorber, shock	Seized	Frozen
Disc (brake)	Rotor/disk	Sidelight	Parking light
Distance piece	Spacer	Silencer	Muffler
Drop arm	Pitman arm	Sill panel (beneath doors)	Rocker panel
Drop head coupe	Convertible	Small end, little end	Piston pin or wrist pin
Dynamo	Generator (DC)	Spanner	Wrench
Earth (electrical)	Ground	Split cotter (for valve spring cap)	Lock (for valve spring retainer)
Engineer's blue	Prussian blue	Split pin	Cotter pin
Estate car	Station wagon	Steering arm	Spindle arm
Exhaust manifold	Header	Sump	Oil pan
Fault finding/diagnosis	Troubleshooting	Swarf	Metal chips or debris
Float chamber	Float bowl	Tab washer	Tang or lock
Free-play	Lash	Tappet	Valve lifter
Freewheel	Coast	Thrust bearing	Throw-out bearing
Gearbox	Transmission	Top gear	High
Gearchange	Shift	Torch	Flashlight
Grub screw	Setscrew, Allen screw	Trackrod (of steering)	Tie-rod (or connecting rod)
Gudgeon pin	Piston pin or wrist pin	Trailing shoe (of brake)	Secondary shoe
Halfshaft	Axleshaft	Transmission	Whole drive line
Handbrake	Parking brake	Tyre	Tire
Hood	Soft top	Van	Panel wagon/van
Hot spot	Heat riser	Vice	Vise
Indicator	Turn signal	Wheel nut	Lug nut
Interior light	Dome lamp	Windscreen	Windshield
Layshaft (of gearbox)	Countershaft	Wing/mudguard	Fender
Leading shoe (of brake)	Primary shoe		

Tools and working facilities

Introduction

A selection of good tools is a fundamental requirement for anyone contemplating the maintenance and repair of a motor vehicle. For the owner who does not possess any, their purchase will prove a considerable expense, offsetting some of the savings made by doing-it-yourself. However, provided that the tools purchased are of good quality, they will last for many years and prove an extremely worthwhile investment.

To help the average owner to decide which tools are needed to carry out the various tasks detailed in this manual, we have compiled three lists of tools under the following headings: *Maintenance and minor repair, Repair and overhaul*, and *Special*. The newcomer to practical mechanics should start off with the *Maintenance and minor repair* tool kit and confine himself to the simpler jobs around the vehicle. Then, as his confidence and experience grows, he can undertake more difficult tasks, buying extra tools as, and when, they are needed. In this way, a *Maintenance and minor repair* tool kit can be built-up into a *Repair and overhaul* tool kit over a considerable period of time without any major cash outlays. The experienced do-it-yourselfer will have a tool kit good enough for most repair and overhaul procedures and will add tools from the *Special* category when he feels the expense is justified by the amount of use these tools will be put to.

It is obviously not possible to cover the subject of tools fully here. For those who wish to learn more about tools and their use there is a book entitled *How to Choose and Use Car Tools* available from the publishers of this manual.

Maintenance and minor repair tool kit

The tools given in this list should be considered as a minimum requirement if routine maintenance, servicing and minor repair operations are to be undertaken. We recommend the purchase of combination spanners (ring one end, open-ended the other); although more expensive than open-ended ones, they do give the advantages of both types of spanner.

> *Combination spanners - 10, 11, 13, 14, 17 mm*
> *Adjustable spanner - 9 inch*
> *Engine sump/gearbox/rear axle drain plug key (where applicable)*
> *Spark plug spanner (with rubber insert)*
> *Spark plug gap adjustment tool*
> *Set of feeler gauges*
> *Brake adjuster spanner (where applicable)*
> *Brake bleed nipple spanner*
> *Screwdriver - 4 in long x $\frac{1}{4}$ in dia (flat blade)*
> *Screwdriver - 4 in long x $\frac{1}{4}$ in dia (cross blade)*
> *Combination pliers - 6 inch*
> *Hacksaw, junior*
> *Tyre pump*
> *Tyre pressure gauge*
> *Grease gun (where applicable)*
> *Oil can*
> *Fine emery cloth (1 sheet)*
> *Wire brush (small)*
> *Funnel (medium size)*

Repair and overhaul tool kit

These tools are virtually essential for anyone undertaking any major repairs to a motor vehicle, and are additional to those given in the *Maintenance and minor repair* list. Included in this list is a comprehensive set of sockets. Although these are expensive they will be found invaluable as they are so versatile - particularly if various drives are included in the set. We recommend the $\frac{1}{2}$ in square-drive type, as this can be used with most proprietary torque wrenches. If you cannot afford a socket set, even bought piecemeal, then inexpensive tubular box spanners are a useful alternative.

The tools in this list will occasionally need to be supplemented by tools from the *Special* list.

> *Sockets (or box spanners) to cover range in previous list*
> *Reversible ratchet drive (for use with sockets)*
> *Extension piece, 10 inch (for use with sockets)*
> *Universal joint (for use with sockets)*
> *Torque wrench (for use with sockets)*
> *'Mole' wrench - 8 inch*
> *Ball pein hammer*
> *Soft-faced hammer, plastic or rubber*
> *Screwdriver - 6 in long x $\frac{5}{16}$ in dia (flat blade)*
> *Screwdriver - 2 in long x $\frac{5}{16}$ in square (flat blade)*
> *Screwdriver - 1$\frac{1}{2}$ in long x $\frac{1}{4}$ in dia (cross blade)*
> *Screwdriver - 3 in long x $\frac{1}{8}$ in dia (electricians)*
> *Pliers - electricians side cutters*
> *Pliers - needle nosed*
> *Pliers - circlip (internal and external)*
> *Cold chisel - $\frac{1}{2}$ inch*
> *Scriber (this can be made by grinding the end of a broken hacksaw blade)*
> *Scraper (this can be made by flattening and sharpening one end of a piece of copper pipe)*
> *Centre punch*
> *Pin punch*
> *Hacksaw*
> *Valve grinding tool*
> *Steel rule/straight edge*
> *Allen keys*
> *Selection of files*
> *Wire brush (large)*
> *Axle stands*
> *Jack (strong scissor or hydraulic type)*

Special tools

The tools in this list are those which are not used regularly, are expensive to buy, or which need to be used in accordance with their manufacturers' instructions. Unless relatively difficult mechanical jobs are undertaken frequently, it will not be economic to buy many of these tools. Where this is the case, you could consider clubbing together with friends (or a motorists' club) to make a joint purchase, or borrowing the tools against a deposit from a local garage or tool hire specialist.

The following list contains only those tools and instruments freely available to the public, and not those special tools produced by the vehicle manufacturer specifically for its dealer network. You will find occasional references to these manufacturers' special tools in the text of this manual. Generally, an alternative method of doing the job without the vehicle manufacturers' special tool is given. However, sometimes, there is no alternative to using them. Where this is the case and the relevant tool cannot be bought or borrowed you will have to entrust the work to a franchised garage

> Valve spring compressor
> Piston ring compressor
> Balljoint separator
> Universal hub/bearing puller
> Impact screwdriver
> Micrometer and/or vernier gauge
> Carburettor flow balancing device (where applicable)
> Dial gauge
> Stroboscopic timing light
> Dwell angle meter/tachometer
> Universal electrical multi-meter
> Cylinder compression gauge
> Lifting tackle
> Trolley jack
> Light with extension lead

Buying tools

For practically all tools, a tool factor is the best source since he will have a very comprehensive range compared with the average garage or accessory shop. Having said that, accessory shops often offer excellent quality tools at discount prices, so it pays to shop around.

Remember, you don't have to buy the most expensive items on the shelf, but it is always advisable to steer clear of the very cheap tools. There are plenty of good tools around at reasonable prices, so ask the proprietor or manager of the shop for advice before making a purchase.

Care and maintenance of tools

Having purchased a reasonable tool kit, it is necessary to keep the tools in a clean serviceable condition. After use, always wipe off any dirt, grease and metal particles using a clean, dry cloth, before putting the tools away. Never leave them lying around after they have been used. A simple tool rack on the garage or workshop wall, for items such as screwdrivers and pliers is a good idea. Store all normal spanners and sockets in a metal box. Any measuring instruments, gauges, meters, etc., must be carefully stored where they cannot be damaged or become rusty.

Take a little care when tools are used. Hammer heads inevitably become marked and screwdrivers lose the keen edge on their blades from time-to-time. A little timely attention with emery cloth or a file will soon restore items like this to a good serviceable finish.

Working facilities

Not to be forgotten when discussing tools, is the workshop itself. If anything more than routine maintenance is to be carried out, some form of suitable working area becomes essential.

It is appreciated that many an owner mechanic is forced by circumstances to remove an engine or similar item, without the benefit of a garage or workshop. Having done this, any repairs should always be done under the cover of a roof.

Wherever possible, any dismantling should be done on a clean flat workbench or table at a suitable working height.

Any workbench needs a vice: one with a jaw opening of 4 in (100 mm) is suitable for most jobs. As mentioned previously, some clean dry storage space is also required for tools, as well as the lubricants, cleaning fluids, touch-up paints and so on which become necessary.

Another item which may be required, and which has a much more general usage, is an electric drill with a chuck capacity of at least $\frac{5}{16}$ in (8 mm). This, together with a good range of twist drills, is virtually essential for fitting accessories such as wing mirrors and reversing lights.

Last, but not least, always keep a supply of old newspapers and clean, lint-free rags available, and try to keep any working area as clean as possible.

Spanner jaw gap comparison table

Jaw gap (in)	Spanner size
0·250	$\frac{1}{4}$ in AF
0·275	7 mm AF
0·312	$\frac{5}{16}$ in AF
0·315	8 mm AF
0·340	11/32 in AF; $\frac{1}{8}$ in Whitworth
0·354	9 mm AF
0·375	$\frac{3}{8}$ in AF
0·393	10 mm AF
0·433	11 mm AF
0·437	$\frac{7}{16}$ in AF
0·445	$\frac{3}{16}$ in Whitworth; $\frac{1}{4}$ in BSF
0·472	12 mm AF
0·500	$\frac{1}{2}$ in AF
0·512	13 mm AF
0·525	$\frac{1}{4}$ in Whitworth; $\frac{5}{16}$ in BSF
0·551	14 mm AF
0·562	$\frac{9}{16}$ in AF
0·590	15 mm AF
0·600	$\frac{5}{16}$ in Whitworth; $\frac{3}{8}$ in BSF
0·625	$\frac{5}{8}$ in AF
0·629	16 mm AF
0·669	17 mm AF
0·687	$\frac{11}{16}$ in AF
0·708	18 mm AF
0·710	$\frac{3}{8}$ in Whitworth; $\frac{7}{16}$ in BSF
0·748	19 mm AF
0·750	$\frac{3}{4}$ in AF
0·812	$\frac{13}{16}$ in AF
0·820	$\frac{7}{16}$ in Whitworth; $\frac{1}{2}$ in BSF
0·866	22 mm AF
0·875	$\frac{7}{8}$ in AF
0·920	$\frac{1}{2}$ in Whitworth; $\frac{9}{16}$ in BSF
0·937	$\frac{15}{16}$ in AF
0·944	24 mm AF
1·000	1 in AF
1·010	$\frac{9}{16}$ in Whitworth; $\frac{5}{8}$ in BSF
1·023	26 mm AF
1·062	$1\frac{1}{16}$ in AF; 27 mm AF
1·100	$\frac{5}{8}$ in Whitworth; $\frac{11}{16}$ in BSF
1·125	$1\frac{1}{8}$ in AF
1·181	30 mm AF
1·200	$\frac{11}{16}$ in Whitworth; $\frac{3}{4}$ in BSF
1·250	$1\frac{1}{4}$ in AF
1·259	32 mm AF
1·300	$\frac{3}{4}$ in Whitworth; $\frac{7}{8}$ in BSF
1·312	$1\frac{5}{16}$ in AF
1·390	$\frac{13}{16}$ in Whitworth; $\frac{15}{16}$ in BSF
1·417	36 mm AF
1·437	$1\frac{7}{16}$ in AF
1·480	$\frac{7}{8}$ in Whitworth; 1 in BSF
1·500	$1\frac{1}{2}$ in AF
1·574	40 mm AF; $\frac{15}{16}$ in Whitworth
1·614	41 mm AF
1·625	$1\frac{5}{8}$ in AF
1·670	1 in Whitworth; $1\frac{1}{8}$ in BSF
1·687	$1\frac{11}{16}$ in AF
1·811	46 mm AF
1·812	$1\frac{13}{16}$ in AF
1·860	$1\frac{1}{8}$ in Whitworth; $1\frac{1}{4}$ in BSF
1·875	$1\frac{7}{8}$ in AF
1·968	50 mm AF
2·000	2 in AF
2·050	$1\frac{1}{4}$ in Whitworth; $1\frac{3}{8}$ in BSF
2·165	55 mm AF
2·362	60 mm AF

Routine maintenance

Introduction

Maintenance is essential for ensuring safety and desirable for the purpose of getting the best in terms of performance and economy from the vehicle. Over the years the need for periodic lubrication – oiling, greasing and so on - has been drastically reduced if not totally eliminated. This has unfortunately tended to lead some owners to think that because no such action is required the items either no longer exist or will last for ever. This is a serious delusion. It follows therefore that the largest initial element of maintenance is visual examination. This may lead to repairs or renewals.

Every 250 miles (400 km) travelled or weekly – whichever comes first

Check the engine oil level and top-up if necessary (photo).
Check the battery electrolyte level and top-up if necessary.
Check the windshield washer fluid level and top-up if necessary.Check the tyre pressures (when cold).
Examine tyres for wear or damage
Check the brake reservoir fluid level and top-up if necessary.
Check the radiator coolant level and top up if necessary.
Check the brake operation is satisfactory.
Check the operation of all lights, instruments, warning devices, accessories, controls etc.

Every 6000 miles (10 000 km) travelled, or 6 months – whichever comes first

1   Engine
Check and adjust the valve/rocker clearance.
Drain the engine oil when warm, and top-up with the correct quantity and grade of oil. Renew oil filter.
Inspect the condition of the alternator/fan belt. Renew the belt or adjust the tension as necessary.

2   Fuel and Evaporative Emission Control System
Adjust the idle speed and mixture.
Clean the air cleaner element.*

Examine the fuel lines and connections for security of attachment, damage and deterioration.

3   Ignition system
Check the contact breaker points; clean and adjust as necessary.
Check the spark plugs; clean and adjust the gaps as necessary.
Check and adjust the ignition timing.

4   Exhaust Emission Control System
Examine the exhaust system for security of fitment and for deterioration of the silencer.
Check the operation of the servo diaphragm and the vacuum control valve.

5   Electrical system
Check the battery specific gravity.
Clean the battery terminals.
Check that all warning lights are working.

6   Chassis and body
Check the condition of the seat belts and their anchorage points; renew any seat belts which are frayed or otherwise damaged.
Check and adjust the clutch pedal free-travel.
Check and adjust the steering wheel play.
Check and adjust the brake shoe to drum clearances.
Check and adjust the brake pedal travel.
Check and adjust the parking brake operation.
Check oil/fluid level in transmission unit; top up if necessary.
Check oil level in rear axle; top up if necessary.
Check disc brake pads and renew if necessary.

In very dusty conditions the air cleaner element should be cleaned every 1000 miles.

Every 12 000 miles (20 000 km) travelled, or 12 months – whichever comes first

1   Engine
Tighten the inlet manifold, exhaust manifold and cylinder head attachment bolts.

The engine oil drain plug

Topping-up the engine oil

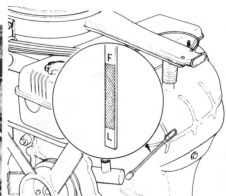
The engine oil dipstick

2   Fuel and Evaporative Emission Control systems

Renew the in-line fuel filter.
Check and adjust the carburettor throttle and choke linkage.
Examine the evaporative emission control system, fuel tank and vapour lines for security of attachment, damage and deterioration.
Check and adjust carburettor float level.

3   Ignition system

Examine the distributor cap, rotor and condenser. Clean or renew parts as necessary.
Check the ignition coil and HT leads. Clean or renew parts as necessary.

4   Crankcase Emission Control System

Check the condition of the cooling system hoses; renew as necessary.
Check the operation of the positive crankcase ventilation valve. Clean or renew as necessary.

5   Cooling system

Check the condition of the cooling system hoses; renew as necessary.

6   Chassis and body

Top-up the oil level in the steering gear.
Examine the brake linings; fitting new ones as necessary.
Apply a few drops of general purpose lubricating oil to the door, hood, pedal, carburettor control, pivot points etc.
Drain the transmission oil when warm and top-up with the correct grade and quantity of oil (photo).
Drain the rear axle oil when warm and top-up with the correct grade and quantity of oil.

Every 24 000 miles (40 000 km) travelled, or 2 years – whichever comes first

1   Engine

Check the engine cylinder compression pressures.

2   Fuel and Evaporative Emission Control Systems

Renew the air cleaner element*.
Renew the carbon (charcoal) canister.
Check the operation of the check valve; renew as necessary.

3   Cooling system

Drain, flush and refill the cooling system, using new antifreeze.

4   Chassis and body

Renew the piston cups (seals) in the brake master cylinder.
Lubricate and adjust the front wheel bearings.

In very dusty conditions the air cleaner element should be replaced every 12 000 miles

Every 48 000 miles (80 000 km) travelled, or 4 years – whichever comes first

1   Chassis and body

Replace the flexible hoses in the braking system.

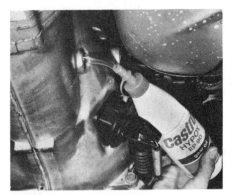

Topping-up the gearbox oil

The brake master cylinder fluid reservoir cap

Renewing the air cleaner element

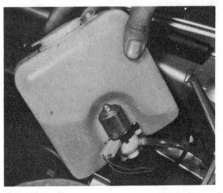

The front windscreen washer bottle is located on the suspension tower

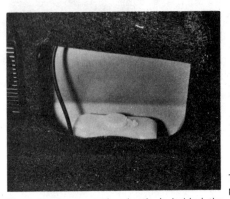

The rear screen washer bottle is behind the side trim in the luggage compartment

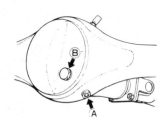

The rear axle. 'A' drain plug and 'B' the filler plug

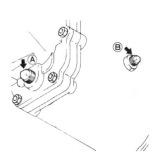

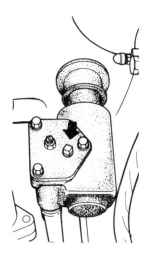

The transmission. 'A' drain plug and 'B' the filler plug

The steering box filler plug

Jacking and towing

Before carrying out any servicing or repair operations, make sure that you know where to position the jack and axle stands. It is most important to use only the specified points in order to prevent accidents and damage to the vehicle itself (see the accompanying illustration).

If the vehicle breaks down or becomes otherwise immobile, a front and rear towing eye is provided to which a tow rope may be attached.

On vehicles equipped with automatic transmission, it is recommended that the propeller shaft is disconnected from the rear axle companion flange, and tied up out of the way, before any towing of the vehicle is carried out.

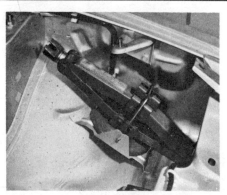

The jack is located under the bonnet

The spare wheel is under the carpet in the luggage area

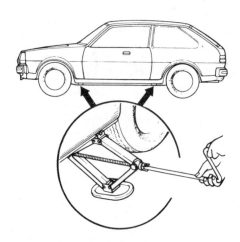

The jacking points

Recommended lubricants

Component	Castrol Product
1 Engine	Castrol GTX
2 Manual gearbox	Castrol Hypoy B (90 EP)
2 Automatic transmission	Castrol TQF
3 Rear axle	Castrol Hypoy B (90 EP)
4 Wheel bearings	Castrol LM Grease
5 Brake master cylinder	Castrol Girling Universal Brake and Clutch fluid
6 Steering box	Castrol Hypoy B (90 EP)

Note: The above are general recommendations only. Lubrication requirements vary from territory to territory and depend on vehicle usage. Consult the operator's handbook supplied with your vehicle.

Chapter 1 Engine

See Chapter 13 for specifications and information applicable to 1979 thru 1983 North American models

Contents

1

Specifications

General

Type	Four cylinder, four stroke, in line, water cooled, overhead camshaft
Bore:	
1000 cc	2·76 in (70 mm)
1300 cc	2·87 in (73 mm)
Stroke:	
1000 cc	2·52 in (64 mm)
1300 cc	2·99 in (76 mm)
Displacement:	
1000 cc	985 cc (60·1 cu in)
1300 cc	1272 cc (77·6 cu in)
Compression ratio:	
1000 cc	8·8 : 1
1300 cc	9·2 : 1
Firing order	1 – 3 – 4 – 2
Oil pressure:	
Hot at 3000 rpm	50 – 64 lbf/in² (3·5 – 4·5 kgf/cm²)
Idle	4·3 lbf/in² (0·3 kgf/cm²)
Valve clearance (warm):	
Valve side:	
Inlet	0·010 in (0·25 mm)
Exhaust	0·012 in (0·30 mm)
Cam side:	
Inlet	0·007 in (0·18 mm)
Exhaust	0·009 in (0·22 mm)

Cylinder head

Material	Aluminium alloy
Maximum permissible distortion of cylinder head surface	0·006 in (0·15 mm)

Valve seat:
    Angle (inlet and exhaust) . . . . 90°
    Width (inlet and exhaust) . . . . 0·055 in (1·4 mm)
Valve guide:
    Length . . . . 1·988 in (50·5 mm)
    Outer diameter . . . . 0·5512 to 0·5529 in (14·0 to 14·044 mm)
    Inner diameter . . . . 0·3150 to 0·3183 in (8·0 to 8·083 mm)
Valve stem to guide clearance (new):
    Inlet . . . . 0·0007 to 0·0021 in (0·018 to 0·053 mm)
    Exhaust . . . . 0·0007 to 0·0023 in (0·018 to 0·058 mm)
Valve stem to guide clearance wear limit . . . . 0·008 in (0·20 mm)

Valves

	1000 cc	1300 cc
Inlet valve:		
Overall length	4·1733 in (106 mm)	4·1536 in (105·5 mm)
Head diameter	1·3780 ± 0·0039 in (35 ± 0·1 mm)	1·4173 ± 0·0039 in (36 ± 0·1 mm)
Exhaust valve		
Overall length	4·0552 in (103 mm)	4·0749 in (103·5 mm)
Head diameter	1·1811 ± 0·0039 in (30 ± 0·1 mm)	1·2205 ± 0·0039 in (31 ± 0·1 mm)

Stem diameter:
    Inlet and exhaust . . . . 0·3150 to 0·3168 in (8·0 to 8·045 mm)
    Wear limit . . . . 0·3140 in (7·975 mm)
Face angle . . . . 90°

Valve springs

Outer:
    Wire diameter . . . . 0·157 in (4·0 mm)
    Outer coil diameter . . . . 1·268 in (32·2 mm)
Free length:
    New . . . . 1·587 in (40·3 mm)
    Limit . . . . 1·539 in (39·1 mm)
Fitted length . . . . 1·319 in (33·5 mm)
Fitted load:
    New . . . . 43·7 lbf (19·8 kgf)
    Limit . . . . 36·6 lbf (16·6 kgf)
Inner:
    Wire diameter . . . . 0·118 in (3·0 mm)
    Outer coil diameter . . . . 0·909 in (23·1 mm)
Free length:
    New . . . . 1·449 in (36·8 mm)
    Limit . . . . 1·406 in (35·7 mm)
Fitted length . . . . 1·260 in (32·0 mm)
Fitted load:
    New . . . . 20·9 lbf (9·5 kgf)
    Limit . . . . 17·9 lbf (8·1 kgf)

Rocker arm

Bore in rocker arm . . . . 0·7480 to 0·7493 in (19·0 to 19·033 mm)

Rocker arm shaft

Outer diameter . . . . 0·7480 to 0·7464 in (19·0 to 18·959 mm)
Length . . . . 13·032 in (331 mm)
Clearance in rocker arm:
    New . . . . 0·0008 to 0·0029 in (0·020 to 0·074 mm)
    Limit . . . . 0·004 in (0·10 mm)

Camshaft

Journal diameter:
    Front . . . . 1·6536 to 1·6516 in (42·0 to 41·949 mm)
    Centre . . . . 1·6536 to 1·6504 in (42·0 to 41·919 mm)
    Rear . . . . 1·6536 to 1·6516 in (42·0 to 41·949 mm)
Wear limit of journal . . . . 0·0020 in (0·05 mm)
Basic circle of cam . . . . 1·4961 ± 0·0020 in (38 ± 0·05 mm)
Cam elevation . . . . 1·7369 in (44·116 mm)
Wear limit of cam elevation . . . . 1·7290 in (43·916 mm)
Camshaft endplay:
    New . . . . 0·001 to 0·007 in (0·02 to 0·18 mm)
    Wear limit . . . . 0·008 in (0·20 mm)
Camshaft run-out limit . . . . 0·0012 in (0·03 mm)
Camshaft support bore bearing clearance:
    Front . . . . 0·0014 to 0·0030 in (0·035 to 0·076 mm)
    Centre . . . . 0·0026 to 0·0042 in (0·065 to 0·106 mm)
    Rear . . . . 0·0014 to 0·0030 in (0·035 to 0·076 mm)

Limit of bearing clearance .... 0·0059 in (0·15 mm)

Camshaft drive
Type .... Chain and sprockets
Number of chain links 90
Number of sprocket teeth:
    Camshaft sprocket 36
    Crankshaft sprocket 18

Valve timing
Mazda Hatchback 323

	1000 cc		1300 cc	
	Europe	*Except Europe*	*Europe*	*Except Europe*
Inlet valve opens	15° BTDC	13° BTDC	15° BTDC	15° BTDC
Inlet valve closes	44° ABDC	50° ABDC	44° ABDC	55° ABDC
Exhaust valve opens	53° BBDC	57° BBDC	53° BBDC	58° BBDC
Exhaust valve closes	6° ATDC	6° ATDC	6° ATDC	12° ATDC

Mazda GLC (USA)
Inlet valve opens .... 13° BTDC
Inlet valve closes .... 50° ABDC
Exhaust valve opens .... 57° BBDC
Exhaust valve closes .... 6° ATDC

Connecting rod
Length (centre to centre):
    1000 cc .... 5·3544 ± 0·0020 in (136·00 ± 0·05 mm)
    1300 cc .... 5·1182 ± 0·0020 in (130·00 ± 0·05 mm)
Permissible bend or twist .... 0·002 in per 5·0 in (0·02 mm per 50·0 mm)
Side clearance .... 0·004 to 0·008 in (0·11 to 0·21 mm)
Small end bush:
    Inner diameter .... 0·7874 to 0·7880 in (20·0 to 20·014 mm)
    Outer diameter .... 0·9055 to 0·9077 in (23·0 to 23·056 mm)
    Bore in connecting rod .... 0·9055 to 0·9063 in (23·0 to 23·021 mm)
    Piston pin and bush clearance .... 0·0004 to 0·0012 in (0·01 to 0·03 mm)
Connecting rod bearing:
    Bearing clearance:
        New .... 0·0011 to 0·0029 in (0·027 to 0·073 mm)
        Limit .... 0·0039 in (0·10 mm)
        Available underside bearings .... 0·0039 in (0·10 mm), 0·010 in (0·25 mm), 0·020 in (0·50 mm), 0·030 in (0·75 mm)

Pistons and piston rings
Piston diameter (measured at 90° to pin bore axis and 0·67 in (17 mm) below oil ring groove):
    1000 cc .... 2·7541 ± 0·0004 in (69·952 ± 0·010 mm)
    1300 cc .... 2·8721 ± 0·0004 in (72·952 ± 0·010 mm)
Gudgeon pin hole bore .... 0·7874 to 0·7869 in (20·0 to 19·988 mm)
Ring groove width:
    Top .... 0·0787 to 0·0803 in (2·0 to 2·040 mm)
    Second .... 0·0787 to 0·0800 in (2·0 to 2·034 mm)
    Oil .... 0·1575 to 0·1588 in (4·0 to 4·032 mm)
Ring groove depth .... 0·1299 in (3·3 mm)
Piston to cylinder clearance:
    New .... 0·0021 to 0·0026 in (0·053 to 0·066 mm)
    Limit .... 0·006 in (0·15 mm)
Available oversize pistons .... 0·006 in (0·15 mm), 0·010 in (0·25 mm), 0·020 in (0·50 mm), 0·030 in (0·75 mm), 0·040 in (1·00 mm)
Piston ring width:
    Top and second .... 0·0787 to 0·0775 in (2·0 to 1·970 mm)
    Oil .... 0·1575 to 0·1563 in (4·0 to 3·970 mm)
Piston ring thickness:
    Top and second .... 0·1220 ± 0·0039 in (3·1 ± 0·1 mm)
    Oil .... 0·0984 ± 0·0039 in (2·5 ± 0·1 mm)
Piston ring side clearance:
    Top .... 0·0014 to 0·0028 in (0·035 to 0·070 mm)
    Second .... 0·0012 to 0·0025 in (0·030 to 0·064 mm)
    Oil .... 0·0012 to 0·0024 in (0·030 to 0·062 mm)
Ring gap .... 0·008 to 0·016 in (0·2 to 0·4 mm)
Available oversize piston rings .... 0·010 in (0·25 mm), 0·020 in (0·50 mm), 0·030 in (0·75 mm), 0·040 in (1·00 mm)

Gudgeon pin
Diameter .... 0·7874 to 0·7868 in (20·0 to 19·984 mm)
Length .... 2·1654 ± 0·0020 in (55·0 ± 0·05 mm)
Clearance between piston and pin .... 0·0006 to 0·0002 in (0·014 to 0·005 mm)

Crankshaft

Main journal diameter
    New .... 2·4804 to 2·4780 in (63·0 to 62·940 mm)
    Limit .... 0·0020 in (0·05 mm)
Crankpin diameter:
    New .... 1·7717 to 1·7693 in (45·0 to 44·940 mm)
    Limit .... 0·0020 in (0·05 mm)
Crankshaft endplay:
    New .... 0·003 to 0·009 in (0·08 to 0·24 mm)
    Limit .... 0·012 in (0·30 mm)
Crankshaft run-out limit .... 0·0012 in (0·03 mm)
Thrust bearing:
    Available oversizes .... 0·010 in (0·25 mm), 0·020 in (0·50 mm), 0·030 in (0·75 mm)
Main bearing:
    Clearance (new) .... 0·0012 to 0·0024 in (0·031 to 0·061 mm)
    Limit .... 0·0031 in (0·08 mm)
    Available undersize bearing .... 0·010 in (0·25 mm), 0·020 in (0·50 mm), 0·030 in (0·75 mm)

Cylinder block

Bore diameter:
    1000 cc .... 2·7559 to 2·7566 in (70·0 to 70·019 mm)
    1300 cc .... 2·8741 to 2·8748 in (73·0 to 73·019 mm)
Wear limit of bore .... 0·0059 in (0·15 mm)
Oversize boring .... 0·0059 in (0·15 mm), 0·010 in (0·25 mm),
                                        0·020 in (0·50 mm), 0·030 in (0·75 mm), 0·040 in (1·00 mm)

Lubrication system

Oil pump type .... Rotor
Feeding capacity (2000 rpm) .... 2·3 Imp gal/min (2·8 US gal/min) (10·6 litres/min)
Oil pump drive .... Chain and sprockets
Number of chain links 44
Number of sprocket teeth 29
Outer rotor and body clearance:
    New .... 0·006 to 0·010 in (0·14 to 0·25 mm)
    Limit .... 0·012 in (0·30 mm)
Clearance between rotor lobes:
    New .... 0·002 to 0·006 in (0·04 to 0·15 mm)
    Limit .... 0·010 in (0·25 mm)
Rotor endfloat:
    New .... 0·002 to 0·004 in (0·04 to 0·10 mm)
    Limit .... 0·006 in (0·15 mm)
Clearance between pump shaft and body:
    New .... 0·0013 to 0·0034 in (0·032 to 0·086 mm)
    Limit .... 0·004 in (0·10 mm)
Oil filter:
    Type .... Full flow, cartridge
    Relief valve opens .... 11 to 17 lbf/in^2 (0·8 to 1·2 kgf/cm^2)
Engine oil capacity .... 2·6 Imp quarts (3·2 US quarts) (3·0 litres)

Torque wrench settings

	lbf ft	kgf m
Main bearing caps	43 to 47	6·0 to 6·5
Connecting rod caps	29 to 33	4·0 to 4·5
Oil pump sprocket	22 to 25	3·0 to 3·5
Oil sump	5 to 7	0·65 to 0·95
Cylinder head:		
Cold engine (initial)	47 to 51	6·5 to 7·0
Warm engine (final)	51 to 54	7·0 to 7·5
Distributor drive gear	51 to 58	7·0 to 8·0
Valve rocker arm cover	1·8 to 2·5	0·25 to 0·35
Crankshaft pulley	80 to 87	11·0 to 12·0
Inlet manifold	14 to 19	1·9 to 2·6
Exhaust manifold	12 to 17	1·6 to 2·3
Spark plugs	11 to 15	1·5 to 2·1
Oil pressure switch	9 to 13	1·2 to 1·8
Temperature gauge unit	4 to 7	0·5 to 4·7
Flywheel	60 to 65	8·0 to 9·0

1 General description

The engine has been designed to provide the vehicle with a lively and economical performance, and is, in most respects, typical of conventional overhead camshaft types. Dependent upon the intended market, there are two engine sizes available; 1000 cc or 1300 cc. The difference in capacity is achieved by a larger bore diameter and longer stroke on the 1300 cc engine. Except for bore size and stroke, the cylinder block and cylinder head are identical on both engines. The cylinder head is of crossflow design, with the inlet and exhaust valves arranged in near-hemispherical combustion chambers. The valves are operated by rockers in contact with a single chain-driven camshaft, mounted centrally between the two banks of valves. Valve adjustment is provided by conventional screw and lock nut fittings in the rockers.

The distributor, which is mounted on the inlet side of the cylinder

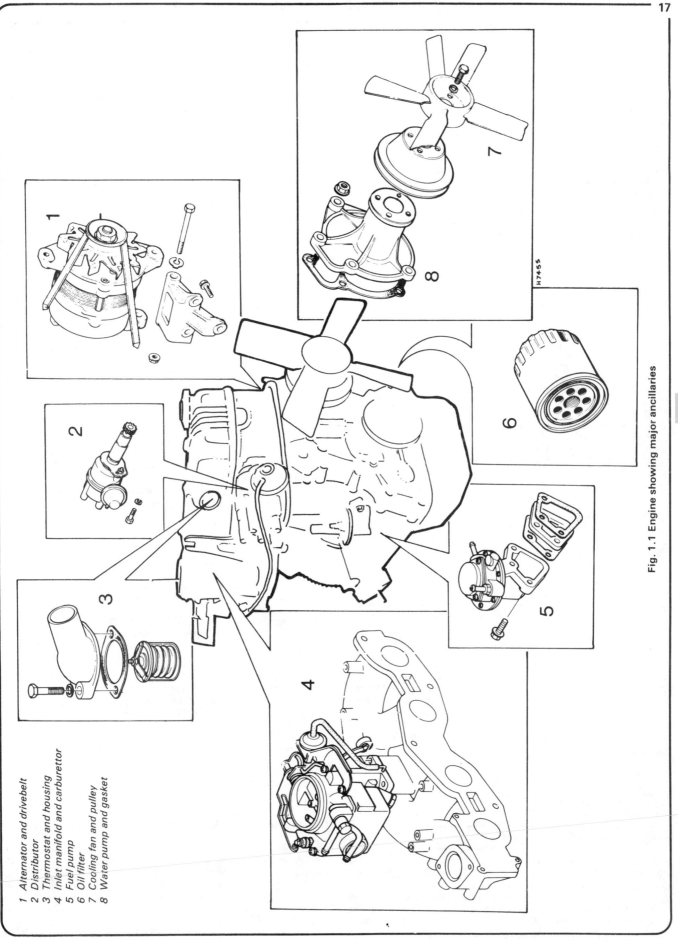

Fig. 1.1 Engine showing major ancillaries

1

1 Alternator and drivebelt
2 Distributor
3 Thermostat and housing
4 Inlet manifold and carburettor
5 Fuel pump
6 Oil filter
7 Cooling fan and pulley
8 Water pump and gasket

head, is driven by a helical gear at the forward end of the camshaft next to the camshaft chain sprocket.

The cylinder block is an iron casting, providing support for the crankshaft by way of five main bearings.

The oil pump is mounted low down on the right-hand side of the engine, driven by a chain from a sprocket on the front end of the crankshaft. Oil is picked up through a strainer housed in the sump and passed directly to a full flow filter mounted at the rear of the cylinder block, from which the oil is passed to the main gallery and the various parts of the engine. The filter incorporates a relief valve which ensures a full supply of oil to the engine in the event of the filter element being blocked, and returns any surplus oil to the sump if pressure becomes excessive.

2 Major operations possible with engine fitted

The following tasks can be performed with the engine fitted. However, the degree of difficulty varies, and for anyone who has lifting tackle available it is recommended that items such as the flywheel, pistons, connecting rods and crankshaft bearings are attended to after removal of the engine from the vehicle.

1 *Removal and refitting of the cylinder head.*
2 *Removal and refitting of the camshaft and bearings.*
3 *Removal and refitting of the engine supports (mounts).*
4 *Removal and refitting of the crankshaft rear oil seal.*
5 *Removal and refitting of the flywheel.*
6 *Removal and refitting of the clutch pilot bearing.*
7 *Removal and refitting of the engine front cover oil seal.*
8 *Removal and refitting of the engine front cover.*
9 *Removal and refitting of the timing chain and sprockets.*
10 *Removal and refitting of the oil sump.*
11 *Removal and refitting of the oil pump.*
12 *Removal and refitting of the main bearings.*
13 *Removal and refitting of the pistons and connecting rods.*
14 *Removal and refitting of the connecting rod big-end bearings.*

3 Major operations which require removal of the engine

The only item which cannot be removed and installed without the engine being removed, is the crankshaft.

4 Methods and equipment for engine removal

1 The engine can either be removed complete with the transmission, or the two units can be removed separately. If the transmission is to be removed, it is preferable to remove it while still coupled to the engine.

2 Essential equipment includes a suitable jack and support(s) so that the vehicle can be raised whilst working underneath, and a hoist or lifting tackle capable of taking the weight of the engine (and transmission, if applicable). The engine and manual transmission weigh in the order of 300 lb (135 kg); if the engine alone is being lifted, allow for approximately three-quarters of this weight; if the lifting tackle demands a reduction of weight, items such as the cylinder head, manifolds and engine ancillaries can be removed before lifting the engine, provided that a suitable sling can be used for lifting purposes. If an inspection pit is available some problems associated with jacking will be alleviated, but at some time during the engine removal procedure a jack will be required beneath the transmission (a trolley jack is very useful for this application).

5 Engine – removal with transmission

1 Before commencing work, it will be necessary to drive the vehicle onto ramps, over an inspection pit, or to jack it up and support it for access to the exhaust and transmission.

2 Open the bonnet to its full extent. Using a felt tipped pen, mark the bonnet hinge positions. Remove the two nuts on each hinge and the screw that retains the bonnet stay. With the assistance of a second person, remove the bonnet and stow it where it cannot be accidentally damaged.

3 Disconnect the battery terminals and stow them back out of the way.

4 Drain the cooling system (refer to Chapter 2 if necessary).

5 Remove the air cleaner (refer to Chapter 3 if necessary).

6 Loosen the hose clips and remove the top and bottom radiator hoses.

7 Where applicable, remove the bolts that retain the fan shroud; position the shroud over the fan.

8 Undo and remove the four bolts that retain the radiator. Remove the radiator from the engine compartment and stow it in a safe place.

9 Where applicable, lift out the fan shroud.

10 Loosen the bolts that retain the alternator and slide it in against the engine. Remove the fan belt.

11 Where applicable, remove the air pump and drivebelt.

12 Remove the four bolts that retain the fan and pulley(s) to the water pump. Lift off the fan and pulley(s).

13 It is now time to disconnect the electrical connections. Before commencing this, it is wise to have a quantity of identification tags handy to attach to the various leads as they are disconnected; also make a note of the various routes and clips that the looms pass through.

14 Disconnect the low tension lead between the distributor and the ignition coil.

15 Pull the HT leads from their respective spark plugs. Remove the HT lead from the ignition coil, and remove the distributor cap and leads from the distributor. Take care not to lose the rubber seal between the cap and the distributor.

16 Pull out the multi-pin output connector from the rear of the alternator. Prise off the rubber boot to disconnect the alternator input wire, which is retained by a brass nut.

17 Disconnect the water temperature sensor wire from the thermostat housing.

18 Disconnect the leads to the starter motor.

19 Remove the earth strap attached to the engine mount and the inner wing.

20 Disconnect the earth strap between the clutch bellhousing and the engine rear bulkhead (photo).

21 At the carburettor, disconnect the accelerator and choke control cables; stow them well out of the way.

22 On vehicles fitted with servo assisted braking systems, remove the vacuum supply pipe to the brake servo.

23 Remove the hoses between the heater and the engine.

24 Disconnect the fuel pipe to the fuel pump; use a piece of clean metal bar or similar the same diameter as the bore of the fuel pipe, to prevent fuel spillage and the ingress of dirt.

25 Undo and remove the three bolts and washers that retain the exhaust pipe to its mounting flange. The exhaust pipe is attached to the clutch bellhousing by a clip. Working underneath the vehicle, detach the clip from its mounting bracket (photo).

26 On vehicles equipped with full emission control systems:

a) *Disconnect the bullet-type connectors from the water thermoswitch located under the inlet manifold.*

b) *Remove the vacuum tube between the EGR control valve and the three-way solenoid, and the vacuum tube between the inlet manifold and the three-way solenoid.*

c) *At the air control valve, disconnect the two vacuum tubes; one tube is connected to the inlet manifold, and the other tube is connected to the check valve.*

d) *Disconnect the air pump hoses between the air control valve and the air injection manifold.*

Note: *Further information on emission control systems can be found in Chapter 3.*

27 Working inside the vehicle, unscrew the gear lever control knob.

28 Lift off the gear lever cover, which is held to the floor aperture surround by elastic sewn into the cover.

29 Remove the four bolts that retain the gear lever housing; lift out the gear lever and housing assembly.

30 Working beneath the vehicle, drain the transmission oil into a container of suitable capacity.

31 On vehicles fitted with automatic transmission, disconnect the

5.20 The clutch bellhousing-to-bulkhead earth strap

5.25 The exhaust pipe mounting at the clutch bellhousing

5.35 The speedometer drive cable

5.41 An engine mounting

5.42 The transmission rear support

5.43 Lifting out the engine and transmission

gear selector shift rod from the inhibitor switch. Disconnect the multi-pin connector between the inhibitor switch and the underside of the vehicle.

32 Disconnect the two leads from the downshift solenoid.

33 Disconnect the vacuum feed pipe to the diaphragm unit.

34 Pull off the two bullet connectors from the reversing light switch.

35 Unscrew and remove the speedometer drive cable (photo).

36 Mark the propeller shaft to ensure correct reassembly, and remove it from the vehicle (refer to Chapter 7 if necessary).

37 Disconnect the clutch release cable from the side of the clutch bellhousing.

38 Before proceeding any further, examine the engine and transmission to ensure that all connections have been detached, stowed and labelled as necessary. It is essential that the space around the transmission and engine is clear of pipes, wires, rods, etc, because any of them could be damaged if knocked when the engine/transmission assembly is being hoisted out of the vehicle.

39 Now bring the hoist over the engine and attach the sling to the lifting eyes on the front of the engine and transmission housing.

40 Support the weight of the transmission with a jack, using a suitable wooden packer between the jack head and the transmission.

41 Take the weight of the engine and transmission on the hoist, then proceed to loosen the engine mounts. The nuts which retain the mounts to the vehicle chassis need only be loosened as they are of the slide-in V-type (photo).

42 Remove the transmission rear support from the transmission and then remove it from the vehicle completely (photo).

43 The hoist should now be taking the weight of the engine, which means that the transmission supporting jack can be carefully lowered and the engine and transmission slowly lifted out. Take great care not to let the engine sway, or damage to accessories, wiring or vehicle paintwork may occur (photo).

44 Now that the engine and transmission assembly is out of the vehicle, the two units may be separated. Undo the nuts and bolts securing the flywheel housing and starter motor to the engine. Remove the starter motor.

45 Carefully ease the transmission away from the engine, taking particular care to prevent the weight of the transmission being taken on the input shaft. **Note**: *With automatic transmissions use a tyre lever between the engine flexplate and the converter, to prevent the converter disengaging from the transmission as it is moved away.*

---

6 Engine – removal without transmission

1 Initially proceed as described in the preceding Section up to, and including, paragraph 26.

2 Remove the two nuts and bolts that retain the starter motor to the clutch bellhousing. Lift out the starter motor.

3 Undo and remove the remaining bolts around the clutch bellhousing.

8.8 The crankshaft rear oil seal in position

4 Support the transmission with a jack, using a suitable wooden packer between the jack head and the transmission.

5 Before proceeding any further, examine the engine to ensure that all connections have been detached, stowed and labelled as necessary. It is essential that the space around the engine is clear of pipes, wires, etc, because any of them could be damaged if knocked when the engine is being hoisted out of the vehicle.

6 Now bring the hoist over the engine and attach the sling to the lifting eye on the front of the engine and around the rear end of the exhaust manifold, or the rear lifting eye.

7 Take the weight of the engine on the hoist, then loosen the nuts on the engine mounts. The engine mounts will need to be removed from the engine block; proceed to unbolt them from each side of the engine.

8 Carefully draw the engine forwards, away from the transmission, taking care that the engine weight is not taken by the transmission input shaft (manual transmission). **Note**: *With automatic transmission, use a tyre lever between the engine flexplate and the converter to prevent the converter disengaging from the transmission as the engine is moved away. Take care not to let the engine sway, or damage to accessories, wiring or vehicle paintwork may occur.*

9 Now that the engine is removed, do not attempt to move the vehicle if the transmission is still fitted, unless some method can be devised of suspending it from the vehicle frame.

---

7 Engine supports (mounts) – removal and refitting (engine in the vehicle)

1 In order to gain access to the underside of the vehicle, it will need to be raised on a hoist or suitable jacks, or alternatively placed over an inspection pit.

Engine front support

2 At the mount-to-chassis bracket, loosen the single nut a couple of turns.

3 Position a jack beneath the oil sump, with a wooden block between the jack head and the oil pan. Raise the engine so that a piece of wood can be inserted between the oil sump and the crossmember. Lower the jack, so allowing the engine to rest on the piece of wood on the crossmember.

4 The mount assembly may now be removed from the cylinder block, by removing the bolts that secure it.

5 Refitting is the reverse of the removal procedure but, where applicable, do not forget to refit the earth strap.

Engine rear support

6 Support the transmission with a jack; place a suitable wooden packer between the jack head and the transmission.

7 Undo and remove the four nuts and washers that secure the crossmember to the body and the transmission. Remove the crossmember.

8 Remove the two bolts that retain the rubber mount to the transmission case. Remove the rubber mount.

9 Refitting is a straightforward reversal of the removal procedure.

---

8 Clutch pilot bearing, flywheel and crankshaft rear oil seal – removal and refitting (engine in the vehicle)

Note: *The procedure given in this Section describes oil seal replacement where it is* **not** *required to renew the rear main bearing. If the rear main bearing requires renewal also, the two jobs must be carried out separately since the engine cannot be conveniently supported if the transmission and oil sump are both removed at the same time.*

1 Remove the transmission as described in Chapter 6.

2 Remove the clutch assembly (refer to Chapter 5 if necessary).

3 If it is required to remove the clutch pilot bearing, it can be done at this stage provided that a suitable extractor is available. If the bearing cannot be extracted do not use too much force, but remove the flywheel first as described in paragraph 4. Then carefully tap the bearing out with a suitable size of bar.

4 Remove the flywheel attaching bolts and carefully draw off the flywheel. Great care should be taken if leverage is required, in order to prevent damage to the ring gear and block end face. Remember that the flywheel is fairly heavy and cumbersome, particularly in view of the limited accessibility.

5   If the rear oil seal requires renewal, use an awl to punch two holes in it on opposite sides of the crankshaft just above the split line of the bearing cap to cylinder block.

6   Screw in two self-tapping screws to the holes previously punched in the oil seal, then lever at each screw in turn to extract the seal. Small blocks of wood placed against the block will provide a fulcrum point for levering. Take great care not to damage the crankshaft seal surface whilst the seal is being removed.

7   Carefully clean the oil seal recess in the cylinder block and bearing cap. Clean the crankshaft seal surface and inspect it for any scoring, etc.

8   When fitting the new seal, lubricate the surfaces of the seal, crankshaft, cylinder block and bearing cap with engine oil. Insert the seal with the lips inwards, carefully pressing it in. If it is stubborn, use a small wooden block and light hammer blows; take care to ensure that the seal is not crooked as it is pressed home (photo).

9   The flywheel can now be examined for damage to the ring gear. If there are any broken or chipped teeth, or if other damage is evident, the old ring gear can be removed from the flywheel. This can be done by heating the ring gear with a blow lamp as evenly as possible to a temperature of 250 to 260°C (480 to 500°F) and driving it off, or cutting a small groove at the root of two adjacent teeth and then splitting the ring with a cold chisel.

10 To fit a new ring gear, it must be heated evenly to a temperature of 250 to 260°C (480 to 500°F) in a domestic oven or an oil bath (a naked flame is not very suitable unless the heat can be applied evenly), then carefully drive it on to the cold flywheel. Note that the ring gear tooth chamfer should be towards the engine.

11 When the flywheel has cooled off, it can be refitted to the engine. Align the O-marked hole on the flywheel with the reamer hole on the crankshaft. Coat the bolt threads with a gasket sealer and fit the reamer bolt and washer in the O-marked hole. Fit the remaining bolts and washers and tighten them in a crosswise sequence to the specified torque (photo).

12 Refit the clutch (refer to Chapter 5 if necessary).

13 Refit the transmission (refer to Chapter 6 if necessary).

9   Engine front cover oil seal – renewal (engine in the vehicle)

1   Remove the bonnet (refer to Chapter 12 if necessary).

2   Drain the cooling system (refer to Chapter 2 if necessary).

3   Remove the radiator (refer to Chapter 2 if necessary).

4   Remove the alternator (and where applicable, the air pump) drivebelt(s).

5   Undo and remove the bolt that retains the cranksaft pulley, then use a suitable extractor to draw the pulley off. Take care that the Woodruff key is not lost (photo).

6   Carefully pry out the oil seal. Take care not to damage the sealing face on the front cover, and check that it is clean once the seal is removed.

7   Lubricate the sealing faces with engine oil and carefully press in a new seal.

8   Refit the pulley and tighten the bolt to the specified torque.

9   Refit the drivebelt(s), radiator, and bonnet in the reverse order to removal. Finally fill the cooling system (refer to Chapter 2 if necessary).

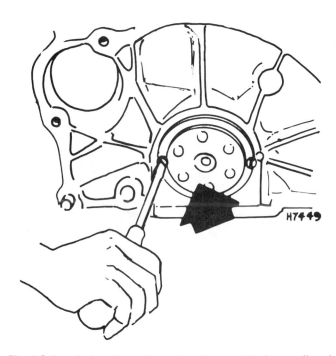

Fig. 1.2 A typical method of removing the crankshaft rear oil seal (Sec 8)

8.11 The O-mark on the flywheel

9.5 The crankshaft pulley retaining bolt

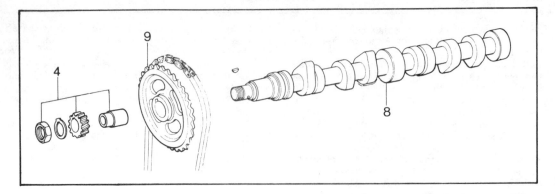

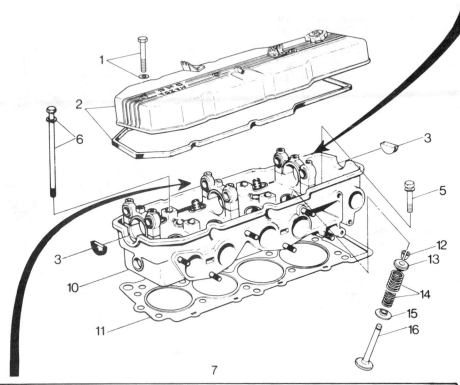

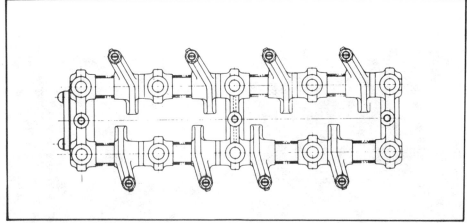

Fig. 1.3 An exploded view of the cylinder head assembly

1 Rocker cover bolt and washer
2 Rocker cover and gasket
3 Oil seal
4 Locknut, washer and distributor drivegear
5 Cylinder head-to-front cover bolt and washer
6 Cylinder head bolt and washer
7 Rocker arm assembly
8 Camshaft
9 Camshaft sprocket
10 Cylinder head
11 Cylinder head gasket
12 Split collets
13 Upper spring seat
14 Valve springs
15 Lower spring seat
16 Valve

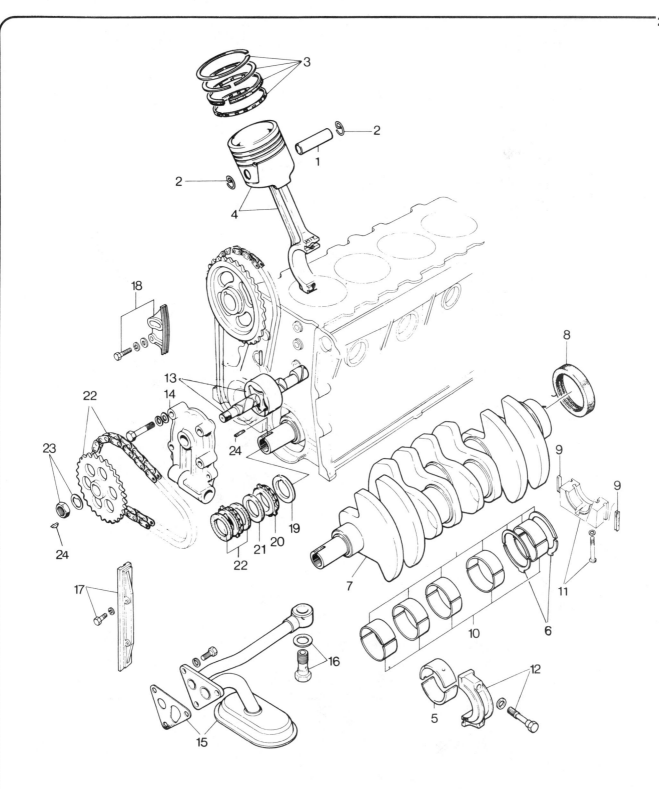

Fig. 1.4 An exploded view of the cylinder block assembly

1 Gudgeon pin
2 Retaining clip
3 Piston rings
4 Piston and connecting rod assembly
5 Connecting rod bearing
6 Thrust washers
7 Crankshaft
8 Oil seal

9 Rear main bearing cap seals
10 The main bearings
11 Main bearing cap, bolt and washer
12 Connecting rod bearing cap, bolt and washer
13 Oil pump shaft and rotor assembly

14 Oil pump cover
15 Oil strainer and outlet pipe assembly and gasket
16 Connection bolt and washer
17 Vibration damper and bolt
18 Chain guide strip and bolt
19 Spacer (when applicable)
20 Crankshaft timing chain

sprocket
21 Spacer
22 Oil pump drive sprockets and chain (and rubber rings, where applicable)
23 Oil pump sprocket retaining nut and lock tab
24 Woodruff key

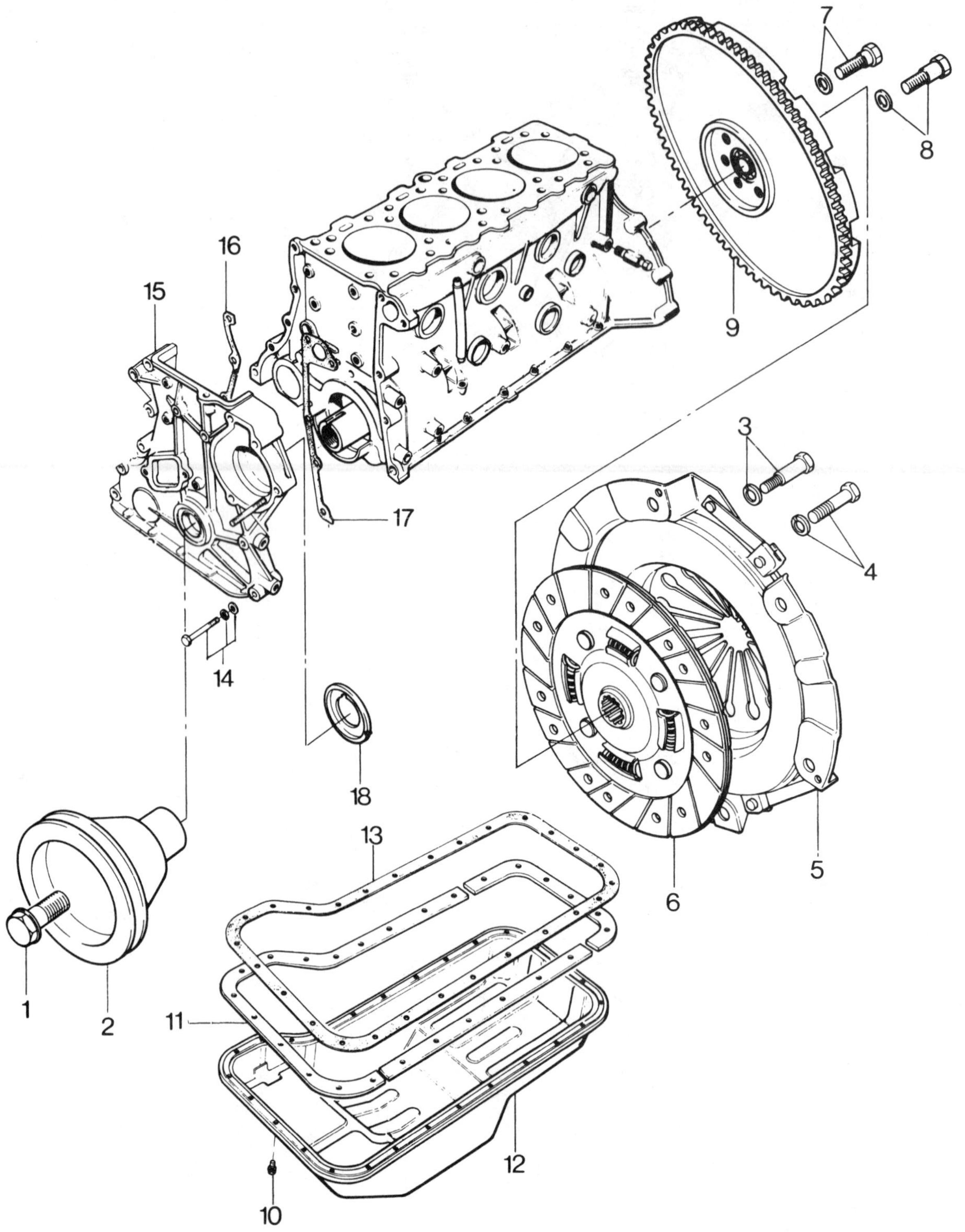

Fig. 1.5 Exploded view of the sump, front cover and clutch assembly

1	Crankshaft pulley bolt		plate assembly	10	Sump bolt	washers
2	Crankshaft pulley	6	Clutch friction disc	11	Stiffener plate	15 Front cover
3	Bolt and washer (reamer	7	Flywheel bolt and washer	12	Sump	16 Front cover gasket
	type)	8	Flywheel bolt and washer	13	Gasket	17 Front cover gasket
4	Bolt and washer		(reamer type)	14	Front cover bolt and	18 Oil thrower
5	Clutch cover and pressure	9	Flywheel			

10 Engine front cover, timing chain, oil pump, oil pump chain and sprockets – removal and refitting (engine in the vehicle)

1   Initially proceed as described in paragraphs 1 to 5, of the previous Section.
2   Remove the water pump (refer to Chapter 2 if necessary).
3   Remove the air cleaner (refer to Chapter 3 if necessary).
4   Remove the spark plug leads and tie them out of the way.
5   Remove the three bolts and washers that secure the rocker arm cover, and remove the rocker arm cover together with its gasket. Take care not to lose the semi-circular rubber oil seals at each end of the cover to cylinder head bolts.
6   Undo and remove the single bolt and washer that retain the front cover to the cylinder head.
7   Remove the oil sump, as described in Section 12.
8   Remove the alternator bracket to block bolts and stow the alternator in a safe place. Where applicable, remove the air pump bracket to block bolts and remove the air pump from the vehicle.
9   Remove the bolts which retain the engine front cover and take the cover off. This will probably be stuck to the front of the block, but a few taps with a soft-faced hammer should free it. Alternatively, provided that care is taken not to damage the sealing faces, a knife blade can be carefully inserted to break the seal.
10   To remove the oil pump sprockets and chain, remove the oil thrower and spacer from their keyway in the crankshaft. Knock back the lock tab and remove the nut and lock tab from the oil pump sprocket. Pull off the oil pump sprocket and the crankshaft sprocket, together with the oil pump drive chain. Take care that the Woodruff key in the oil pump spindle is not lost. Undo and remove the four bolts and washers that secure the oil pump to the engine block. Remove the oil pump, and withdraw the spindle shaft and rotors from the engine block. If the oil pump is suspect or requires attention refer to Section 17.
11   If the timing chain is to be renewed or removed, remove the cylinder head, as described in Section 21. (The inlet and exhaust manifolds need not be removed from the head unless considered necessary.)
12   Remove the oil pump and crankshaft sprockets and chain as described in paragraph 10.
13   Remove the timing chain tensioner (two bolts) taking care not to let it spring apart.
14   Loosen the screws which retain the timing chain guide strip.
15   Where applicable, remove the rubber spacer from the crankshaft.
16   Remove the spacer from the face of the crankshaft timing chain sprocket.
17   With a suitable puller, draw off the crankshaft timing chain sprocket together with the timing chain. If the Woodruff key is loose, remove it and stow it in a safe place. Remove the chain together with the crankshaft sprocket.
18   If applicable, remove the spacer that locates between the crankshaft timing chain sprocket and the crankshaft shoulder.
19   To refit the timing chain, align the nickel plated link in the timing chain with the dot on the crankshaft sprocket tooth.
20   Rotate the engine until number one piston is at the top dead centre position; this can be done by engaging a gear and rocking the car back or forth. The keyway slot in the crankshaft should now be facing upwards at 90° to the sump face.
21   If the Woodruff key has been removed, it must now be refitted.
22   Where applicable, refit the spacer to the crankshaft shoulder.
23   Taking care to ensure that the timing chain and crankshaft sprocket are still aligned as previously mentioned, refit the sprocket and chain to the crankshaft. Feed the chain upwards, along the vibration damper on one side, and the chain guide strip at the other side. Keeping tension on the chain to prevent it from coming off the crankshaft sprocket, secure it tightly to a convenient anchorage point ready for the cylinder head assembly.
24   Refit the spacer (and, where applicable, a rubber spacer) to the face of the crankshaft sprocket.
25   Assemble the oil pump inner rotor and shaft to the outer rotor, and refit them into their bore in the engine block.
26   Locate the oil pump body over the oil pump spindle, and secure it with the four bolts and washers.
27   If the oil pump spindle key has been removed, it must now be

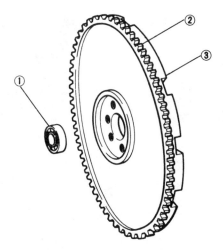

Fig. 1.6 The flywheel and ring gear assembly

1 Clutch pilot bearing
2 Ring gear
3 Flywheel

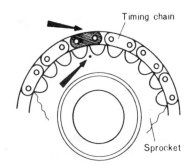

Fig. 1.7 The nickel plated link and the sprocket tooth dot

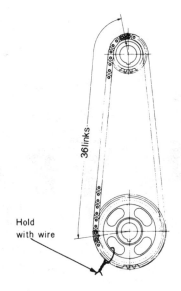

Fig. 1.8 The correct timing chain alignment (Sec 10)

11.10 Adjusting the timing chain tension

11.13 The inspection cover in position

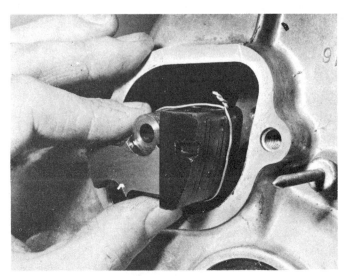
11.21 Fitting the timing chain tensioner

refitted.

28 Locate the oil pump sprocket and the crankshaft oil pump drive sprocket to the oil pump drive chain.

29 Align the keyways in the two sprockets to match the Woodruff keys on the crankshaft and the oil pump spindle.

30 Refit the two sprockets, together with the drive chain; ensure that the sprockets are fully seated on their respective shafts.

31 Assemble the lock tab and nut to the oil pump spindle. —Tighten the sprocket nut to the specified torque and bend over the lock tab.

32 Refit the spacer to the front face of the crankshaft oil pump drive sprocket, followed by the flat oil thrower (or if applicable, the dished oil thrower).

33 Refit the cylinder head and camshaft as described in Section 40. When assembling the camshaft sprocket to the camshaft, ensure that the nickel plated link in the timing chain and the dot on the sprocket tooth are aligned, before the camshaft is located into the sprocket.

34 Assemble the timing chain tensioner and secure it with the two bolts; in order to carry out timing chain adjustment, press the slipper head into the tensioner housing and secure it in this position with locking wire.

35 To adjust the tension of the chain, press the top of the chain guide strip towards the chain then tighten the retaining screws.

36 Remove the locking wire from the tensioner, allowing the slipper head to contact the timing chain.

37 Before refitting the engine front cover, clean the gasket surface of the cover, cylinder block and water pump.

38 Position the front cover gaskets to the block using a non-setting gasket sealant. Refit the front cover and tighten the bolts in a crosswise order.

39 Refit the alternator bracket, followed by the alternator.

40 If applicable, refit the air pump and bracket.

41 Refit the water pump (refer to Chapter 2 if necessary).

42 Check that the crankshaft pulley is clean and smear engine oil on the sealing surface.Refit the pulley and tighten the retaining bolt to the specified torque.

43 The remainder of the refitting procedure is the reverse of the removal procedure. On completion, refill the cooling system and adjust the drivebelt(s) tension (refer to Chapter 2 if necessary). Finally, adjust the valve clearances, as described in Section 41.

11 Timing chain tensioner – adjustment, removal and refitting (engine in the vehicle)

Adjustment

1   Timing chain adjustment is not part of any routine maintenance programme and is normally taken care of automatically. However, after removal and refitting of the cylinder head, it is recommended that the adjustment procedure in this Section is carried out.

2   Loosen the alternator (and if applicable, the air pump) adjustment bolts and remove the drivebelt(s).

3   Remove the cooling fan and pulley (refer to Chapter 2 if necessary).

4   Undo and remove the two bolts that retain the timing chain tensioner inspection plate. Remove the inspection plate together with its gasket.

5   Using a piece of thin gauge metal approximately 0·25 in (6·4 mm) wide and 6 in (150 mm) long, bend over about 0·25 in (6·4 mm) at one end of the metal strip to 90°. Measure the dimension across the inspection aperture and form another 90° bend in the strip, this time opposite to the first bend, the dimension between the two bends being slightly less than the inspection aperture width (See Fig. 19). Use this tool to pull the slipper head away from the timing chain; by inserting the tool into the aperture, and locating the formed lip onto the slipper head face, the slipper head can then be pulled back away from the timing chain. Retain the tool in this position with a suitable piece of wood, wedged between the side of the inspection aperture and the tool.

6   If applicable, remove the spark plug leads, and remove the rocker cover as described in Section 10, paragraph 5.

7   Remove the two blind bolts from the engine front cover, to gain access to the chain guide retaining screws. Loosen the two screws to allow the chain guide to slide back and forth on its elongated slot.

8   Slightly rotate the crankshaft in the forward direction of engine

revolution; this can be carried out with a suitably sized ring spanner applied to the crankshaft pulley retaining bolt.

9 Press the top of the chain guide strip with a lever through the opening of the cylinder head. Lever the chain guide strip until it is pressing onto the timing chain, but do not lever with excessive force.

10 Whilst still holding the lever, tighten the chain guide retaining screws (photo).

11 Replace the two blind plugs to the front cover.

12 Remove the tool, which was used to compress the slipper head, from the inspection aperture.

13 Replace the inspection cover and gasket, refit the retaining bolts and tighten them (photo).

14 Assemble the pulley and fan to the water pump and secure with the four bolts and washers.

15 Refit the drivebelt(s) and adjust the tension as described in Chapter 2.

16 If applicable, replace the rocker cover, using a new gasket and semi-circular rubber oil seals. Assemble the three retaining bolts and washers; tighten them to the specified torque.

17 Replace the spark plug leads.

Removal and refitting

18 To remove the tensioner, proceed as described in paragraphs 2, 3 and 4.

19 If applicable, remove the spark plug leads and detach the rocker cover, as described in Section 10, paragraph 5.

20 Remove the two attaching bolts from the tensioner, then withdraw the tensioner through the aperture in the front cover. **Note**: *Take great care not to drop the retaining screws inside the front cover. When lifting out the tensioner, invert the slipper head upwards to prevent the assembly coming apart. If any components are accidentally dropped into the front cover it will mean removal of the sump and front cover.*

21 When refitting the tensioner, compress the slipper head, and using locking wire, secure it in this position (photo).

22 Refit the tensioner through the aperture in the front cover, and retain it in position with the two bolts.

23 Adjust the timing chain tension by following the procedure of paragraphs 7 to 11.

24 Using long nosed pliers, carefully remove the locking wire to release the slipper head.

25 Complete the refitting procedure by following paragraphs 13 to 17.

12 Sump – removal and refitting (engine in the vehicle)

1 In order to gain access beneath the vehicle, it will need to be raised on a hoist or suitable jacks and supports, or alternatively placed over an inspection pit.

2 Drain the crankcase oil into a container of suitable capacity. Remove the oil dipstick.

3 Remove the retaining bolts and lower the sump onto the crossmember. If the sump is sticking to the block, a few careful blows with a soft faced hammer may remove it, but take care that the sump is not damaged. If this does not move it, a thin bladed knife may be very carefully inserted between the sealing faces, but extreme care must be taken not to damage this sealing face.

4 Remove the oil pump pick-up tube and strainer by removing the three bolts and washers that retain it to the oil pump, then undo and remove the connection bolt that retains the oil feed pipe to the block. Lift out the sump, together with the oil pump pick-up tube and feed pipe.

5 When refitting the sump, first ensure that all the gasket surfaces are clean and undamaged. Clean the pick-up tube filter, and the sump.

6 Fit a new gasket to the block, using a non-setting gasket sealant.

7 Rest the sump on the crossmember, and whilst holding it slightly downwards at the front end, fit the oil pump pick-up tube and strainer, together with a new gasket to the oil pump. Refit the connection bolt to the block taking care that a washer is fitted either side of union faces.

8 Lift up the sump, assemble the stiffener plates, and screw in all the retaining bolts. The sump bolts should be tightened in a crosswise order to prevent any distortion.

9 On completion, top-up the engine with the correct quantity and grade of oil. Start the engine and check for any leakage.

13 Main bearings – removal and refitting (engine in the vehicle)

Note: *It is of the utmost importance that this operation is carried out in conditions which are as clean as possible, to prevent ingress of dirt to the bearings.*

1 Initially remove the oil sump, and the oil pick-up tube and strainer, as described in the previous Section.

2 Remove one bearing at a time by removing the retaining bolts and washers from the bearing cap. Take the bearing shell out of the cap. **Note**: *When removing the rear main bearing cap, fit two bolts into the tapped holes so that leverage can be applied directly downwards. Do not rock or twist the bearing cap.*

3 Now insert a small bolt into the oil hole in the crankshaft. The bolt head must be larger than the oil hole but must not project a greater distance than the thickness of the bearing shell.

4 Rotate the crankshaft in the normal direction of rotation to force the bearing out of the block.

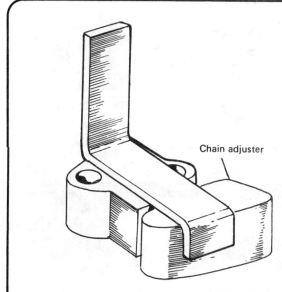

Chain adjuster

Fig. 1.9 Tool to pull slipper head away from timing chain

Fig. 1.10 Adjusting the timing chain tension (Sec 11)

Fig. 1.11 The main bearings (Sec 13)

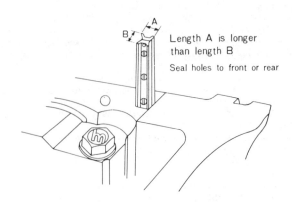

Length A is longer than length B

Seal holes to front or rear

Fig. 1.12 Correct fitment of the rear main bearing cap side seals (Sec 13)

Fig. 1.13 Fitting a thrust washer to the rear main bearing cap (Sec 13)

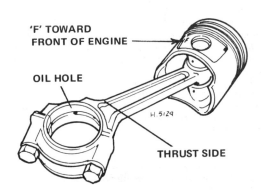

'F' TOWARD FRONT OF ENGINE

OIL HOLE

THRUST SIDE

Fig. 1.14 Positioning of the piston and connecting rod (Sec 15 and 16)

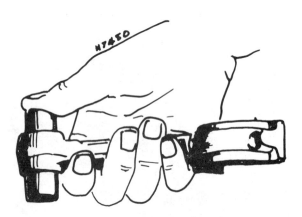

Fig. 1.15 Checking the sliding fit of a gudgeon pin (Sec 16)

16.4 A gudgeon pin retaining clip

5   Clean up the crankshaft journals and inspect for nicks, burrs and bearing pick-up.
6   When refitting an upper bearing shell, place the plain end over the shaft on the locking tang side of the block. Partially fit the bearing so that the bolt (see paragraph 3) can be inserted in the crankshaft oil hole; rotate the crankshaft in the opposite direction to normal rotation until the bearing seats. Remove the bolt.
7   The bearing cap and shell can now be refitted, after lubricating the bearing surface with engine oil. Tighten the bolts to the specified torque. **Note**: *When refitting the rear main bearing cap, it is good practice to renew the crankshaft thrust washers and the main bearing cap side seals. Note that the thrust washers are fitted to the block with the oil grooved surface facing the crankshaft thrust side.*
8   Finally, refit the oil pick-up tube and strainer, and refit the sump as described in Section 12.

14 Connecting rod (big-end) bearings – removal and refitting (engine in the vehicle)

Note: *It is of the utmost importance that this operation is carried out in conditions which are as clean as possible to prevent ingress of dirt to the bearings.*
1   Initially remove the sump, and the oil pick-up tube and strainer, as described in Section 12.
2   Remove one bearing at a time by rotating the crankshaft as necessary, to gain access to the cap bolts. Remove the cap bolts, take off the cap, then remove the bearing shell from the cap and connecting rod.
3   Clean the crankshaft journals, and inspect for nicks, burrs and bearing pick-up. Check that the bearing shell seatings in the cap and connecting rod are clean.
4   Refit the bearing shells in the connecting rod and bearing cap, with the tangs located in the slots.
5   Lubricate the bearing surfaces with engine oil, refit the cap and tighten the bolts to the specified torque.
6   Refit the oil pick-up tube and strainer, and the sump as described in Section 12.

15 Pistons and connecting rods – removal and refitting (engine in the vehicle)

1   Remove the cylinder head, as described in Section 21.
2   Remove the sump and oil pick-up tube and strainer, as described in Section 12.
3   Turn the crankshaft until the piston to be removed is at bottom dead centre (BDC). Place a cloth on top of the piston crown to collect any carbon deposits, etc.
4   Carefully scrape off any deposits at the top of the cylinder bore, remove the bearing cap and carefully drive the piston and connecting rod up out of the bore using a hammer handle. Ensure that the bearing cap and connecting rod are marked as to their relative cylinder positions. If the piston will eventually be removed from the connecting rod, this too must be marked.
5   When refitting, lubricate the pistons, rings and cylinder bores with engine oil.
6   Ensure that the rings are positioned correctly (see Section 30) and fit a piston ring compressor. Press the piston and connecting rod into the correct bore; ensure that the 'F' mark on the piston is towards the front of the engine (see Fig. 1.14). Guide the connecting rod towards the journal, lubricate the bearing shells with engine oil, then refit the bearing cap. Tighten the bolts to the specified torque.
7   Refit the oil pick-up tube and strainer, and the sump as described in Section 12.
8   Refit the cylinder head, as described in Section 40.

16 Pistons and connecting rods – dismantling and reassembly

1   If the engine block is to be rebored, or the pistons are going to be renewed, it will be necessary to remove the pistons from the connecting rods.
2   Mark the pistons and gudgeon pins to identify them with their respective connecting rods.
3   Carefully remove the piston rings by opening them out at their

ends and sliding them up over the piston crown.
4   Using suitable pliers, remove the gudgeon pin retaining clips. Provided that they are undamaged, they can be saved for re-use (photo).
5   Hold the piston in hot water for about fifteen seconds, then press out the pin. This should not be difficult as the piston will expand, but provided that great care is taken to support the piston, light hammer blows against a suitable drift will do the job.
6   Inspection of the piston, piston rings and cylinder bores, and assembly of the rings to the pistons are dealt with in Sections 29 and 30 later in this Chapter.
7   Check that the gudgeon pin slides through the small end bush, but is not loose.
8   Lubricate the gudgeon pin with engine oil. Fit it through the piston and small end bush and fit the retaining clips. Note that the 'F' mark on the piston will eventually point towards the front of the engine. See Fig. 1.14 for the correct relationship to the connecting rod.
9   Carefully refit the piston rings to their grooves.

17 Oil pump – dismantling, inspection and reassembly

1   Remove the split pin that retains the spring seat, spring and regulator valve in the relief valve chamber.
2   Wash all the parts in petrol (gasoline) or paraffin (kerosene), and dry them with compressed air. Use a small brush to clean out the pump housing and the relief valve chamber to ensure that all dirt is removed.
3   Check the pump cover for wear and scoring.
4   Refit the outer rotor, and inner rotor and shaft, to the block. Using

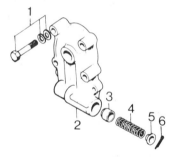

Fig. 1.16 The oil pump cover components (Sec 17)

1 Bolt and washers   *4 Spring*
2 Oil pump cover   *5 Spring seat*
3 Regulator valve   *6 Split pin*

Fig. 1.17 Checking the rotor lobes clearance (Sec 17)

Fig. 1.18 Checking the outer rotor-to-cylinder block bore clearance (Sec 17)

feeler gauges, check the clearance between the lobes of the rotors against the figure specified. If the clearance is larger than specified, replace both rotors.

5   Check the clearance between the outer rotor and the cylinder block bore. If the clearance exceeds the specified limit, replace the rotor, or in extreme cases, the cylinder block.

6   Check the endfloat of the rotors by placing a straight edge across the cylinder block and measuring the clearance between the rotor and the straight edge. Then place the straight edge across the pump cover, and measure the clearance between the straight edge and the cover. The two dimensions added together will indicate the endfloat of the rotors. If the endfloat exceeds the figure specified, the pump cover can be machined to obtain the correct endfloat.

7   Lubricate all the parts of the pump with engine oil.

8   Refit the regulator valve, spring and spring seat, to the relief valve chamber, retaining them in position with the split pin.

9   The oil pump is now ready for refitting as described in Section 10, paragraphs 25 to 43.

18 Oil filter – removal and refitting

1   In order to gain access to the underside of the vehicle, it will need to be raised on a hoist or suitable jacks and supports, or alternatively placed over an inspection pit.

2   Place a drip pan beneath the oil filter, then unscrew the filter. If it is stubborn, wrap a strip of emery cloth around it to provide a better grip.

3   When refitting, coat the seal with engine oil and fit the filter firmly by hand.

4   Lower the vehicle, run the engine at a fast idle and check for leaks.

5   Top-up the oil level.

19 Engine dismantling – general

1   It is best to mount the engine on a dismantling stand, but if one is not available stand the engine on a strong bench with wooden blocks so that it is at a comfortable working height. It can be dismantled on the floor but this is not as convenient.

2   During the dismantling process, great care should be taken to keep the exposed parts free from dirt. As an aid to achieving this, thoroughly clean down the outside of the engine, removing all traces of oil and dirt.

3   Use paraffin (kerosene) or one of the proprietary cleaning solvents. These solvents will make the job much easier, for after the solvent has been applied and allowed to stand for a time, a vigorous jet of water will wash off the solvent, together with all the grease and dirt. If the

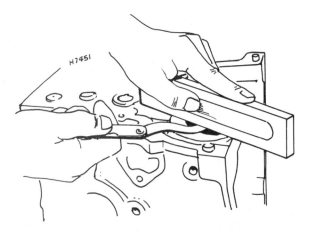

Fig. 1.19 Checking the rotor endfloat (Sec 17)

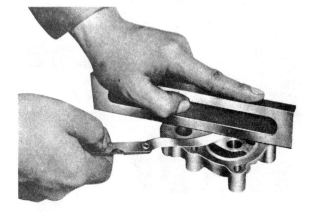

Fig. 1.20 Checking the rotor endfloat (Sec 17)

Fig. 1.21 Lubricating the oil filter rubber seal (Sec 18)

dirt is thick and deeply embedded, work the solvent into it with a wire brush.

4 Finally wipe down the exterior of the engine with a rag and only then, when it is quite clean, should the dismantling process begin. As the engine is stripped, clean each part in a bath of paraffin (kerosene) or petrol (gasoline).

5 Never immerse parts with oilways (for example the crankshaft) in solvents, but to clean wipe down carefully with a petrol dampened cloth. Oilways can be cleaned out with nylon pipe cleaners. If an air line is available, all parts can be blown dry and the oilways blown through as an added precaution.

6 Re-use of the old gaskets is false economy and will lead to oil and water leaks, if nothing worse. Always use new gaskets throughout.

7 Do not throw the old gasket away for it sometimes happens that an immediate replacement cannot be found and the old gasket is then very useful as a template. Hang up old gaskets as they are removed.

8 To strip the engine it is best to work from the top down. The underside of the crankcase, when supported on wooden blocks, acts as a firm base. When the stage is reached where the crankshaft and connecting rods have to be removed, the engine can be upturned and all other work carried out in this position.

9 Whenever possible refit nuts, bolts and washers finger-tight from wherever they were removed. This helps avoid loss and muddle later. If they cannot be refitted, lay them out in such a fashion that it is clear from where they came.

20 Engine ancillary components – removal

Before basic engine dismantling begins, it is necessary to strip it of ancillary components. Typically, these will be as follows:

a) *Alternator*
b) *Water pump*
c) *Oil filter*
d) *Distributor*
e) *Inlet manifold and carburettor*
f) *Exhaust manifold*
g) *Engine mount brackets*

1 Remove the alternator, together with its mounting bracket (photo).
2 Remove the water pump (refer to Chapter 2 if necessary) (photo).
3 Unscrew the oil filter from the cylinder block.
4 Remove the distributor from its mounting on the side of the cylinder head (refer to Chapter 4 if necessary).
5 Remove the inlet manifold complete with the carburettor (refer to Chapter 3 if necessary).
6 Remove the exhaust manifold (refer to Chapter 3 if necessary).
7 Remove the engine mounts from the block.

21 Camshaft and cylinder head – removal (engine in the vehicle)

1 Begin by removing the battery leads and stowing them out of the way to prevent short-circuiting.
2 Drain the cooling system (refer to Chapter 2 if necessary).
3 Remove the top hose from the radiator and thermostat housing on the cylinder head.
4 Remove the air cleaner (refer to Chapter 3 if necessary).
5 Rotate the engine crankshaft so that number 1 piston is at top dead centre (TDC) on its firing stroke; this is most easily done by selecting 3rd gear and moving the vehicle a little until the top dead centre (TDC) mark on the crankshaft pulley aligns with the timing pointer, whilst at the same time checking that the distributor rotor aligns (approximately) with number 1 spark plug lead in the distributor cap. (See Chapter 4 for further information). For vehicles with automatic transmission, remove the spark plugs and rotate the crankshaft, using the fanbelt, to obtain TDC.
6 Apply the handbrake with 3rd gear still engaged as the refitting procedure is simplified a little if this position is not disturbed.
7 Index mark the distributor body and cylinder head, and the rotor and the body, to ensure that correct alignment can be obtained when reassembling. Disconnect the distributor leads, noting their relative positions, then remove the clamp and withdraw the distributor.
8 The next step is to remove the inlet manifold and carburettor from the cylinder head which will require removal of the hoses, carburettor

20.1 Removing the alternator

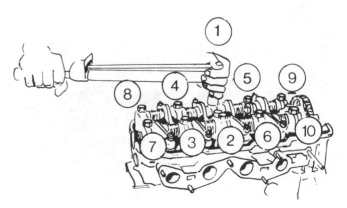
20.2 Removing the water pump

Fig. 1.22 The correct sequence for tightening the cylinder head bolts (Sec 21, 39, 40 and 41)

21.13 The dot and the bright link

21.18 Lifting off the rocker arm assembly

controls, fuel pipes and all pneumatic and electrical connections associated with the emission control equipment. Ensure that all the connections are suitably identified so that they cannot be mixed up when reconnecting. **Note:** *Removal of the inlet manifold is described in Chapter 3, but there is no need to remove the assembly completely from the engine compartment, provided that it can be adequately supported. This will mean that a little work may be saved in not having to remove some of the connections.*

9 Remove the exhaust manifold from the cylinder head (refer to Chapter 3 if necessary).

10 Remove the radiator and water pump (refer to Chapter 2 if necessary). Ensure that all the hose connections and electrical leads are suitably identified, then remove them from the cylinder head.

11 Undo the crankshaft pulley bolt and draw out the pulley, remove the cover from the timing chain tensioner.

12 Slacken the rocker arm cover bolts and remove them. Lift off the cover; if it is found to be sticking, either tap around the joint with a soft-faced hammer, or insert a knife between the cover and the cylinder head to break the seal. Take great care not to damage the sealing faces.

13 Once the rocker cover has been removed, check the position of the camshaft and sprocket wheel. If the engine number 1 piston is at TDC on its firing stroke, the two cam lobes operating the number 1 cylinder valves should be both pointing downwards and the rockers should be slack. Examination of the sprocket should show a reference dot being adjacent to a bright link in the timing chain – both mark and link should be in the plane of the cylinder head top surface and axis of the camshaft (photo).

14 It is now necessary to relieve the tension on the timing chain. Ideally this will mean that the crankshaft needs to be rotated slightly in the direction of rotation, but it is important that it is restored to its original position if the timing datum is not to be lost. To relieve the tension, remove the two blind bolts in the front cover, to gain access to the chain guide retaining screws; loosen the screws, and, through the aperture in the cylinder head, use a screwdriver to lever the chain guide away from the chain. Remove the timing chain tensioner through the aperture in the front cover. For further information refer to Section 11.

15 Undo and remove the two screws that secure the shim plate to the rocker shaft half clamp; the shim plate controls the camshaft endplay. To gain access to the retaining screws, the camshaft sprocket will need to be rotated slightly in order to pass a screwdriver through the holes in the camshaft sprocket.

16 Fold back the lock tab, and undo and remove the nut that retains the distributor drive gear, and the camshaft sprocket. Remove the lock tab, distributor drive gear and spacer.

17 It is now time to undo the cylinder head bolts. The same bolts also fix the rocker assembly and camshaft in position on the cylinder head.

21.20 The single cylinder head-to-front cover bolt

Fig. 1.23 The cylinder head-to-front cover bolt (Sec 21)

18 Once all the bolts have been undone and removed in the reverse order to that shown in Fig. 1.22, carefully lift the rocker arm assembly from the cylinder head. The assembly does not hold together by itself, because there is nothing to restrain the movement of the rocker pedestals along the rocker shafts. Pay particular attention to how this assembly is arranged — it is easier to ensure that it does not accidentally fall apart! Also mark the fitted position of the camshaft bearing half clamps (photo).

19 Having removed the rockers, lift the camshaft out of the bearings in the cylinder head. Extract the camshaft from the sprocket. Carefully, keeping tension on the timing chain, remove the sprocket, and, using wire, tie the chain to the chain guide strip.

20 Undo the single bolt which secures the cylinder head to the top of the engine front cover (photo).

21 Carefully lift the cylinder head off the cylinder block. The weight of the head is not unreasonable for one person to manage, but in view of the posture to be adopted to reach the cylinder head, it would be as well to have assistance for this task.

22 If the crankshaft head appears to be stuck to the engine block, **do not** try to prise it off with a screwdriver, but tap the cylinder head firmly with a plastic or wooden-headed hammer. The mild shocks should break any bond between the gasket, the head and the engine block.

24.2A Using a spring compressor to remove the split collets

22 Camshaft and cylinder head –removal and refitting (engine out of the vehicle)

1   Remove the ancillary components, and proceed as directed in Section 21. **Note**: *To obtain more leverage when removing the cylinder head, the manifolds can be left on the head until after its removal from the block.*

2   Refitting is the reverse of this procedure (see Section 40).

23 Rocker arm assembly – dismantling and reassembly

1   The rocker arm assembly will have been removed by necessity before the removal of the cylinder head from the engine, and the components may now be separated, paying particular attention to their position and orientation.

2   There are no pins or fasteners which hold the rocker assembly together, and once free from the cylinder head it can very easily be dismantled. Be careful that it does not catch you unawares and fall apart accidentally.

3   If necessary, unscrew the camshaft shim plate from the front shaft bearing half clamp.

4   All components should now be carefully laid out ready for inspection.

5   Assemble the component parts to the rocker shafts in the reverse order of removal. Pay particular attention to the oil hole in the centre bearing half clamp; this must face to the inlet side of the rocker shaft (Fig. 1.25).

24.2B Lifting off the upper spring seat

24 Cylinder head assembly – dismantling and reassembly

1   Remove the cylinder head to a clean bench and stand it on wooden blocks.

2   The valves can be removed from the cylinder head by the following

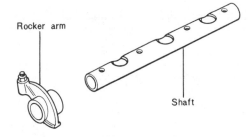

Fig. 1.24 A rocker arm and shaft (Sec 23)

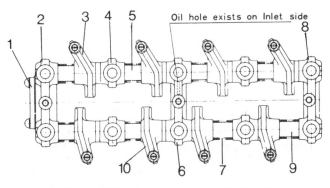

Fig. 1.25 The rocker shaft assembly (Sec 23)

1 The shim plate	6 Centre bearing cap
2 Front bearing cap	7 Spring
3 Rocker arm (inlet)	8 Rear bearing cap
4 Supporter	9 Rocker shaft (exhaust)
5 Rocker arm shaft (inlet)	10 Rocker arm (exhaust)

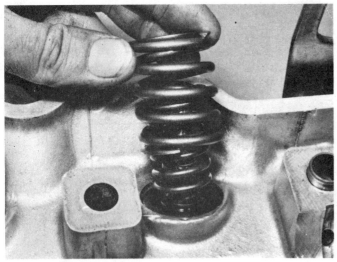
24.2C Removing the valve springs

24.2D Removing a valve

method. Using a valve spring compressor, compress each set of valve springs in turn, until the valve spring split collets (locks) near the top of the valve stem, can be removed. Release the compressor and then remove the upper spring seat, springs, lower spring seat (if applicable) and finally the valve itself. **Note:** *If, when the spring compressor is screwed down the spring seat refuses to budge and expose the collets, do not continue to screw the compressor down as there is a likelihood of damaging it.* Gently tap the top of the tool directly over the collets with a light hammer; this should free the seat. Hold the compressor to prevent it from kicking aside when the seat is released (photos).

3   It is essential that the valves and their associated components are kept to their respective places in the cylinder head. Therefore, if they are being kept and used again, place them in a piece of card with slots cut in it, marked 1 to 4 exhaust and 1 to 4 inlet. Keep the valve springs, seats and collets in order as well.

4   Finally remove the spark plugs.

5   Having dismantled the cylinder head to this stage, the next step is inspection, as described in Sections 32 to 36.

6   When reassembling the cylinder head, the procedure is basically the reverse of that used when dismantling. Ensure that the valves are refitted in their original positions, if they are being re-used; if the valves have been reground to the seats these must also be refitted in their correct positions. Lubricate each valve stem with a little engine oil as it is being refitted to prevent any chance of possible binding in the valve guides.

25 Cylinder block – dismantling (engine out of the vehicle)

1   With the engine removed from the vehicle, and the cylinder head removed from the block, as described in Section 21, prepare to dismantle the engine as described in the following Sections. Since many of the operations can be carried out with the engine fitted, cross-references are made to later Sections for full details of many of the steps, in order to prevent repetition of procedures.

2   On manual transmission models, remove the clutch, as described in Chapter 5.

3   Remove the retaining bolts, and take off the sump. Take great care that the sump is not distorted or the sealing faces damaged, particularly if it is stubborn to remove. For further information see Section 12.

4   Remove the crankshaft pulley, and the engine front cover. With the front cover removed, it is now time to remove the oil pump and timing chain. For details of these operations refer to Section 10. When extracting the crankshaft sprockets and spacers, if the Woodruff key is loose remove it, and stow it in a safe place. If the crankshaft is to be renewed or machined, the key must be removed regardless.

5   Remove the flywheel attaching bolts and carefully ease the flywheel off.

6   Loosen the connecting rod (big-end) bearing caps on each connecting rod in turn. Remove one cap, and its associated connecting rod and piston, one at a time, ensuring that the relative position of the cap, bearing and connecting rod is not lost. If there are no identification markings (cylinder numbers) on the caps and connecting rods, mark them now using a centre punch. For further information on removal of the connecting rods and bearings, refer to Section 14.

7   Loosen the crankshaft main bearing caps then remove them one at a time. If there are no identification markings as to their correct fitted positions, mark them now using a centre punch. The crankshaft can now be lifted out, followed by the bearing shells and thrust washers. Ensure that the bearing shells are suitably marked to ensure that they can be refitted in the correct positions (if they are to be used again).

8   If any machining of the block is required, or an exchange block is going to be obtained, remove any remaining items such as dowels, coolant drain plugs, oil separator, lifting eyes, etc.

26 Engine examination and renovation – general

1   With the engine stripped and all parts thoroughly cleaned, every component should be examined for wear. The items listed in the Sections following should receive particular attention, and where necessary be renewed or renovated.

2   So many measurements of engine components require accuracies down to tenths of a thousandth of an inch. It is advisable therefore to either check your micrometer against a standard gauge occasionally to ensure that the instrument zero is set correctly, or use the micrometer as a comparative instrument. This last method, however, necessitates that a comprehensive set of slip and bore gauges is available.

3   Before any inspection of parts is commenced, clean them using petrol (gasoline) or paraffin (kerosene), then blow them dry with compressed air.

27 Crankshaft – examination and renovation

1   Examine the crankpin and main journal surfaces for signs of scoring or scratches, and check the ovality and taper of each journal in turn. If the bearing surface dimensions do not fall within the tolerance ranges given in the Specifications at the beginning of this Chapter, the crankpins and/or main journals will have to be reground. Big-end bearing and crankpin wear is accompanied by distinct metallic knocking, particularly noticeable when the engine is pulled from low revs. Main bearing and main journal wear is accompanied by severe engine vibration/rumble, getting progressively worse as engine revs increase. If the crankshaft is reground the workshop should supply the necessary undersize bearing shells.

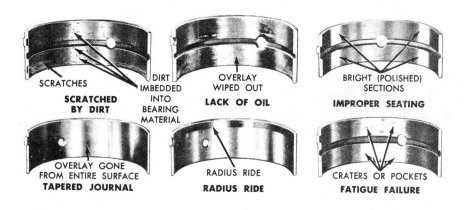

Fig. 1.26 Typical bearing condition (Sec 28)

2   Ensure that the crankshaft oilways are unobstructed.

28 Big-end and main bearing shells – examination and renovation

1   Big-end failure is accompanied by a noisy knocking from the crankcase and a slight drop in oil pressure. Main bearing failure is accompanied by a drop in oil pressure and by vibration which can be quite severe as the engine speed rises and falls (see previous Section).

2   Bearings which have not broken up, but are badly worn will also give rise to low oil pressure and some vibration. Inspect the big-ends, main bearings and thrust washers for signs of general wear, scoring, pitting and scratches. The bearings should be matt grey in colour. Renew the bearings if they are not in this condition or if there is any sign of scoring or pitting. **You are strongly advised to renew the bearings – regardless of their condition. Refitting used bearings is a false economy.**

3   The undersizes available are designed to correspond with crankshaft regrind sizes, ie, 0.020 in (0.50 mm) bearings are correct for a crankshaft reground 0.020 in (0.50 mm) undersize. The bearings are in fact, slightly more than the stated undersize, as running clearances have been allowed for during their manufacture.

29 Cylinder bores – examination and renovation

1   The cylinder bores must be examined for taper, ovality, scoring and scratches. Start by carefully examining the top of the cylinder bores. If they are at all worn a very slight ridge will be found on the thrust side. This marks the top of the piston travel. The owner will have a good indication of the bore wear prior to dismantling the engine, or removing the cylinder head. Excessive oil consumption accompanied by blue smoke from the exhaust is a sure sign of worn cylinder bores and piston rings.

2   Measure the bore diameter just under the ridge with a micrometer or cylinder bore gauge and compare it with the diameter at the bottom of the bore, which is not subject to wear. If the difference between the two measurements is more than 0.006 in (0.15 mm) then it will be necessary to fit oversize pistons and rings. If no micrometer or cylinder bore gauge is available, remove the rings from a piston and place the piston in each bore in turn about three quarters of an inch below the top of the bore. If a 0.010 inch (0.25 mm) feeler gauge can be slid between the piston and the cylinder wall on the thrust side of the bore then remedial action must be taken. Oversize pistons are available as stated in the Specifications.

3   These are accurately machined to just below the rebore measurements so as to provide correct running clearances in bores bored out to the exact oversize dimensions.

4   If the bores are slightly worn but not so badly worn as to justify reboring them, special oil control rings can be fitted to the existing

Fig. 1.27 Measuring the bore diameter with a cylinder bore gauge (Sec 29)

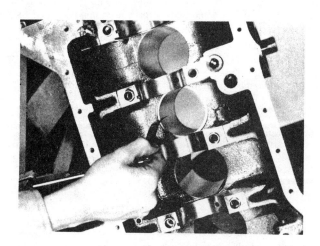

Fig. 1.28 Measuring piston ring gap (Sec 30)

Fig. 1.29 Checking the piston ring side clearance (Sec 30)

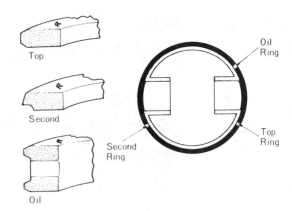

Fig. 1.30 The correct radial positions of the piston ring gaps (Sec 30)

pistons, which will restore compression and stop the engine burning oil. Several different types are available and the manufacturer's instructions concerning their fitting must be followed closely.

30 Pistons and piston rings – examination and renovation

1   If the old pistons are to be refitted, carefully remove the piston rings and thoroughly clean them. Take particular care to clean out the piston ring grooves. At the same time do not scratch the aluminium. If new rings are to be fitted to the old pistons, then the top ring should be stepped to clear the ridge left above the previous top ring. If a normal but oversize new ring is fitted, it will hit the ridge and break, because the new ring will not have worn in the same way as the old, (which will have worn in unison with the ridge).
2   Before fitting the rings on the pistons, each should be inserted approximately 3 inches (8 cm) down the cylinder bore and the gap measured with a feeler gauge as shown in Fig. 1.28. This should be as detailed in the Specifications at the beginning of this Chapter. It is essential that the gap is measured at the bottom of the ring travel. If it is measured at the top of a worn bore and gives a perfect fit it could easily seize at the bottom. If the ring gap is too small rub down the ends of the ring with a fine file, until the gap, when fitted, is correct. To keep the rings square in the bore for measurement, line each up in turn with an old piston down about 3 inches (8cm). Remove the piston and measure the piston ring gap.
3   Check the side clearance of the piston rings using a feeler gauge as shown in Fig. 1.29. Compare the clearance with that given in the Specifications, and renew the rings or pistons as appropriate, if they are outside the specified limits.
4   When fitting new piston and rings to a rebored engine the ring gap can be measured at the top of the bore as the bore will now not taper. It is not necessary to measure the side clearance in the piston ring groove with rings fitted, as the groove dimensions are accurately machined during manufacture. When fitting new oil control rings to old pistons it may be necessary to have the groove widened by machining to accept the new wider rings. In this instance the manufacturer's representative will make this quite clear and will supply the address to which the pistons must be sent for machining.
5   When fitting the rings on the pistons, make sure that the ring gaps are properly spaced around the piston circumference. They should be positioned about 120 degrees apart so that the gap is not located on the thrust side or the piston pin side (Fig. 1.30). Apply a little engine oil to the rings and make sure that they are not binding in the piston grooves.

31 Connecting rods – examination

1   The connecting rod is a fairly robust item, but its loading is very

harsh and therefore the examination required is directed at finding deep scratches or notches. Generally if the engine is being overhauled, simply because it has worn out, then it may be found that the connecting rods are unserviceable.
2   However, if overhaul was necessitated by a failure of a particular component - piston or valve, then the rod should be wiped clean and its surface thoroughly inspected for damage. If there are sizeable dents or if a crack is found, the rod should be renewed.

32 Camshaft – examination and renovation

1   Inspect each half clamp bearing for scoring, pitting and wear marks. If any such marks are evident in the half clamps or the cylinder head bearing areas, a new cylinder head and half clamps will have to be purchased.
2   Examine the camshaft lobes for scoring and signs of abnormal wear. Slight dressing of the cam lobes is permissible using an oilstone - provided that the overall cam lobe lift (elevation) is not reduced by more than 0.008 in (0.20 mm). This can be checked by running the camshaft on lathe centres or V-blocks (or with the camshaft fitted on the cylinder head). If the wear pattern is unsatisfactory, the camshaft must be renewed.

33 Rocker arm and shaft assembly – inspection and renovation

1   Ensure that all the oilways are unobstructed.
2   Inspect the ends of the rocker arms for wear, indentations and scoring. Light wear is permissible, but if there is evidence of grooving, the rocker arms must be replaced. They should not be refaced by grinding.
3   Check the rocker arms for wear of the bushes. This can be done by holding the rocker arm in place on the shaft and, by gripping the rocker arm tip, noting if there is any lateral rocker arm shake. If any shake is present, and the arm is very loose on the shaft, renew the arm.
4   Check each shaft for straightness by rolling it on a flat surface, such as a piece of plate glass. The surface of the shaft should be free from any wear ridges caused by the rocker arms. If any wear is present renew the shaft. Wear will normally only take place if the oilways have become obstructed.

34 Valves and valve seats – examination and renovation

1   Examine the heads of the valves for pitting and burning, especially the heads of the exhaust valves. The valve seatings should be examined at the same time. If the pitting on the valves and seats is very slight the marks can be removed by grinding the seats and valves together with coarse, and then fine, valve grinding paste. Where bad

35.2 A valve guide seal

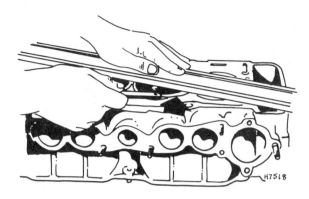

Fig. 1.31 Checking a cylinder head for distortion (Sec 36)

pitting has occurred to the valve seats it will be necessary to recut them and to fit new valves. If the valve seats are so worn that they cannot be recut, then it will be necessary to fit valve seat inserts. These latter two jobs should be entrusted to a vehicle main dealer or engineering works. In practice it is very seldom that the seats are so badly worn that they require renewal. Normally, it is the valve that is too badly worn, and the owner can easily purchase a new set of valves and match them to the seats by valve grinding.

2   Valve grinding is carried out as follows: Place the cylinder head upside down on a bench with a block of wood at each end to give clearance for the valve stems. Alternatively, place the head at 45 degrees to a wall with the combustion chambers away from the wall.

3   Smear a trace of coarse carborundum paste on the seat face and apply a suction grinder tool to the valve heads. With a semi-rotary action, grind the valve head to its seat, lifting the valve occasionally to redistribute the grinding paste. When a dull matt surface finish is produced on both the valve seat and the valve, then wipe off the paste and repeat the process with fine carborundum paste, lifting and turning the valve to redistribute the paste as before. A light spring placed under the valve head will greatly ease this operation. When a smooth unbroken ring of light grey matt finish is produced, on both valve and the valve seat faces, the grinding operation is complete.

4   Scrape away all carbon from the valve head and the valve stem. Carefully clean away every trace of grinding compound, taking great care to leave none in the ports or in the valve guides. Clean the valves and valve seats with a paraffin (kerosene) soaked rag, then with a clean rag. Finally, if an air line is available, blow the valve, valve guides and valve ports clean.

35 Valve guides – examination and renovation

1   If it is found that the valve guides require renewal, where the cylinder head is in an otherwise serviceable condition (or can easily be made serviceable), it is not recommended that any attempt is made to remove them as they are an interference fit. This is a job for a specialist and should either be carried out by a vehicle main dealer or by an automotive engineer with the necessary equipment.

2   If it is found necessary to renew the seals on the ends of the valve guides, these can be pulled off and replacements pressed on. A tool is available from the vehicle manufacturers for this purpose (photo).

36 Cylinder head – decarbonisation and inspection

1   It is very unlikely that with modern fuels and oils, decarbonisation will be necessary at anything shorter then 60 000 mile (100 000 km) intervals.

2   This operation can be carried out with the engine either in or out of the vehicle, but the procedure given here is described as though the engine is fitted. The spark plugs can remain fitted as this helps to prevent carbon deposits from entering the screw threads.

3   With the cylinder head off, carefully remove with a wire brush and blunt scraper all traces of carbon deposits from the combustion spaces and ports. The valve stems and valve guides should be also freed from any carbon deposits. Wash the combustion spaces and ports down with petrol and scrape the cylinder head surface of any foreign matter with the side of a steel rule or a similar article. Take care not to scratch the surface.

4   Clean the pistons and top of the cylinder bores. If the pistons are still in the cylinder bores then it is essential that great care is taken to ensure that no carbon gets into the cylinder bores as this could scratch the cylinder walls or cause damage to the piston and rings. To ensure that this does not happen, first turn the crankshaft so that two of the pistons are at the top of the bores. Place a clean lint-free cloth into the other two bores or seal them off with paper and masking tape. The waterways and oilways should also be covered with a small piece of masking tape to prevent particles of carbon entering the cooling system and damaging the water pump, or entering the lubrication system and damaging the oil pump or bearing surfaces.

5   There are two schools of thought as to how much carbon ought to be removed from the piston crown. One is that a ring of carbon should be left around the edge of the piston and on the cylinder bore wall as an aid to keeping oil consumption low. Although this is probably true for older engines with heavily worn bores, on newer engines it is best to remove all traces of carbon during decarbonisation.

6   If all traces of carbon are to be removed, press a little grease into the gap between the cylinder walls and the two pistons which are to be worked on. With a blunt scraper carefully scrape away the carbon from the piston crown, taking care not to scratch the aluminium. Also scrape away the carbon from the surrounding lip of the cylinder wall. When all carbon has been removed, scrape away the grease which will now be contaminated with carbon particles, taking care not to press any into the bores. To assist prevention of carbon build up the piston crown can be polished with a metal polish, but take care that none seeps between the piston and the cylinder wall or the rings may seize in their grooves. Remove the cloth or masking tape from the other two cylinders and turn the crankshaft so that the two pistons which were at the bottom are now at the top. Place a lint-free cloth into the other two bores or seal them off with paper and masking tape. Do not forget the waterways and oilways as well. Proceed as previously described.

7   If a ring of carbon is going to be left around the piston then this can be helped by inserting an old piston ring into the top of the bore to rest on the piston and ensure that carbon is not accidentally removed. Check that there are no particles of carbon in the cylinder bores. Decarbonisation is now complete.

8   Using a straight-edge and feeler gauge, check the cylinder head for distortion and corrosion. The maximum permissible distortion is given in the Specifications. Renew any cylinder head which is outside the acceptable limits.

37 Miscellaneous items – inspection and renovation

Timing chain and sprockets

1   Renew a timing chain and sprockets if there is obvious wear or damage, or in the case of major engine overhaul.

Distributor drive gear

2   Renew a drive gear which is damaged or has wear marks on the teeth.

Manifolds

3   Manifolds are dealt with in Chapter 3, but in the case of engine overhaul they should be examined for cracks, and cleaned of any carbon deposits and traces of gasket material.

Chain tensioner and guide strip

4   Wear on the chain tensioner slipper head or guide strip will entail replacement items being fitted as spare parts are not supplied.

Sump

5   The oil sump should be cleaned, and traces of gasket material removed. Check for cracks and damaged sealing faces. If damage cannot be rectified by localized dressing or repair, a replacement item should be obtained.

Crankshaft pulley

6   Clean the oil seal contact surface to remove sludge, corrosion or varnish deposits. Light dressing with crocus cloth is permitted, but do not polish the surface as this can produce an inferior seal and lead to premature seal wear.

Cylinder block core plugs

7   Leaking core plugs can be removed by drilling a hole in the centre and then prying them out. Before fitting a new core plug, ensure that the bore is clean and dry. Smear a gasket sealing compound around the bore before fitting the plug, with the flanged edge inward. Do not strike the crowned portion of the plug as this will damage it, but use a proper fitting tool.

Spark plugs

8   Spark plugs should normally be renewed at the time of a major engine overhaul, but provided that their condition is satisfactory, cleaning will suffice. Full details of spark plugs are given in Chapter 4.

38 Engine reassembly – general

1   To ensure maximum life with minimum trouble from a rebuilt engine, not only must every part be correctly reassembled, but everything must be spotlessly clean, all the oilways must be clear, locking washers and spring washers must always be fitted where indicated and all bearings and other working surfaces must be thoroughly lubricated during reassembly. Before reassembly begins, renew any bolts or studs whose threads are in any way damaged; whenever possible use new spring washers and cotter pins.

2   Apart from your normal tools, a supply of lint-free cloths, an oil can filled with engine oil (an empty washing-up fluid plastic bottle thoroughly clean and washed out will invariably do just as well), a supply of new spring washers, a set of new gaskets and a torque wrench should be collected together.

3   The order of reassembly for the engine is as follows:

a) Refit the crankshaft and main bearings.
b) Refit the pistons, connecting rods and bearings.
c) Refit the oil pump.
d) Refit the timing chain an- oil pump drive chain.
e) Refit the engine front cover and oil sump.
f) Refit the cylinder head and valves.
g) Reassemble the rocker arm assembly.
h) Refit the timing chain tensioner.
j) Refit the water pump and crankcase pulley.
k) Adjust the valve clearances.
l) Refit the flywheel and clutch.
m)Refit the remaining ancillaries.

4   Since many of the reassembly operations can be carried out on an already fitted engine as described previously in this Chapter, only an outline procedure, together with any necessary precautions, is now given to avoid repetition.

39 Cylinder block – reassembly (engine out of the vehicle)

1   Ensure that the crankcase is thoroughly clean and that all oilways are clear. A thin twist drill is useful for cleaning the oilways, or if possible they may be blown out with compressed air. Treat the crankshaft in the same fashion, and then inject engine oil into the oilways.

2   Never re-use old bearing shells; wipe the shell seats in the crankcase clean and then fit the upper halves of the new main bearing shells into their seats.

3   Fit the main bearing shells in the cylinder block and lubricate them with engine oil (photos).

4   Fit the crankshaft using the original thrust washers in the cylinder block with the oil grooved surface towards the crankshaft thrust side (photos).

5   Fit a new oil seal on the rear end of the crankshaft after applying engine oil to the seal lip (photo).

6   Insert the side seals into the grooves in the rear main bearing cap (see Fig. 1.12).

7   Position the bearing shells and original thrust washers in the caps, lubricate the crankcase journals, then fit and torque tighten the bearing caps. Take great care to ensure that they are fitted in their original positions (photos).

8   Using a wooden block and hammer, take up the crankcase endplay in one direction, then measure the endplay between the crankshaft thrust face and thrust washer using a feeler gauge. If necessary, remove the crankshaft and fit replacement thrust washers to bring the endplay within the specified limits (Fig. 1.32).

9   With the crankshaft fitted, position the bearing shells in the connecting rods and bearing caps, lubricate the running surfaces with engine oil, then fit them in their correct positions. Ensure that they are in the correct relationship with the cylinder block (front to rear). A piston ring compressor will be required to retain the rings while the piston is being fitted. The piston can then be positioned into the lubricated bore by pressing down on the crown with a hammer handle or similar, until the connecting rod mates with the crankshaft journal (crankpin). For further information see Section 15 (photos).

10   Rotate the crankshaft to ensure that there is no binding or undue friction (high spots).

11   Fit the oil pump shaft and inner rotor and the outer rotor to their cylinder block bore. Assemble the oil pump cover and secure it in position with its four bolts (photos).

12   If applicable, fit the oil pump feed pipe and strainer to the oil pump, and secure the feed pipe to the cylinder block (photo).

13   Where applicable, fit the spacer to the crankshaft.

14   Fit the timing chain and crankshaft sprocket to the crankshaft, aligning the crankshaft sprocket and crankshaft keyways with the key in position. Refer to Figs. 1.7 and 1.8 for the position of the bright links (photos).

15   Fit the spacer (and where applicable, the rubber ring), that locates between the timing chain and the oil pump drive sprockets, onto the crankshaft (photo).

16   Assemble the crankshaft oil pump drive sprocket, and the oil pump drive shaft sprocket, to the drive chain. Assemble the crankshaft sprocket to the crankshaft, and the oil pump drive sprocket to the oil pump shaft, aligning the keyways to both the crankshaft and the oil pump drive shaft. Assemble the locking tab and nut to the oil pump drive sprocket and tighten the nut to the specified torque. Bend over the locking tab (photos).

17   If the chain guide strip and the vibration damper were removed, now is the time to refit them. The retaining screws for the chain guide strip should be left loose at this stage. Using locking wire, secure the timing chain to the tops of the chain guide strip and the vibration damper. This will ensure that when the cylinder block is inverted, the timing chain will be unable to drop from the crankshaft sprocket (photos).

18   Fit the second spacer to the face of the oil pump drive sprocket on the crankshaft, followed by the flat oil thrower (or where applicable,

39.3 Fitting the main bearing shells in the block

39.4A Don't forget the thrust washers ...

39.4B ... before fitting the crankshaft

39.5 Fit the oil seal at the rear end of the crankshaft

39.7A Lubricate the journals

39.7B Tightening a main bearing cap

39.9A Fit the bearing shells ...

39.9B ... push the piston down the bore ...

39.9C ... fit the bearing cap ...

39.9D ... and tighten it

39.11A Fit the oil pump ...

39.11B ... and the oil pump cover

39.12 Securing the feed pipe to the cylinder block

39.14A Fit the Woodruff key …

39.14B … and, with the dot and bright link aligned, assemble the crankshaft sprocket and chain

39.15 Don't forget the spacer

39.16A The oil pump sprockets and chain in position

39.16B Bend over the locking tab

39.17A The vibration damper in position

39.17B Fit the chain guide, but do not tighten it yet

39.17C Secure the chain to the tops of the chain guide strip and the vibration damper

39.18A Fit a new oil seal to the front cover …

39.18B … and fit it to the engine block

39.21 Fit the cylinder head

39.22 Position the camshaft into the sprocket

39.24A Fit the camshaft shim plate ...

39.24B ... and tighten through the holes in the sprocket

39.25 Fitting the timing chain tensioner

39.28A Assemble the spacer, distributor drive gear, locking tab washer and nut, to the camshaft ...

39.28B ... then torque tighten the nut

the dished oil thrower, concave side outwards). Fit a new oil seal in the engine front cover, and lubricate the lips with engine oil. Do not forget the front cover gasket which can be retained in position, using a non-setting gasket sealant. Fit the engine front cover and secure it in position (photos).

19 Position a new gasket on the sump using a non-setting gasket sealant. Fit the sump, torque-tightening the bolts in a crosswise order.

20 Invert the engine.

21 Fit the cylinder head which has already been assembled as described previously, using a new cylinder head gasket (photo).

22 If the camshaft sprocket is not in the timing chain, position it as shown in Fig. 1.8, position the camshaft onto the cylinder head and locate it through the camshaft sprocket. Ensure that the camshaft journals and bearing areas are lubricated with engine oil. For further information, see Section 40 (photo).

23 Fit the rocker arm assembly and torque-tighten the bolts in the order shown in Fig. 1.22. Refer to Section 40, for further details of this procedure.

24 Fit the shim plate to the front camshaft half clamp bearing, and retain it in position with the two retaining screws. In order to gain screwdriver access to these screws, rotate the engine a little to line up the holes in the camshaft sprocket with the screws. Using feeler gauges, measure the camshaft endfloat, which must be as specified; oversize shim plates are available to correct endfloat (Fig. 1.33) (photos).

25 The timing chain tensioner may now be fitted. Using locking wire, fully compress the slipper head into the tensioner assembly, and retain it in this position. The locking wire should be left in this position until timing chain adjustment has been carried out (photo).

26 Adjust the timing chain tensioner as described in Section 11. Do not forget to remove the locking wire from the tensioner assembly after satisfactory chain tension has been achieved. Fit a new gasket and fit the inspection cover.

27 Fit the crankshaft pulley, and tighten the retaining bolt to the specified torque.

28 Assemble the spacer, distributor drive gear, locking tab washer and nut, to the front end of the camshaft. Tighten the nut to the

Fig. 1.32 Checking the crankshaft endplay (Sec 39)

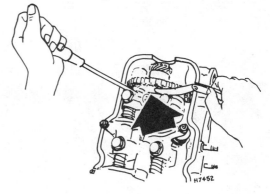

Fig. 1.33 Checking camshaft endfloat (Sec 39)

specified torque and bend over the locking tab (photos).

29 Fit the distributor, aligning the index marks made when it was removed.

30 Fit the water pump (refer to Chapter 2 if necessary).

31 Adjust the valve clearances as described in Section 41. Do not forget the semi-circular oil seals before fitting the rocker arm cover.

32 Fit the flywheel. Refer to Section 8 for further information.

33 Fit the clutch. Refer to Chapter 5 for further information.

34 All that is now required is fitting of the ancillaries, which can be done in the reverse order to removal (see Section 20). Note that it is considered preferable to fit the oil filter cartridge after fitting of the engine, since there is a possibility of damage occurring to it as the engine is lowered into the vehicle engine bay. When fitting the intake and exhaust manifolds, ensure that new gaskets are used.

40 Camshaft and cylinder head – refitting (engine in the vehicle)

1 Ensure that the cylinder block and cylinder head mating faces are clean.

2 Position a new cylinder head gasket on the block, then feed the timing chain through the aperture in the cylinder head, as it is positioned on the block. Assemble the camshaft sprocket to the timing chain, ensuring that the bright link aligns with the dot on the sprocket (see Fig. 1.7).

3 Assemble the camshaft to the cylinder head, and aligning the Woodruff key in the camshaft with the keyway in the sprocket, locate the camshaft into the sprocket. Ensure that the camshaft journals and bearing areas are lubricated with engine oil.

4 The assembled rocker arm assembly (see Section 23) can now be positioned on the cylinder head. Refer to Fig. 1.22 for the bolt tightening sequence and lightly tighten each bolt. Now move each rocker arm supporter to offset each rocker arm 0.04 in (1 mm) from the valve centre (Fig. 1.34), then fully tighten the bolts to the specified cold (initial) torque.

5 Fit the shim plate to the front camshaft half clamp bearing, and retain it in position with two screws. In order to gain screwdriver access to these screws, rotate the engine a little to line up the holes in the camshaft sprocket with the screws. Using feeler gauges, measure the camshaft endfloat, which must be as specified. Oversize shim plates are available to correct endfloat (Fig. 1.33).

6 Fit and tighten the single bolt which secures the cylinder head to the engine front cover.

7 Assemble the spacer, distributor drive gear, locking tab washer and nut, to the front end of the camshaft. Tighten the nut to the specified torque and bend over the locking tab.

8 Fit the distributor, aligning the index marks made when it was removed.

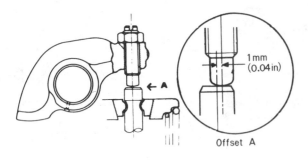

Fig. 1.34 The correct valve rocker offset position (Sec 40)

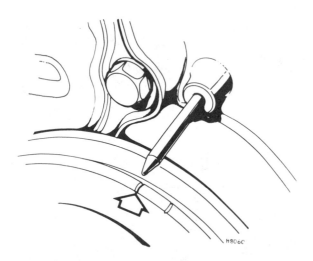

Fig. 1.35 The TDC mark aligned with the pointer

41.6 Adjusting a valve clearance

42.2 Lowering the engine and transmission into position

9   Adjust the timing chain tensioner, as described in Section 11. Fit the inspection cover using the new gasket.

10   The remainder of the fitting procedure is the reverse of the removal procedure, but do not fit the rocker arm cover until valve clearance adjustment has been carried out, as described in Section 41.

11   On completion, fill the cooling system and run the engine at a fast idle until warm. Switch off and re-torque the cylinder head bolts to the specified final figure.

41 Valve clearance – adjustment

1   If the cylinder head has been dismantled, it will be necessary to adjust the valve clearances initially with the engine cold. The figures given in the Specifications are suitable for this purpose, provided that the engine is warmed up properly after refitting and the clearances rechecked. Care must be taken during this warming up period to ensure that excessive engine speeds and temperatures are not attained.

2   Having warmed up the engine, shut off the engine and remove the rocker arm cover.

3   Refer to the cylinder head torque-tightening sequence (Fig. 1.22). Loosen No. 1 bolt about $\frac{1}{4}$-turn and retighten to the specified torque. Repeat this procedure for the other bolts in the correct order.

4   Rotate the crankshaft so that No. 1 piston is at top dead centre (TDC) on the compression stroke. This position can be checked by aligning the TDC mark on the pulley with the timing pointer (Fig. 1.35) when the distributor rotor is pointing to No. 1 spark plug terminal in the distributor cap (or where the valve rockers of No. 1 cylinder can be moved slightly up and down). A three notch crank pulley is fitted to some models. Refer to Chapter 4 for details of the timing marks.

5   Using a feeler gauge between the rocker and the camshaft, or rocker and valve stem, check for the specified clearance.

6   If adjustment is required, loosen the adjusting screw locknut and turn the adjusting screw until the correct clearance is obtained. Tighten the locknut afterwards and recheck the clearance (photo).

7   Repeat the procedure for the remaining cylinders with the piston at TDC on the compression stroke, in the firing order 1-3-4-2.

8   On completion ensure that the semi-circular oil seals are fitted at each end of the rocker arm cover. Using a new rocker cover gasket, fit the rocker cover and tighten the cover retaining bolts to the specified torque.

42 Engine – refitting

1   Whether the engine is to be fitted on its own, or together with the transmission, the procedure is basically the reverse of the removal procedure.

2   Raise the engine (and transmission, where appropriate), using a hoist, and lower it into position in the vehicle engine bay. A jack will be required to support the transmission until the rear support is fitted if the transmission is connected to the engine (photo).

3   Fit the engine and transmission mounts.

4   Where applicable, fit the starter motor and the flywheel housing bolts.

5   Fit the exhaust pipe.

6   Fit the oil filter (where applicable).

7   Connect the electrical and earth wires, and fuel and air hoses to the engine.

8   Fit the carburettor control cables.

9   Fit the alternator (if not already fitted), and where applicable the air pump. Fit the drivebelt(s).

10   Connect the heater hoses, (and where applicable, the air pump hoses).

11   Fit the fan, radiator and hoses.

12   If applicable, connect the gearshift linkage, speedometer drive cable and propeller shaft.

13   Top up the engine oil and coolant (and transmission oil, where applicable).

14   Fit the air cleaner.

15   Fit the engine compartment bonnet.

16   Fit the battery connections.

43 Engine – initial start-up after overhaul

1   Make sure that the battery is fully charged and that all lubricants, coolant and fuel are replenished.

2   If the fuel system has been dismantled it will require several revolutions of the engine on the starter motor to pump the petrol up to the carburettor.

3   As soon as the engine fires and runs, keep it going at a fast idle only, (no faster) and bring it up to the normal working temperature.

4   As the engine warms up there will be odd smells and some smoke from parts getting hot and burning off oil deposits. The signs to look for are leaks of water or oil which will be obvious if serious. Check also the exhaust pipe and manifold connections, as these do not always find their exact gas tight position until the warmth and vibration have acted on them, and it is almost certain that they will need tightening further. This should be done of course, with the engine stopped.

5   When normal running temperature has been reached adjust the engine idling speed as described in Chapter 3, then set the valve clearances as described in Section 41.

6   Stop the engine and wait a few minutes to see if any lubricant or coolant is dripping out when the engine is stationary.

7   Road test the car to check that the timing is correct and that the engine is giving the necessary smoothness and power. Do not race the engine - if new bearings and/or pistons have been fitted it should be treated as a new engine and run in at a reduced speed for the first 300 miles (500 km).

44 Fault diagnosis – engine

Symptom	Reasons	Remedy
Engine fails to turn over when starter control operated		
No current at starter motor	Flat or defective battery	Change or renew battery. Push start car
	Loose battery leads	Tighten both terminals and earth ends of earth leads
	Defective starter solenoid or switch or broken wiring	Run a wire direct from the battery to the starter motor or bypass the solenoid
	Engine earth strap disconnected	Check and retighten strap
Current at starter motor	Jammed starter motor drive pinion	Place car in gear and rock to and fro
	Defective starter motor	Remove and recondition
Engine turns over but will not start		
No spark at spark plug	Ignition damp or wet	Wipe dry the distributor cap and ignition leads
	Ignition leads to spark plugs loose	Check and tighten at both spark plug and distributor cap ends
	Shorted or disconnected low tension leads	Check the wiring on the coil and to the distributor
	Dirty, incorrectly set, or pitted contact	Clean, file smooth, and adjust

	breaker points	
	Faulty condenser	Check contact breaker points for arcing, remove and fit new
	Defective ignition switch	Bypass switch with wire
	Ignition leads connected wrong way round	Remove and refit leads to spark plugs in correct order
	Faulty coil	Remove and fit new coil
	Contact breaker point spring earthed or broken	Check spring is not touching metal part of distributor. Check insulator washers are correctly placed. Renew points if the spring is broken
No fuel at carburettor float chamber or at jets	No petrol in petrol tank	Refill tank!
	Vapour lock in fuel line (in hot conditions or at high altitude)	Blow into petrol tank, allow engine to cool, or apply a cold wet rag to the fuel line
	Blocked float chamber needle valve	Remove, clean and refit
	Fuel pump filter blocked	Remove, clean and refit
	Choked or blocked carburettor jets	Dismantle, and clean
	Faulty fuel pump	Remove, overhaul and refit

Engine stalls and will not start

Excess of petrol in cylinder or carburettor flooding	Too much choke allowing too rich a mixture to wet plugs	Remove and dry spark plugs or with wide open throttle, push start the car
	Float damaged or leaking or needle not seating	Remove, examine, clean and refit float and needle valve as necessary
	Float level incorrectly adjusted	Remove and adjust correctly
No spark at spark plug	Ignition failure – sudden	Check over low and high tension circuits for breaks in wiring
	Ignition failure – misfiring precludes total stoppage	Check contact breaker points, clean and adjust Renew condenser if faulty
	Ignition failure – in severe rain or after traversing water splash	Dry out ignition leads and distributor cap
No fuel at jets	No petrol in petrol tank	Refill tank!
	Sudden obstruction in carburettor	Check jets, filter, and needle valve in float chamber for blockage
	Water in fuel system	Drain tank and blow out fuel lines

Engine misfires or idles unevenly

Intermittent spark at spark plug	Ignition leads loose	Check and tighten as necessary at spark plug and distributor cap ends
	Battery leads loose on terminals	Check and tighten terminal leads
	Battery earth strap loose on body attachment point	Check and tighten earth lead to body attachment point
	Engine earth lead loose	Tighten lead
	Low tension leads to coil loose	Check and tighten leads if found loose
	Dirty, or incorrectly gapped plugs	Remove, clean and regap
	Dirty, incorrectly set, or pitted contact breaker points	Clean, file smooth and adjust
	Tracking across inside of distributor cover	Remove and fit new cover
	Ignition too retarded	Check and adjust ignition timing
	Faulty coil	Remove and fit new coil
Fuel shortage at engine	Mixture too weak	Check jets, float chamber needle valve, and filter for obstruction. Clean as necessary
	Carburettor incorrectly adjusted	Remove and overhaul carburettor
	Air leak in carburettor	Remove and overhaul carburettor
	Air leak at inlet manifold to cylinder head, or inlet manifold to carburettor	Test by pouring oil along joints. Bubbles indicate leak. Renew manifold gasket as appropriate

Lack of power and poor compression

Mechanical wear	Incorrect valve clearances	Adjust rocker arms to take up wear
	Burnt out exhaust valves	Remove cylinder head and renew defective valves
	Sticking or leaking valves	Remove cylinder head, clean, check and renew valves as necessary
	Weak or broken valve springs	Check and renew as necessary
	Worn valve guides or stems	Renew valve guides and valves
	Worn pistons and piston rings	Dismantle engine, renew pistons and rings
Fuel/air mixture leaking from cylinder	Burnt out exhaust valves	Remove cylinder head, renew defective valves
	Sticking or leaking valves	Remove cylinder head, clean, check, and renew valves as necessary
	Worn valve guides and stems	Remove cylinder head and renew valves and valve guides
	Weak or broken valve springs	Remove cylinder head, renew defective springs
	Blown cylinder head gasket (accompanied	Remove cylinder head and fit new gasket

	by increase in noise)	
	Worn pistons and piston rings	Dismantle engine, renew pistons and rings
	Worn or scored cylinder bores	Dismantle engine, rebore, renew pistons and rings
Incorrect adjustments	Ignition timing wrongly set. Too advanced or retarded	Check and reset ignition timing
	Contact breaker points incorrectly gapped	Check and reset contact breaker points
	Incorrect valve clearances	Check and reset rocker arm to valve stem gap
	Incorrectly set spark plugs	Remove, clean and regap
	Carburation too rich or too weak	Tune carburettor for optimum performance
Carburation and ignition faults	Dirty contact breaker points	Remove, clean and refit
	Fuel filters blocked causing poor top end performance through fuel starvation	Dismantle, inspect, clean and refit all fuel filters
	Distributor automatic balance weights or vacuum advance and retard mechanisms not functioning correctly	Overhaul distributor
	Faulty fuel pump giving top end fuel starvation	Remove, overhaul, or fit exchange reconditioned fuel pump
Excessive oil consumption	Excessively worn valve stems and valve guides	Remove cylinder head and fit new valves and valve guides
	Worn piston rings	Fit oil control rings to existing pistons or purchase new pistons
	Worn pistons and cylinder bores	Fit new pistons and rings, rebore cylinders
	Excessive piston ring gap allowing blow-up	Fit new piston rings and set gap correctly
	Piston oil return holes choked	Decarbonise engine and pistons
Oil being lost due to leaks	Leaking oil filter gasket	Inspect and fit new gasket as necessary
	Leaking rocker cover gasket	Inspect and fit new gasket as necessary
	Leaking timing gear cover gasket	Inspect and fit new gasket as necessary
	Leaking sump gasket	Inspect and fit new gasket as necessary
	Loose sump plug	Tighten and fit new gasket as necessary
Unusual noises from engine Excessive clearances due to mechanical wear	Worn valve gear (noisy tapping from rocker box)	Inspect and renew rocker shaft, rocker arms, and ball pins as necessary
	Worn big-end bearing (regular heavy knocking)	Drop sump, if bearings broken up clean out oil pump and oilways, fit new bearings. If bearings not broken but worn, renew bearings
	Worn timing chain and gears (rattling from front of engine)	Remove timing cover, fit new timing wheels and timing chain
	Worn main bearings (rumbling and vibration)	Drop sump, remove crankshaft, if bearings worn but not broken up, renew. If broken up strip oil pump and clean out oilways, renew bearings
	Worn crankshaft (knocking, rumbling and vibration)	Regrind crankshaft, fit new main and big-end bearings

1

Chapter 2 Cooling system

Contents

Specifications

Type . . . . Pressurized and sealed, with centrifugal pump, thermostat, fan and radiator

Radiator
Type . . . . Corrugated fin
Pressure cap setting . . . . 13 lbf/in^2 (0·91 kgf/cm^2)

Thermostat
Type . . . . Wax pellet
Starts to open . . . . 82°C (180°F)
Fully open . . . . 95°C (203°F)
Lift height (nominal) . . . . 0·315 in (8 mm)

Water pump type . . . . Impeller (centrifugal)

Fan/alternator belt tension
New belt . . . . 0·31 to 0·39 in (8 to 10 mm)
Used belt . . . . 0·43 to 0·51 in (11 to 13 mm)

Coolant capacity
With heater . . . . 9·6 Imp pints/11·6 US pints/5·5 litres
Without heater . . . . 8·8 Imp pints/10·6 US pints/5·0 litres

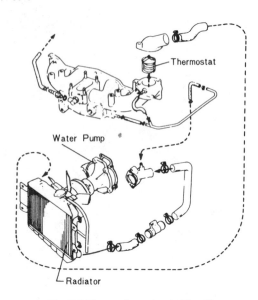

Fig. 2.1 The coolant circuit (Sec 1)

1 General description

The cooling system comprises a radiator, fan, thermostat, pressure cap and centrifugal water pump. The fan and water pump are driven by a V-belt from the crankshaft pulley (this belt also drives the alternator). When the engine is cold, coolant circulates in the engine and bypass hoses since the thermostat is closed and does not allow circulation through the radiator. As the engine warms up, the thermostat starts to open the outlet to the radiator which provides cooling for the engine. Coolant now circulates through the radiator, and into the cylinder and block water jackets and then to the outside surfaces of the combustion chambers which are the hottest parts of the engine. After cooling the combustion chambers, the coolant circulates back through the block, through the thermostat and back into the radiator. The coolant is cooled in the radiator by the combined effects of the cooling fan and the forward motion of the vehicle as air passes through the cooling fins. An internal combustion engine runs most efficiently when hot and in order to increase the temperature above the boiling point of water, but at the same time control the temperature within suitable limits, a pressurized system is used. This is accomplished by the use of a radiator pressure cap which contains two valves. The main valve will relieve the system pressure when it increases to 13 lbf/in^2 (0·91 kgf cm^2), which occurs as the temperature rises to the maximum permitted for the vehicle. When the system cools down, the coolant contracts and in order to prevent low pressures which cause the radiator and hoses to collapse, another relief valve comes into operation. This allows air to enter the system and thus balance the pressure.

2 Cooling system – routine maintenance

1   The cooling system requires very little routine maintenance but in view of its important nature, this maintenance must not be neglected.
2   The maintenance intervals are given in the Routine Maintenance Section at the beginning of this manual.
3   Apart from regular checking of the coolant level, and inspection for leaks and deterioration of hose connections, the only other items of major importance are the use of antifreeze solutions or rust inhibitors suitable for aluminium engines, and renewal of the coolant. These items are covered separately in this Chapter.
4   It must be remembered that the cooling system is pressurized. This means that when the engine is hot, the coolant will be at a temperature in excess of 100°C (212°F). Great care must therefore be taken if the radiator pressure cap has to be removed when the engine is hot since steam and boiling water will be ejected. If possible, let the engine cool down before removing the pressure cap. If this is not possible, place a cloth on the cap and turn it slowly anti-clockwise to the first notch. Keep it in this position until all the steam has escaped, then turn it further until it can be removed.

3 Cooling system – draining

1   The vehicle should be placed on level ground. If the coolant is to be re-used, place suitable containers under the drain cocks for its collection.
2   Remove the radiator cap. If the engine is hot see paragraph 4, of the previous Section.
3   Move the heater temperature control to the 'hot' position, to ensure that the heater matrix coolant is drained.
4   Open the radiator and engine block drain cocks and drain off all the coolant (photo).
5   When coolant ceases to flow from the drain cocks, probe them with a piece of stiff wire to ensure that there is no sediment blocking the drain orifices.
6   On completion, refer to Section 4 or 5, as required.

4 Cooling system – flushing

1   With the passage of time, deposits can build up in an engine, which can lead to engine overheating and possible serious damage.
2   It is a good policy, whenever the cooling system is drained, to flush the system with cold water from a hosepipe. This can conveniently be done by leaving a hosepipe in the radiator filler orifice for about 15 minutes while water is allowed to run through. This will usually be sufficient to clear any sediment which may be present.
3   If there appears to be a restriction, first try reverse flushing; this is the application of the hose to the drain orifices, forcing water back through the radiator tubes and out of the filler orifice.
4   If the radiator flow is restricted by something other than loose sediment, then no amount of flushing will shift it, and it is then that a proprietary chemical cleaner, suitable for aluminium engines, is needed. Use this according to the manufacturer's directions and make sure that the residue is fully flushed out afterwards. If leaks develop after using a chemical cleaner, a proprietary radiator sealer may cure them, but the signs are that the radiator has suffered considerable chemical corrosion and that the metal is obviously getting very thin in places.

5 Cooling system – filling

1   When draining (and flushing if applicable) has been carried out, close the drain cocks and top-up the cooling system with water which contains the correct proportion of antifreeze or inhibitor (see Section 6). Ensure that the heater temperature control is in the 'hot' position to prevent any airlocks from occurring in the heater matrix.
2   When the radiator is full, run the engine for 5 to 10 minutes at a fast idle. As the water circulates, and the thermostat opens, the

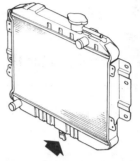

Fig. 2.2 The radiator drain cock (Sec 3)

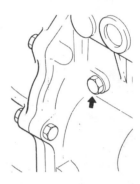

2

Fig. 2.3 The engine block coolant drain cock (Sec 3)

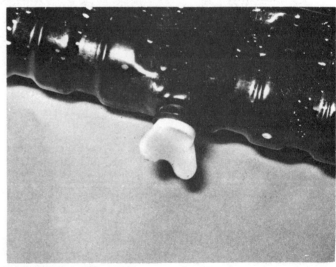

3.4 The radiator drain cock

coolant level will be seen to fall. Top-up the radiator to about halfway up the filler elbow, refit the cap, then run the engine for a few more minutes and check carefully for water leaks.
3   Allow the system to cool, then recheck the coolant level. Top-up if necessary until halfway up the filler elbow, then refit the cap firmly.

8.2A The top hose

8.2B The bottom hose

8.4A Remove the radiator mounting bolts ...

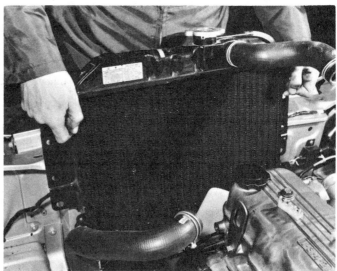

8.4B ... and lift out the radiator

9.3 Removing the thermostat cover

10.4 The fan and pulley ready for removal

6 Antifreeze and rust inhibitors

1 Tap water alone should not be used in an aluminium engine except in an emergency. If it has to be used, it should be drained off at the earliest opportunity and the correct coolant mixture used instead.

2 Generally speaking, the basis of the coolant mixture can be tap water, except where this has a high alkali content or is exceptionally hard. If these conditions exist, clean rainwater or distilled water should be used.

3 Antifreeze must be of a type suitable for use with aluminium engines (ethylene glycol based antifreeze is suitable) and many proprietary products will be available for use. The fact that all products tend to be expensive should not deter you from using them, since they are a good insurance against fraying and corrosion.

4 The following table gives suitable concentrations of antifreeze. Do not use concentrations in excess of 55% except where protection to below –37°C (–35°F) is required.

Coolant freezing point	Mixture percentage (volume)		Specific gravity of mixture at 20°C (68°F)
	Antifreeze	Water	
–30°C (–22°F)	45	55	1·066
–45°C (–49°F)	55	45	1·078

5 Antifreeze mixtures are normally only suitable for use for a period of two years (even so-called permanent antifreeze), after which they should be discarded and a fresh mixture used.

6 Antifreeze normally contains suitable corrosion inhibitors for protection of the engine. However, if antifreeze is not used for some reason, the vehicle manufacturers market suitable inhibitors which will give satisfactory protection to the engine against corrosion.

7 Fan/alternator drivebelt – tension adjustment

1 In order to obtain satisfactory engine cooling and alternator charge rate, the drivebelt must be tensioned correctly.

2 To check the belt tension, apply a pressure of approximately 22 lbf (10 kgf) to the belt midway between the fan pulley and alternator pulley, and check for a deflection as given in the Specifications (see Fig. 2.4).

3 If adjustment is required, loosen the alternator mounting bolt and pivot bolt and move the alternator as necessary. Avoid leverage against the side of the alternator or irreparable damage may result.

4 Tighten the bolts after adjustment has been made, then recheck the tension. After running the vehicle for about 150 miles (250 km), recheck the belt tension and adjust if necessary.

8 Radiator – removal and refitting

1 Drain the cooling system as described in Section 3.

2 Loosen the hose clips, then disconnect the top and bottom radiator hoses (photos).

3 If applicable, undo and remove the four bolts that secure the radiator shroud in position and move the shroud back over the fan.

4 Remove the radiator mounting bolts then carefully lift the radiator upwards (photos). The radiator shroud (if applicable) can then be removed.

5 If necessary, clean the radiator internally as described in Section 4. If the exterior cooling fins are blocked with dirt, flies, grease, etc, they should be cleared using compressed air on a water jet.

6 Refitting is a reversal of the removal sequence but ensure that the fan shroud (where applicable) is positioned over the fan before fitting the radiator.

7 Fill the cooling system, referring to Sections 5 and 6 for further information.

9 Thermostat – removal, testing and refitting

1 Drain the coolant until it is below the thermostat housing.

2 Loosen the hose clips and detach the top hose.

3 Undo and remove the two bolts and washers that retain the

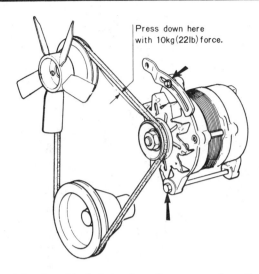

Press down here with 10kg (22lb) force.

Fig. 2.4 Adjusting drivebelt tension; the arrows show the alternator mounting bolt and pivot bolt (Sec 7)

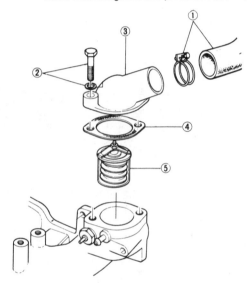

Fig. 2.5 The thermostat and cover (Sec 9)

1 Top hose and clip
2 Cover retaining bolt and washer
3 Thermostat cover
4 Gasket
5 Thermostat

Fig. 2.6 Testing a thermostat (Sec 9)

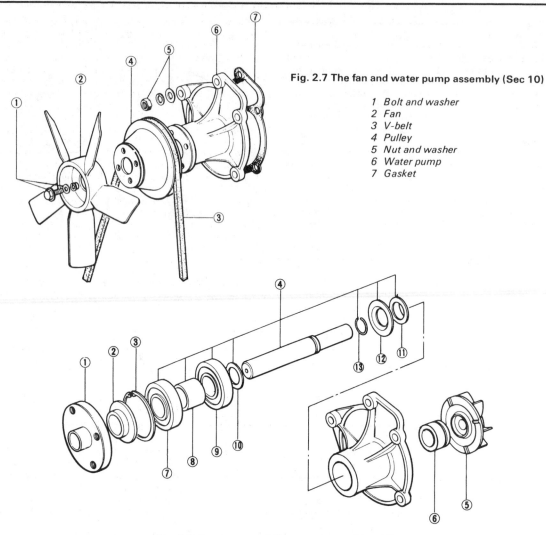

Fig. 2.7 The fan and water pump assembly (Sec 10)

1 Bolt and washer
2 Fan
3 V-belt
4 Pulley
5 Nut and washer
6 Water pump
7 Gasket

Fig. 2.8 Components of the water pump (Sec 11)

1 Pulley boss	assembly	8 Spacer	11 Baffle plate
2 Dust seal plate	5 Impeller	9 Bearing	12 Dust seal plate
3 Circlip	6 Seal assembly	10 Washer	13 Stop ring
4 Shaft, spacer and bearing	7 Bearing		

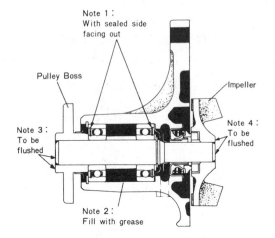

Note 1 :
With sealed side facing out

Pulley Boss

Impeller

Note 3 :
To be flushed

Note 4 :
To be flushed

Note 2 :
Fill with grease

Fig. 2.9 Sectional view of the water pump (Sec 11)

thermostat cover in position. Remove the cover and its gasket from the engine. The thermostat can then be lifted from its housing (photo).

4   To test the thermostat for correct operation, suspend it in a container of hot water. Using a thermometer to measure the water temperature, heat the water and note the temperature at which the thermostat begins to open. Also measure the lift and compare the figures obtained with those specified.

5   Refitting of the thermostat is a straightforward reversal of the removal procedure, but always use a new cover gasket with non-setting gasket sealant on the sealing faces. Don't forget to top-up the system on completion.

10 Water pump – removal and refitting

1   Drain the cooling system as previously described, then remove the lower hose from the water pump.

2   Disconnect the lower radiator hose and the upper engine hose

which leads to the radiator. Remove the radiator as described in Section 8.

3   Loosen the alternator mounting and pivot bolts and remove the drivebelt.

4   Remove the fan and pulley (photo).

5   Undo and remove the nuts and washers that retain the water pump, and remove the pump and its gasket from the block.

6   When refitting, ensure that the mating faces are clean and fit the pump to the block using a new gasket, following the reverse of the removal procedure.

7   Adjust the alternator drivebelt.

8   Fill the cooling system and check for leaks.

11 Water pump – dismantling, overhaul and reassembly

1   If the water pump is leaking, has excessive endplay or looseness of the shaft, or is unduly noisy in operation, it should be dismantled for overhaul. However, before commencing, ascertain the availability of spare parts; if these are not available, your only course of action is to obtain an exchange pump. This will obviously be more expensive but will save a certain amount of work.

2   Having removed the water pump from the engine, remove the impeller from the rear of the assembly using a suitable extractor.

3   Using a suitable extractor, remove the pulley boss from the shaft.

4   Remove the front dust seal plate.

5   Remove the circlip from its groove in the water pump housing.

6   Support the pump body and then press out the shaft, spacer and bearings assembly (from the rear) through the front of the body.

7   Remove the seal assembly from the pump body.

8   Using a suitable puller, draw the bearings off the shaft, then remove the baffle plate, rear dust seal plate washer and stop ring.

9   If the bearings run smoothly and do not show signs of corrosion they should be suitable for re-use. However, where other parts of the pump are unserviceable, which will normally occur after a considerable period of time, it is worthwhile fitting new bearings. The seal assembly should be renewed regardless of its condition. The circlip and dust seal plates should be renewed if their condition warrants it.

10   Commence reassembly by fitting the stop ring into the groove on the shaft, followed by the rear dust seal plate.

11   Drive the baffle plate onto the shaft taper then fit the shaft into the body.

12   Fit the washer on the shaft and press in the bearing with the seal rearwards (towards the impeller).

13   Position the spacer against the fitted bearing and fill the cavity with a general purpose grease.

14   Fit the remaining bearing with the seal side forwards (towards the fan) then fit the circlip.

15   Position the dust seal plate on the bearing.

16   Press the pulley boss onto the shaft until it is flush with the shaft endface.

17   Fit the seal assembly into the impeller side of the body.

18   Press the impeller onto the shaft until it is flush with the shaft endface.

12 Fault diagnosis – cooling system

Symptom	Reason/s	Remedy
Loss of coolant	Leak in system	Examine all hoses, hose connections, drain taps and the radiator and heater for signs of leakage when the engine is cold, then when hot and under pressure. Tighten clips, renew hoses and repair radiator as necessary.
	Defective radiator pressure cap	Examine cap for defective seal and renew if necessary.
	Overheating causing too rapid evaporation due to excessive pressure in system	Check reasons for overheating.
	Blown cylinder head gasket causing excess pressure in cooling system forcing coolant out	Remove cylinder head for examination.
	Cracked block or head due to freezing	Strip engine and examine. Repair as required.
Overheating	Insufficient coolant in system	Top-up.
	Water pump not turning properly, due to slack fan belt	Tighten fan belt.
	Kinked or collapsed water hoses causing restriction to circulation of coolant	Renew hoses as required.
	Faulty thermostat (not opening properly)	Check and renew as necessary.
	Engine out of tune	Check ignition setting and carburettor adjustments.
	Blocked radiator either internally or externally	Flush out cooling system and clean out cooling fins.
	Cylinder head gasket blown forcing coolant out of system	Remove head and renew gasket.
	New engine not run-in	Adjust engine speed until run-in.
Engine running too cool	Missing or faulty thermostat	Check and renew as necessary.

2

Chapter 3 Fuel, exhaust and emission control systems

See Chapter 13 for specifications and information applicable to 1979 thru 1983 North American models

Contents

Specifications

Air cleaner type .... Renewable paper element

Carburettor (1000 cc, Europe)

Type .... Two barrel, downdraught
Throat diameter:
    Primary .... 1·020 in (26 mm)
    Secondary .... 1·10 in (28 mm)
Venturi diameter:
    Primary .... 0·75 x 0·28 in (19 x 7 mm)
    Secondary .... 0·79 x 0·55 x 0·28 in (20 x 14 x 7 mm)
Main nozzle:
    Primary .... 0·0748 in (1·9 mm)
    Secondary .... 0·0748 in (1·9 mm)
Main jet:
    Primary .... No 88
    Secondary .... No 110
Main air bleed:
    Primary .... No 60
    Secondary .... No 80
Slow jet:
    Primary .... No 44
    Secondary .... No 100
Slow air bleed:
    Primary .... No 130
    Secondary .... No 150
Power jet .... No 40
Fast idle adjustment (clearance between primary throttle valve and
bore when choke valve is fully closed) .... 0·055 to 0·067 in (1·4 to 1·7 mm)
Secondary throttle valve adjustment (clearance between primary
throttle valve and bore when secondary throttle valve starts to
open) .... 0·232 to 0·256 in (5·9 to 6·5 mm)
Idle speed .... 650 to 700 rpm
CO connection at idle .... 1·5 ± 0·5%

Carburettor (1000 cc, except Europe)

Type .... Two barrel, downdraught
Throat diameter:
    Primary .... 1·020 in (26 mm)
    Secondary .... 1·10 in (28 mm)
Venturi diameter:
    Primary .... 0·75 to 0·28 in (19 x 7 mm)
    Secondary .... 0·94 x 0·55 x 0·28 in (24 x 14 x 7 mm)
Main nozzle:
    Primary .... 0·0748 in (1·9 mm)
    Secondary .... 0·0748 in (1·9 mm)
Main jet:
    Primary .... No 84
    Secondary .... No 116

Main air bleed:
    Primary . . . . No 60
    Secondary . . . . No 80
Slow jet:
    Primary . . . . No 44
    Secondary . . . . No 70
Slow air bleed:
    Primary . . . . No 130
    Secondary . . . . No 170
Power jet . . . . No 70
Fast idle adjustment (clearance between primary throttle valve and
bore when choke valve is fully closed) . . . . 0·055 to 0·067 in (1·4 to 1·7 mm)
Secondary throttle valve adjustment (clearance between primary
throttle valve and bore when secondary throttle
valve starts to open) . . . . 0·232 to 0·256 in (5·9 to 6·5 mm)
Idle speed . . . . 650 to 700 rpm

Carburettor (1300 cc, except North America)

Type . . . . Two barrel, downdraught
Throat diameter:
    Primary . . . . 1·020 in (26 mm)
    Secondary . . . . 1·180 in (30 mm)
Venturi diameter:
    Primary . . . . 0·87 x 0·31 in (22 x 8 mm)
    Secondary . . . . 1·06 x 0·51 x 0·28 in (27 x 13 x 7 mm)
Main nozzle:
    Primary . . . . 0·0827 in (2·1 mm)
    Secondary . . . . 0·1102 in (2·8 mm)
Main jet:
    Primary . . . . No 105
    Secondary . . . . No 160
Main air bleed:
    Primary . . . . No 80
    Secondary . . . . No 130
Slow jet:
    Primary . . . . No 46
    Secondary . . . . No 130
Slow air bleed:
    Primary . . . . No 190
    Secondary . . . . No 110
Power jet . . . . No 50
Fast idle adjustment (clearance between primary throttle valve and
bore when choke valve is fully closed) . . . . 0·052 in (1·32 mm)
Secondary throttle valve adjustment (clearance between primary
throttle valve and bore when secondary throttle valve starts to
open) . . . . 0·213 to 0·260 in (5·4 to 6·6 mm)
Idle speed . . . . 600 to 650 rpm manual, and automatic transmission (in 'D' range)
CO connection at idle . . . . 2·0 ± 0·5% (Europe only)

Carburettor (Canada)

Type . . . . Two barrel, downdraught
Throat diameter:
    Primary . . . . 1·020 in (26 mm)
    Secondary . . . . 1·18 in (30 mm)
Venturi diameter:
    Primary . . . . 0·87 x 0·31 in (22 x 8 mm)
    Secondary . . . . 1·06 x 0·51 x 0·28 in (27 x 13 x 7 mm)
Main nozzle:
    Primary . . . . 0·0827 in (2·1 mm)
    Secondary . . . . 0·1102 in (2·8 mm)
Main jet:
    Primary . . . . No 106
    Secondary . . . . No 150
Main air bleed:
    Primary . . . . No 80
    Secondary . . . . No 140
Slow jet:
    Primary . . . . No 190
    Secondary . . . . No 180
Slow air bleed:
    Primary . . . . No 90
    Secondary . . . . No 180
Power jet . . . . No 40
Fast idle adjustment (clearance between primary throttle valve and
bore when choke valve is fully closed) . . . . 0·054 in (1·37 mm)
Secondary throttle valve adjustment (clearance between primary

throttle valve and bore when secondary throttle valve starts to

open	0·213 to 0·260 in (5·4 to 6·6 mm)
Idle speed	700 to 750 rpm (manual transmission); 600 to 650 rpm (automatic transmission in 'D' range)
CO connection at idle	2·0 ± 0·5% (without air injection)

Carburettor (USA)

Type	Two barrel, downdraught
Throat diameter:	
Primary	1·020 in (26 mm)
Secondary	1·18 in (30 mm)
Venturi diameter:	
Primary	0·87 x 0·31 in (22 x 8 mm)
Secondary	1·06 x 0·51 x 0·28 in (27 x 13 x 7 mm)
Main nozzle:	
Primary	0·0827 in (2·1 mm)
Secondary	0·1102 in (2·8 mm)
Main jet:	
California:	
Primary	No 112 (manual transmission(; No 114 (automatic transmission)
Secondary	No 150
Except California:	
Primary	No 106 (low altitude); No 102 (high altitude)
Secondary	No 150 (low altitude); No 130 (high altitude)
Main air bleed:	
Primary	No 80
Secondary	No 140
Slow jet:	
Primary	No 48
Secondary	No 130
Slow air bleed:	
Primary	No 90
Secondary	No 180
Power jet	No 40
Fast idle adjustment (clearance between primary throttle valve and bore when choke valve is fully closed)	0·054 in (1·37 mm)
Secondary throttle valve adjustment (clearance between primary throttle valve and bore when secondary throttle valve starts to open)	0·213 to 0·260 in (5·4 to 6·6 mm)
Idle speed	700 to 750 rpm (manual transmission); 600 to 650 rpm (automatic transmission in 'D' range)
CO concentration at idle	2·0 ± 0·5% (without air injection)

Fuel pump

Type	Mechanical diaphragm
Fuel pressure (except North America)	1·4 to 2·1 lbf/m² (0·10 to 0·15 kgf/cm²)
Fuel pressure (Canada and USA)	2·84 to 3·84 lbf/m² (0·20 to 0·27 kgf/cm²)
Fuel feeding capacity (except North America)	0·6 Imp quarts/min (0·7 US quarts/min) (700 cc/min) at 1600 rpm of engine
Fuel feeding capacity (Canada)	0·6 Imp quarts/min (0·7 US quarts/min (700 cc/min) at engine idle
Fuel feeding capacity (USA)	0·9 Imp quarts/min (1·0 US quarts/min) (1000 cc/min) at engine idle
Fuel tank capacity	8·8 Imp gallons (10·6 US gallons) (40 litres)

Torque wrench settings

	lbf ft	kgf m
Inlet manifold stud nuts	14 to 19	1·9 to 2·6
Exhaust manifold stud nuts	12 to 17	1·6 to 2·3

1 General description

The fuel system for all models comprises a rear-mounted fuel tank, an in-line filter, a mechanical fuel pump, a carburettor and air cleaner. The air cleaner is fitted directly on the carburettor whereas the remaining items are connected via metal and flexible pipes.

Fuel is drawn from the tank, through an in-line filter by a mechanical fuel pump which delivers the fuel to the carburettor float chamber. According to the particular vehicle and its intended market, there may be an evaporative fuel line, a check and cut valve, an in-line check valve and a carbon canister between the pump and the carburettor, and an excess fuel return line from the carburettor to the fuel tank. These items are referred to later in the Chapter as part of the Emission Control system.

For all versions, the carburettor is a two-stage, two venturi, down-draught type, although there are differences according to the intended market. These differences are primarily associated with the emission control requirements in order to obtain satsifactory combustion allied to clean exhaust gases. The primary carburettor stage incorporates a curb idle system, accelerator pump system, idle transfer system, main metering system and power enrichment system. The secondary stage incorporates an idle transfer system, main metering system and power enrichment system. Choking action is by means of a cable operated manual control; the throttle is controlled by a foot pedal and a cable linkage.

The air cleaner varies according to the model, but all types incorporate a Summer/Winter control position which permits cool air to be drawn from the engine compartment or warm air from a heat stove attached to the exhaust manifold. Due to the complexities of the emission control system, and the variations for different markets, these are dealt with separately later in this Chapter.

2 Fuel system – routine maintenance

Air cleaner element – cleaning

1 Unscrew the wing nut from the air cleaner cover. Remove the cover and lift out the element. Taking care that no dirt enters the carburettor, dust out the air cleaner case with a lint-free cloth or a dry paintbrush. Tap the element sharply on a hard, clean, surface or use a low pressure air line to remove any traces of dust, etc. Do not rub the element as it may absorb some dust which will eventually enter the carburettor. When refitting the element ensure that it is seated correctly, then fit the cover (with new gaskets if applicable) and secure it with the wing nut.

Idle speed, idle mixture and CO emissions

2 Idle speed and idle mixture adjustments are of a fairly critical nature to maintain the correct exhaust emission standards. For this reason, a separate Section is devoted to them in this Chapter and reference should be made to this when adjustment is required.

Fuel filter

3 Loosen and remove the hose clips at each end of the filter assembly. Remove the filter assembly from its mounting clamp. The filter is serviced as an assembly only (photo).

Carburettor controls and linkages

4 Inspect the choke and throttle pedal linkages for lost motion and general wear and tear, particularly for fraying cables. Where necessary, take up any slackness then apply two or three drops of engine oil or general purpose lubricating oil to all the running surfaces and pivot points.

Fuel lines and connections

5 Examine all the fuel lines and connections for signs of leakage or looseness which may lead to early failure. Tighten connections and/or renew parts as necessary, but only use genuine replacements for your particular vehicle.

3 Air cleaner – removal and refitting

1 The air cleaner is a round flat type, seated above the carburettor, with a cold and a hot intake duct. Vehicles with emission control systems are fitted with a thermostatically controlled air cleaner (photo).
2 Removal of the air cleaner assembly can be done as a complete unit, removal of the element not being necessary. If it is found desirable to remove the element for some reason, this is dealt with in Section 2.
3 Loosen the single bolt securing the cleaner body to the mounting

2.3 The fuel filter

3.1 The Summer/Winter setting control

3

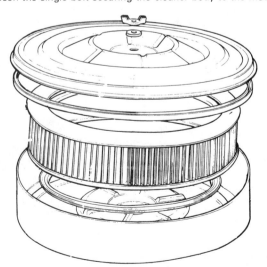

Fig. 3.1 The air cleaner assembly (Sec 2)

3.3 The air cleaner retaining bolt

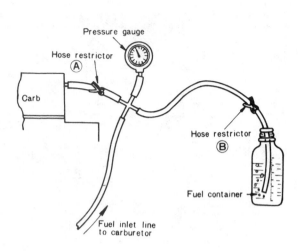

Fig. 3.2 The fuel pressure and volume test set-up (Sec 4)

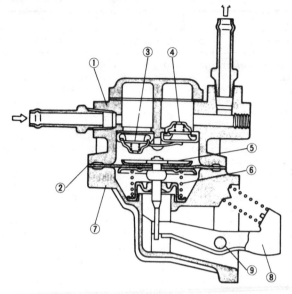

Fig. 3.3 A sectional view of the fuel pump (Sec 5)

1 Upper body   6 Spring
2 Diaphragm   7 Lower body
3 Inlet valve   8 Rocker arm
4 Outlet valve   9 Rocker arm pin
5 Diaphragm plate

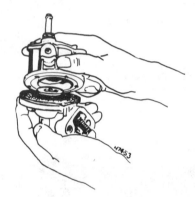

Fig. 3.4 Removing the upper body (Sec 5)

Fig. 3.5 The upper body with its valves, gaskets and retainer (Sec 5)

Fig. 3.6 Removing the diaphragm (Sec 5)

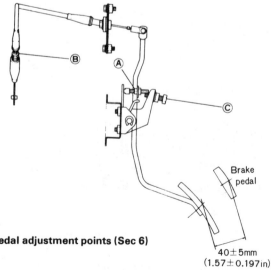

Fig. 3.7 The accelerator cable and pedal adjustment points (Sec 6)

Brake pedal

40±5mm
(1.57±0.197in)

bracket (photo).

4 Pull off the air cleaner, and after it is free of the carburettor, disconnect the hot air inlet pipe and the rocker cover pipe. Where applicable, remove any emission control hoses from the air cleaner.

5 Refitting is the reverse of the removal procedure.

---

4 Fuel pump – removal, refitting and testing

1 The fuel pump, which is actuated by a piston which travels up and down on an eccentric cam at the rear end of the oil pump operating spindle, is located below the thermostat housing.

2 To remove the pump, first disconnect the inlet and outlet hoses.

3 Remove the two bolts attaching the pump to the crankcase and lift away the fuel pump (photo).

4 The operating piston can be removed (if necessary) by removing the single bolt and washer directly above, and in the middle of, the attaching bolt holes. When the bolt and washer have been removed, lift the operating piston up through the bolt hole. It is best removed using long nose pliers or tweezers inserted through the pump mounting aperture. Use them to feed the piston upwards until it protrudes from the bolt hole, and can be gripped with fingers.

5 Refitting is the reverse of the removal procedure, but ensure that new gaskets are used either side of the fibre spacer plate.

6 To test the fuel pump, disconnect the fuel inlet line at the carburettor. Connect a pressure gauge, restrictors and a flexible hose between the fuel inlet line and the carburettor (Fig. 3.2).

7 Position the flexible hose so that the fuel can be discharged into a suitable graduated container.

8 Before taking a pressure reading, start the engine and vent the system into the container by opening the hose restrictor 'B' momentarily.

9 Run the engine at idle, and, closing the hose restrictors 'A' and 'B', allow the pressure to stabilize, and note the reading.

10 If the reading, obtained in paragraph 9, is not within the specified figures, and the fuel lines and filter are in a satisfactory condition, the pump is faulty and should be overhauled.

11 If the pump pressure is correct, use the same equipment to test for the correct delivery volume.

12 Allow the engine to idle (or as specified), then close the hose restrictor 'A' for exactly one minute and remove the amount of fuel collected in the container.

13 If the engine stalls during these tests, obviously the fuel in the float chamber has been used up. Open the restrictor 'A' to allow fuel to reach the carburettor, then proceed as previously mentioned.

---

5 Fuel pump – dismantling, servicing and reassembly

1 Remove the fuel pump as previously described in Section 4.

2 Lightly scratch identification marks on the upper and lower body for convenience in reassembly.

3 Remove the five screws and washers securing the upper body to the lower body. Take care in this operation, so that the pressure from the diaphragm spring does not force the assembly apart until you are ready for it.

4 From inside the upper cover, remove the two screws that secure the valve retainer. Take out the valve retainer, valves and valve gaskets.

5 Remove the diaphragm and spring from the lower body.

6 Using a suitable pin punch, drive out the rocker arm pin, then remove the rocker arm and spring.

7 Clean all the mechanical parts in a suitable cleaning solvent or petrol. Blow out all the passages with compressed air. Check the pump body for cracks, breakage or distorted flanges. Inspect the inlet and outlet valves; replace them if they do not perform properly. Examine the diaphragm for deterioration or any damage. Check the rocker arm, link and pin for wear.

8 Reassembly of the fuel pump is a straightforward reversal of the dismantling procedure.

---

6 Carburettor linkages

1 The choke and accelerator linkages are both operated by cables. It is important that any lost motion in the cables is eliminated (photo).

2 Lost motion in the choke cable is eliminated by loosening the clamp screw retaining the inner cable to the choke operating lever, and pulling the inner cable tight whilst tightening the clamp screw. After adjustment ensure that the choke operates correctly.

3 Any lost motion in the accelerator linkage should be first investigated at the accelerator pedal. The accelerator pedal height should be 1.6 $\pm$ 0.2 in (40 $\pm$ 5 mm) lower than the brake pedal. If necessary, refer to Fig. 3.7, and adjust the nut 'A' to obtain the correct position. Now check the cable free play at the carburettor; it should be 0.04 to 0.12 in (1 to 3 mm). If it is incorrect, rectify it by loosening the locknut and turning the nut 'B' (at the carburettor cable mounting bracket) in the required direction.

4 With all adjustments satisfactorily completed, depress the accelerator to the floor and check to see that the throttle valves are wide open. If necessary, adjust the stopper bolt 'C'.

---

7 Carburettor – removal and refitting

1 Remove the air cleaner as described in Section 3.

2 Release the choke cable from its clamp mounting, loosen the inner cable clamp screw and pull the choke cable clear of the carburettor.

4.3 Removing the fuel pump

6.1 The choke and accelerator cables

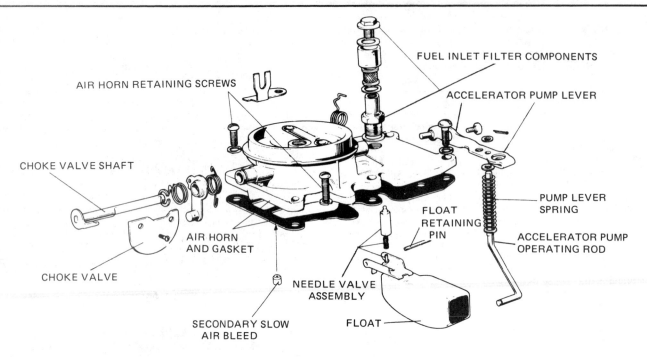

Fig. 3.8 Component parts of the air horn (Sec 8)

Fig. 3.9 The secondary slow air bleed (Sec 8)

Fig. 3.11 Removing the main body from the throttle body (1000 cc) (Sec 8)

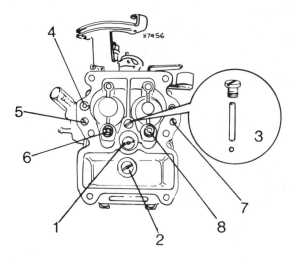

Fig. 3.10 The locations of the jets and air bleeds (1000 cc) (Sec 8)

1 Power valve	5 Primary slow jet
2 Inlet check valve	6 Primary main air bleed
3 Discharge ball and weight	7 Secondary slow jet
4 Primary slow air bleed	8 Secondary main air bleed

3   Open the throttle valve manually and disconnect the accelerator cable from the quadrant linkage. Where applicable, disconnect the rod from the servo diaphragm to the carburettor. Remove the servo diaphragm by disconnecting the vacuum tube, loosening the mounting bracket nut and sliding the unit from its bracket.

4   Remove the vacuum tube between the carburettor and the distributor. Where applicable, remove any vacuum tubes that supply the emission control system.

5   Disconnect the fuel line to the carburettor (and return line, if applicable) (photo).

6   Using a suitable cranked spanner, remove the carburettor attaching nuts from the manifold studs. Lift off the carburettor.

7   When refitting, position a new gasket on the inlet manifold studs, position the carburettor and fit the attaching nuts. The remainder of the refitting procedure is the reverse of the removal.

8   On completion, start the engine and check for fuel leaks.

8 Carburettor (1000 cc) – dismantling, servicing and reassembly

1 Disconnect and remove the throttle return spring.
2 Remove the split cotter pin and washer from the end of the accelerator pump lever and separate the rod from the lever. Remove the washer and spring from the rod and disengage the rod from the carburettor linkage.
3 Remove the circlip retaining the choke operating rod to choke linkage.
4 Undo and remove the screws and washers securing the air horn to the main body. Carefully remove the air horn, taking great care not to damage the float assembly.
5 Carefully tap out the retaining pins to release the float from the air horn. Remove the needle valve assembly.
6 Unscrew and remove the fuel inlet filter assembly from the top of the air horn. If necessary, remove the choke butterfly plate and shaft.
7 Remove the secondary slow air bleed (Fig. 3.9).
8 From the main body, lift out the accelerator pump piston and spring. This will reveal the power jet. With a suitable screwdriver, remove the power jet and its washer.
9 Remove the accelerator pump inlet check valve.
10 Unscrew the plug retaining the accelerator pump and discharge ball and weight, invert the body, and catch the ball and weight.
11 Unscrew and remove the secondary main air bleed. Then, at the main body joint face, remove the secondary slow jet. Moving to the other side of the carburettor (primary side), remove the primary slow air bleed and the primary slow jet; these components are located in the body joint face. Now remove the primary main air bleed from the centre of the primary chamber.
12 Remove the primary main jet and the secondary main jet plugs and washers to reveal the main jets. Now remove the primary main jet and washer, then remove the secondary main jet and washer.
13 Separate the main body from the throttle body by removing the attaching screw.
14 Commence the dismantling of the throttle body (if required) by removing the nut and washer retaining the operating quadrant to the butterfly plate shaft. Lift off the operating quadrant followed by the washer and the secondary chamber operating lever.
15 Remove the screw, washer and lockplate from the end of the choke operating shaft. Pull out the choke operating shaft.
16 Remove the screws securing the butterfly plates to their respective shafts. Pull out the shaft, releasing the butterfly plates from their bores. Keep the plates with their correct shafts.
17 Wash all the parts in clean petrol and dry by shaking and wiping with a lint-free cloth or paper tissue. Alternatively, dry using compressed air.
18 Blow through all the jets and passages to ensure that they are unobstructed. Never probe the jets and passages with needles or wire or irreparable damage may result.
19 Inspect the air horn, body and body joining flanges for cracks, nicks and burrs on their respective sealing faces.
20 Inspect the float for leaking, deformation, a damaged tab and worn retaining pin bore.
21 Check the float needle valve for wear and correct seating. Invert the air horn, assemble the needle valve, then suck at the main fuel passage. The seating is unsatisfactory if any leakage is present.
22 Inspect the fuel strainer for corrosion and damage.
23 Check the choke valve for excessive play. Ensure that it operates smoothly and closes fully.
24 Check for satisfactory operation of the primary and secondary throttle valves and for wear of the shafts.
25 Assembly of the carburettor is essentially the reverse of the removal procedure. Always use new gaskets and ensure that the primary and secondary system parts are not mixed up. Ensure that the primary and secondary throttle valves close correctly when in the closed position.
26 When the float has been refitted, invert the air horn, allow the float seat lip to rest on the needle valve, then measure the distance between the float and the air horn. The correct dimension should be 0.43 in (11 mm). To adjust the dimension, bend the float seat lip in the required direction.
27 Now lift the float up to the full extent of its travel and measure the clearance between the float stopper and the needle valve. The correct clearance should be 0.051 to 0.067 in (1.3 to 1.7 mm). To obtain the correct clearance, bend the float stopper in the required direction.

7.5 The carburettor fuel inlet connection

9.1 The throttle return spring

28 To check the fast idle adjustment, ensure that the choke valve is fully closed. Measure the clearance between the primary throttle valve and the wall of the throttle bore. The correct clearance should be as given in the Specifications. To obtain the correct clearance, screw the adjustment screw in the required direction.
29 The secondary throttle starts to open when the primary throttle valve has opened a pre-determined number of degrees. It is therefore important to obtain the correct setting in paragraph 28, before adjusting the secondary throttle valve.
30 Open the primary throttle valve slowly, until the secondary valve just starts to open. Measure the clearance between the primary throttle valve and the wall of the bore. The clearance should be as given in the Specifications. To correct the clearance, bend the connecting rod, between the primary and secondary shafts, in the required direction.
31 Refit the carburettor as described in Section 7. Start the engine and check for leaks. With the engine operating, check the fuel level in the carburettor. This should be at the specified mark in the sight glass. If all is well, carry out the idle speed and mixture adjustments as described in Sections 10 and 11.

9 Carburettor (1300 cc) – dismantling, servicing and reassembly

1 Disconnect and remove the throttle return spring (photo).

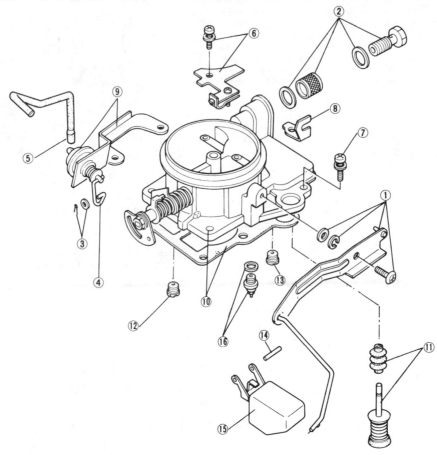

Fig. 3.12 Components of the air horn (Sec 9)

1	The accelerator pump lever assembly		only)	9	Vacuum diaphragm and mounting bracket (USA only)	12	Secondary slow air bleed

1   The accelerator pump lever
    assembly
2   Fuel inlet
3   Circlip and washer (USA
    only)
4   Vacuum servo linkage (USA

   only)
5   Vacuum hose (USA only)
6   Screw and bracket
7   Air horn retaining screw
8   Bracket

9   Vacuum diaphragm and
    mounting bracket (USA only)
10 Air horn and gasket
11 Accelerator pump piston and
    rubber boot

12 Secondary slow air bleed
13 Primary slow air bleed
14 Float retaining pin
15 Float
16 Needle valve assembly

9.3 The accelerator pump lever

2   Disengage the choke operating rod from its location in the actuating lever, then disconnect it from the lever at the end of the choke shaft. Place the rod in a safe place.
3   Unscrew and remove the single fulcrum screw securing the accelerator pump lever to the air horn. When removing the screw, take care not to lose the two washers between the lever and the air horn body. Slide the end of the lever out of the end of the pump piston, then disengage the lever's operating rod from the quadrant mounted on the end of the primary throttle shaft. Stow the rod and lever, together with its screw and washers, in a safe place (photo).
4   Where applicable, remove the split pin and washer attaching the vacuum servo diaphragm unit linkage to the choke operating lever, on the end of the choke valve shaft. Remove the screws securing the servo unit bracket to the carburettor. Pull off the vacuum pipe to the servo unit and lift away the unit and its mounting bracket.
5   Lift off the rubber boot from the accelerator pump piston. Remove the screws and washers attaching the air horn to the main body. Carefully remove the air horn, taking great care not to damage the float. Where applicable, remove the rubber boot from the accelerator pump piston, after removing the air horn.
6   Remove the accelerator pump piston from the main body.
7   Using a suitable pin punch, lightly tap the float retaining pin from its bores; remove the pin and the float.
8   Remove the needle valve assembly and washer.
9   From the air horn jointing face, unscrew, and remove the primary and secondary slow air bleeds.
10   If required, remove the choke valve and shaft. By removing the two valve plate retaining screws, the shaft can be pulled from its locat-

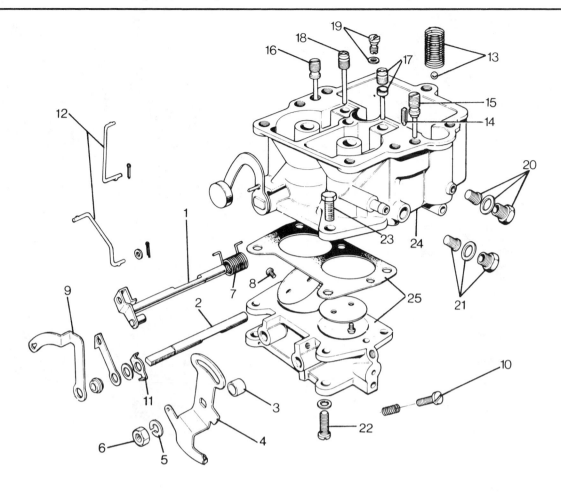

Fig. 3.13 The component parts of the main body and the throttle body (Sec 9)

1	Secondary throttle shaft	15	Primary slow jet
2	Primary throttle shaft	16	Secondary slow jet
3	Spacer	17	Primary main air bleed and emulsion tube
4	Operating quadrant	18	Secondary main air bleed
5	Split washer	19	Power valve
6	Nut	20	Primary main jet, washer and plug
7	Spring		
8	Throttle retaining screw	21	Secondary main jet, washer and plug
9	Choke operating lever	22	Retaining screw (throttle body)
10	Mixture screw	23	Retaining screw (main body)
11	Lockwasher	24	Main body
12	Choke and secondary operating rods	25	Gasket and throttle body
13	Accelerator pump inlet check ball and spring		
14	Accelerator pump discharge weight		

ing bores in the air horn. It is as well to leave the operating lever and its spring assembly on the shaft (uness they require attention). When the shaft has been removed from the air horn, the valve plate is free to be removed.

11 Proceed to dismantle the main body by inverting it to remove the accelerator pump return spring and inlet check ball.

12 From the primary chamber, lift out the accelerator pump discharge weight (outlet valve).

13 Remove the primary slow jet from inside the jointing face; then, at the secondary side of the carburettor, remove the secondary slow jet from the jointing face.

14 Now remove the primary main air bleed and lift out the emulsion tube.

15 In the secondary chamber, unscrew and remove the secondary main air bleed.

16 From the bottom of the fuel chamber, use a suitable screwdriver to remove the power valve and washer.

17 From the underside of the fuel chamber, remove the plugs and washers to reveal the primary and the secondary main jets. Remove

the jets.

18 To remove the main body from the throttle body, remove the securing bolts from the main body, then invert the assembly to remove one screw and washer from inside the throttle body.

19 If required, the throttle body can now be dismantled. As the accelerator linkages are removed, it is advisable to note their fitted position in order to aid reassembly. Always keep the throttle valves with their respective shafts.

20 Wash all the parts in clean petrol (gasoline) and dry by shaking and wiping with a lint-free cloth or paper tissue. Alternatively, dry using compressed air.

21 Blow through all the jets and passages to ensure that they are unobstructed. Never probe the jets and passages with needles or wire or irreparable damage may result.

22 Inspect the air horn, body and body flange for cracks, nicks and burrs on their respective sealing faces.

23 Inspect the float for leaking, deformation, a damaged tab and worn retaining pin bore.

24 Check the needle valve assembly for wear and correct seating.

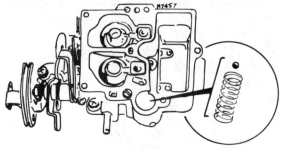

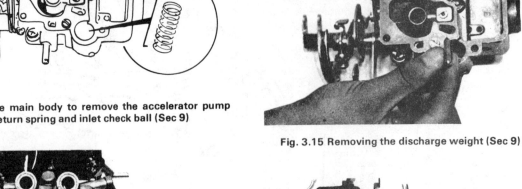

Fig. 3.14 Invert the main body to remove the accelerator pump return spring and inlet check ball (Sec 9)

Fig. 3.15 Removing the discharge weight (Sec 9)

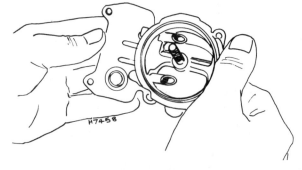

Secondary main jet

Primary main jet

Fig. 3.16 The main jets (Sec 8 and 9)

Fig. 3.17 The air horn, main body and throttle body ready for inspection (Sec 8 and 9)

Fig. 3.18 Checking the choke plate and shaft for smooth action (Sec 8 and 9)

Fig. 3.19 The accelerator pump piston and spring (Sec 8 and 9)

Primary side

Secondary side

Fig. 3.20 The primary and secondary sides (Sec 8 and 9)

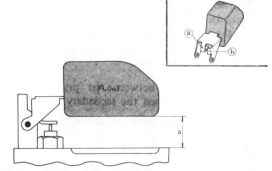

Fig. 3.21 Adjust the dimension 'A' by bending the float seat lip 'a' (Sec 8 and 9)

Blow through the inlet of the valve whilst lifting the needle up and down. With the needle held lightly in, no air should pass through the valve. Try the same test whilst sucking through the valve. If there is any doubt of the efficiency of the valve, renew it.

25 Inspect the fuel strainer for corrosion and damage.

26 Check the choke valve for excessive play. Ensure that it operates smoothly and closes fully.

27 Check for wear on the sliding portion of the pump piston and for smooth action. Check the spring for corrosion, wear and damage.

28 Check for satisfactory operation of the primary and secondary throttle valves and for wear of the shaft.

29 To commence reassembly, build up the throttle body (if it was dismantled), ensuring that the accelerator linkages are fitted in the reverse order of removal as noted in paragraph 19. Check for a smooth and satisfactory operation of the throttle plates in their respective bores.

30 Fit a new gasket between the throttle body and the main body. Secure the two units together, not forgetting the screw and washer that has to be fitted through the throttle body.

31 Screw the secondary and primary main jets into their respective bores below the fuel chamber. Fit a washer and plug to each bore.

32 Fit the power valve and washer to the bottom of the fuel chamber.

33 At the secondary chamber, screw in the secondary main air bleed.

34 Fit the emulsion tube and screw in the primary main air bleed.

35 At the secondary side of the carburettor, screw the secondary slow jet into the jointing face. At the primary side, screw the primary slow jet into the jointing face.

36 Fit the accelerator pump discharge weight (outlet valve) to the primary chamber.

37 Locate the inlet check ball and the accelerator pump return spring into their bore.

38 If the choke valve was removed from the air horn now is the time to reassemble it.

39 Screw the primary and secondary slow air bleeds into their bores in the jointing face of the air horn.

40 Fit the needle valve assembly and tighten it firmly.

41 Hold the float mounting bracket holes in line with the holes in the air horn mounting; fit the retaining pin, ensuring that it is secure.

42 With the float fitted, allow the weight of the float to press the seat lip onto the end of the needle valve. Measure the distance between the air horn and the float face. The dimension obtained should be 0.43 in (11 mm). The air horn gasket must not be in position when measuring this distance. If the dimension is incorrect, rectify it by bending the float seat lip in the required direction.

43 With the correct gap achieved in paragraph 42, proceed to lift the float up as far as it will travel, then measure the gap between the float stop and the end of the needle valve. The gap should be 0.051 to 0.067 in (1.3 to 1.7 mm). To obtain a correct gap bend the float stopper in the required direction.

44 Fit the accelerator pump piston to its bore in the carburettor. Where applicable, locate the pump piston and its rubber boot to the carburettor.

45 Carefully lower the air horn assembly onto the main body. Secure the air horn with the screws and washers. Ensure that the various mounting brackets that are held in place by the securing screws are in their correct positions. Always use a new air horn gasket.

46 Where applicable, fit the rubber boot to the pump piston.

47 Locate the accelerator pump lever to the end of the pump piston. Pass the fulcrum screw through the lever, fit the two washers, then screw the lever assembly to the side of the air horn. With the lever secured, check the action of the accelerator pump, by lifting the lever up and down. The piston should travel freely up and down. Having checked for satisfactory operation, connect the lever's actuating rod to the primary shaft operating quadrant.

48 Connect the choke operating rod to the actuating lever and the lever at the end of the choke shaft.

49 Connect the throttle return spring.

50 Where applicable, fit the vacuum servo diaphragm unit and its bracket. Connect the linkage between the servo unit and the choke shaft operating lever.

51 To carry out the fast idle adjustment, proceed as follows: ensure that the choke valve is fully closed. Invert the carburettor and measure the clearance between the primary throttle valve and the wall of the throttle bore. The clearance should be as given in the Specifications. If the clearance is not as previously mentioned, bend the choke connecting rod in the required direction until the clearance is correct. Always

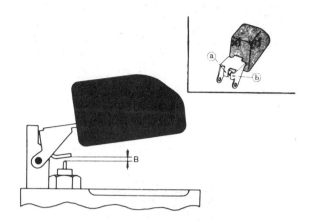

Fig. 3.22 Adjust the clearance 'B' by bending the float stopper 'b' (Sec 8 and 9)

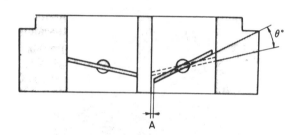

Fig. 3.23 The primary throttle valve-to-bore clearance indicated by 'A' (Sec 8 and 9)

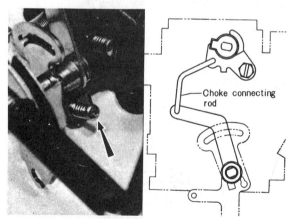

Fig. 3.24 To obtain the correct primary throttle valve-to-bore clearance, adjust the screw (1000 cc) or bend the rod (1300 cc) (Sec 8 and 9)

ensure that the choke valve is fully closed.

52 The secondary throttle valve starts to open after the primary throttle valve has opened to a pre-determined number of degrees. It is therefore very important to ensure that the adjustments described in paragraph 51 are correct before proceeding to the secondary throttle valve.

53 Open the primary throttle valve until the secondary throttle valve just starts to open. Now measure the clearance between the primary throttle valve and the wall of the throttle bore. The clearance should be

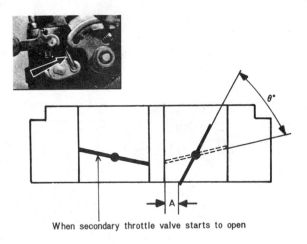

When secondary throttle valve starts to open

Fig. 3.25 Measure the clearance 'A' when the secondary throttle valve starts to open (Sec 8 and 9)

Fig. 3.26 The fuel level sight glass (Sec 8 and 9)

Fig. 3.27 Apply the specified vacuum and measure the gap between the air horn and the choke valve (Sec 9)

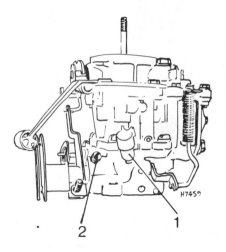

Fig. 3.29 The idle adjustment points (Sec 11)

1 Mixture adjusting screw
2 Throttle adjusting screw

Fig. 3.28 To obtain the correct choke valve-to-air horn clearance, bend the operating rod (Sec 9)

as given in the Specifications. If the clearance is not as previously mentioned, bend the rod, that connects the secondary throttle valve shaft to the primary throttle valve shaft, in the required direction.

54 *Note that the following test can only be carried out provided that a vacuum source and suitable vacuum gauge are available.* On vehicles fitted with a vacuum servo operated linkage, ensure that the choke valve is fully closed. Gradually apply vacuum to the servo unit. When the diaphragm starts to move, take a note of the vacuum applied. On vehicles destined for California this should be 7.5 ± 0.8 in Hg (190 ± 20 mm Hg). For vehicles destined for other than California this should be 8.7 ± 0.8 in Hg (220 ± 20 mm Hg). If this reading is incorrect, the servo unit is faulty and should be inspected, as described later in the emission control system. If the reading obtained is as previously specified, proceed as follows: On vehicles destined for California, apply 12.6 ± 1.2 in Hg (320 ± 30 mm Hg) of vacuum to the servo unit, hold this vacuum, and check the clearance between the choke valve and the air horn. The clearance should be 0.070 ± 0.007 in (1.78 ± 0.17 mm). To correct the clearance, bend the choke connecting rod in the required direction. On vehicles destined for other than California, the vacuum applied should be 13.6 ± 1.2 in Hg (345 ± 30 mm Hg), and the clearance between the choke valve and the air horn should be 0.050 ± 0.006 in (1.3 ± 0.15 mm). Adjustment should be made by bending the choke connecting rod.

55 Assemble the carburettor to the inlet manifold as described in Section 7. Start the engine and check for fuel leaks. With the engine operating, check the fuel level in the carburettor through the sight glass. The fuel level should be at the specified mark in the sight glass. If all is well, carry out the idle speed and mixture adjustments as described in Sections 10 and 11.

10 Accelerator pump lever – adjustment

1 On some carburettors, at the end of the accelerator pump lever there are three holes for the connecting rod which provide different settings for the amount of fuel injected.

2 The hole which is furthest from the accelerator pump piston will give the smallest injection, and the hole nearest the piston the largest; it is up to the owner to decide which gives the best result for his or her own operating conditions.

11 Idle speed and mixture – adjustments

1 It must be appreciated that in order to obtain satisfactory idle speed, idle mixture and exhaust emissions, careful adjustment of the carburettor is required. In all cases, it is important that before any adjustment is commenced, the distributor contact breaker point gap or dwell angle must be correct, and the timing must have been set for the particular vehicle. These items are dealt with in Chapter 4. Additionally, all items of the emission control system must be in a satisfactory state of tune; these items are dealt with later in this Chapter.

2 Dependent on market destination, some carburettors will be found to have a white polythene blind plug fitted over the mixture adjusting screw. Before commencing any adjustment on a carburettor with a blind plug fitted, ensure that no local regulations are being con-travened. It is quite easy to remove a blind plug to carry out mixture adjustments, but it is recommended that after setting as described in the following paragraphs, a qualified engineer checks these settings with exhaust gas analysing equipment at the earliest possible date. Failure to do this may result in excessive pollutants in the exhaust gases.

3 If possible, obtain an engine speed tachometer and connect it in accordance with the manufacturer's instructions.

4 Set the transmission in neutral then run the engine to attain the normal running temperature. Ensure that the choke is not in operation.

5 Set the engine idle speed to that specified, by means of the throttle adjusting screw.

6 If an exhaust gas analyser is available, connect it in accordance with the manufacturer's instructions.

7 Adjust the idle (mixture) screw to obtain the specified idle CO content. If no exhaust gas analyser is available, adjust the idle (mixture) adjusting screw to obtain the fastest steady idle speed. Screwing the screw outwards will tend to give a 'lumpy' running and a sooty exhaust gas which indicates a rich mixture; screwing the screw inwards will cause the engine to increase in speed a little then slow down and will give a 'hollow' exhaust note, indicating a lean (weak) mixture. After obtaining the highest steady speed, re-adjust the throttle adjusting screw to obtain an idle speed of 20 rpm above that specified. Now turn the idle (mixture) adjustment screw in (lean) until the engine speed reduces to that specified. Adjustment is now complete.

12 Inlet manifold – removal and refitting

1 The procedure given in this Section, describes removal and refitting of the manifold complete with carburettor. If considered necessary, the carburettor may first be removed by reference to Section 7 of this Chapter.

2 Drain the cooling system (refer to Chapter 2, if necessary).

3 Remove the air cleaner and accelerator linkage as described previously in this Chapter.

4 Disconnect the choke cable and fuel line to the carburettor.

5 Disconnect the water hoses from the manifold.

6 As appropriate, disconnect the air injection hoses, crankcase ventilation hose and distributor vacuum line. Take care to label the hoses if there is any fear of them being inadvertently mixed up when reconnecting.

7 Remove the inlet manifold to cylinder head nuts and washers, then lift the manifold off the studs on the cylinder head. Remove and discard the gasket (photo).

8 When refitting, clean the mating gasket surfaces thoroughly then position a new gasket on the manifold studs. A little non-setting joint-ing compound is permissible but is not essential.

9 Position the manifold then fit and torque tighten the washers and nuts., The remainder of refitting is the reverse of the removal proce-

12.7 Removing the inlet manifold

13.3 Separating the exhaust pipe from the exhaust manifold

13.4 Removing the exhaust manifold

3

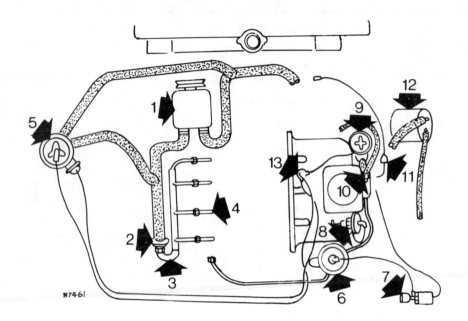

Fig. 3.30 Component parts of the emission control system (California) (Sec 14)

1	Air pump	5	Air control valve
2	Manifold check valve	6	EGR control valve
3	Air injection manifold	7	Three-way solenoid valve
4	Air injection nozzle		

8	Servo diaphragm	11	Spark delay valve
9	Anti-afterburn valve	12	Carbon canister
10	PCV valve	13	Air control check valve

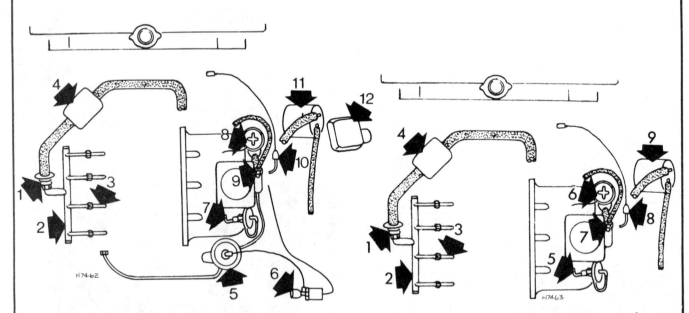

Fig. 3.31 Component parts of the emission control system (except California vehicles) (Sec 14)

1	Reed valve	7	Servo diaphragm
2	Air injection manifold	8	Anti-afterburn valve
3	Air injection nozzles	9	PCV valve
4	Air silencer	10	Spark delay valve
5	EGR control valve	11	Carbon canister
6	Three-way solenoid valve	12	Igniter (transistor ignition)

Fig. 3.32 Component parts of the emission control system (Canada) (Sec 14)

1	Reed valve	6	Anti-afterburn valve
2	Air injection manifold	7	PCV valve
3	Air injection nozzles	8	Spark delay valve
4	Air silencer	9	Carbon canister
5	Servo diaphragm		

dure, but when filling the cooling system refer to Chapter 2 for details.

13 Exhaust manifold – removal and refitting

1   Remove the air cleaner as described in Section 3.
2   Disconnect the exhaust pipe from its clutch bellhousing mounting clip.
3   Undo and remove the three nuts and washers securing the exhaust pipe flange to the exhaust manifold flange. Allow the exhaust pipe to drop off the manifold studs (photo).
4   Remove the manifold attaching nuts and washers then lift away the manifold. Remove and discard the gasket(s) (photo).
5   When refitting, clean the mating gasket surfaces thoroughly, then position new gasket(s) on the manifold studs.
6   Lightly lubricate the mating flange of the manifold to cylinder head with graphite grease, then position the manifold and fit the washers and nuts.
7   Tighten the nuts to the specified torque, then assemble the exhaust pipe flange to the manifold flange. Connect the exhaust pipe to its clutch bellhousing mounting clip.
8   Refit the air cleaner as described in Section 3.

14 Emission control system – general description

1   The emission control system is divided into four basic parts as described in the following paragraphs, these being: the exhaust emission control system (air injection system and deceleration control system) to reduce harmful emissions in the exhaust gases; the exhaust gas recirculation (EGR) system, where a small part of the exhaust gas is introduced into the combustion chamber, which lowers the combustion temperature to reduce the amount of noxious gases produced; the positive crankcase ventilation (PCV) system which channels the blow-by gases from the crankcase into the combustion chamber to be burned; and the evaporative emission control system which stores the fuel vapour from the fuel system and leads it to the combustion chamber. In addition to the systems mentioned, vehicles for certain markets incorporate a catalytic converter, which is fitted in the exhaust system.

Air injection system (California)
2   The air injection system is designed to reduce the hydrocarbons, carbon monoxide and oxides of nitrogen in the exhaust gases. This is accomplished by injecting air under pressure into the exhaust ports near the exhaust valves, where the added oxygen (in the air) plus the exhaust heat, induces combustion during the piston exhaust stroke.
3   The system comprises the engine driven air pump, check valve, air injection nozzles, air injection manifold, air control valve, air control check valve and interconnecting pipes and hoses.
4   When operating, air is pumped via an air control valve and check valve to the air manifold and injection nozzles, where it oxidizes the unburned portion of the exhaust gases. A relief valve located in the air pump relieves excessive pump pressures, and the check valve prevents back-flow of exhaust gases into the air manifold at times when the exhaust gas pressure is higher than the pump pressure (eg high engine speeds or in the event of pump drivebelt failure). The check valve, mounted adjacent to the inlet manifold, is in effect a non-return valve, delivering vacuum to the air control valve. Dependent on inlet manifold conditions, the air control valve allows more or less air, as required, to flow from the air pump to the air injection manifold.

Air injection system (except California and Canada)
5   The principles of this system are the same as described in paragraph 2, but air is delivered from the air cleaner, through an air silencer, to a reed valve and into the air injection manifold.

Deceleration control system (Canada and USA)
6   The deceleration control system is designed to maintain a balanced fuel/air mixture during deceleration. It comprises an anti-afterburn valve, to prevent fuel detonation in the exhaust due to an over-rich mixture during the early part of deceleration, and a throttle opener system to prevent lean mixtures during the latter part of deceleration.
7   The anti-afterburn valve is operated by inlet manifold vacuum. It opens when inlet manifold vacuum suddenly increases during dece-

leration, and remains open for a period of time in proportion to the amount of pressure change sensed by the valve diaphragm.
8   As soon as the anti-afterburn valve completes its operation, the throttle opener system acts to add additional fuel to the lean mixture created in the inlet manifold by the deceleration action, thus ensuring more complete combustion, and cleaner exhaust gases. The throttle opener system is controlled by a vacuum control valve supplying a servo diaphragm valve, and this vacuum then operates the servo diaphragm. The servo diaphragm is connected to the primary throttle shaft; a bellows inside this assembly acts as an altitude compensator by changing its datum length in accordance with prevailing atmospheric pressure, and thus maintaining the correct spring tension on the diaphragm return spring. During deceleration, the vacuum control valve senses the higher inlet manifold vacuum and transmits this vacuum to the servo diaphragm. The diaphragm linkage then opens the primary throttle plate slightly, to supply additional fuel to the lean air/fuel mixture.

Exhaust gas recirculation system (USA only)
9   In this system, a small part of the exhaust gas is introduced into the combustion chamber to lower the combustion flame temperature, to reduce the nitrogen oxide content of the exhaust gases. The system comprises an EGR control valve, a three-way solenoid valve and a water temperature switch. When water passing through the inlet manifold galleries reaches a pre-determined temperature, the water temperature switch activates the three-way solenoid, allowing vacuum to be pulled through the solenoid and the EGR control valve. The EGR control valve is now open to allow the passage of exhaust gases between the exhaust manifold and the inlet manifold.

Positive crankcase ventilation system (Canada and USA)
10 A positive crankcase ventilation (PCV) system is used to divert blow-by gases into the inlet manifold to be burned in the cylinders. The system comprises a positive crankcase ventilation valve and its associated hoses, with ventilating air being routed into the rocker cover from the air cleaner and out again into the inlet manifold.
11 The PCV valve is operated by the pressure difference between the inlet manifold and the rocker cover. When there is no pressure difference (engine not rotating), or the inlet manifold pressure is higher than the rocker cover pressure (backfire), the valve closes under spring action. If there is a large pressure difference (engine idling or decelerating), the high vacuum in the inlet manifold opens the valve against the action of the spring and allows a restricted flow of air through the valve. At small pressure differences (normal operation), the valve is balanced and now the maximum flow of ventilating air occurs.

Evaporative emission control system (Canada and USA)
12 The evaporative emission control system is designed to prevent the emission of fuel vapours into the atmosphere, and comprises an evaporative fuel line, check and cut valve and a carbon canister.
13 Fuel vapours rising from the fuel tank enter the evaporative fuel line and, if the engine is not operating, collect in the carbon canister. With the engine operating, the fuel vapours are routed through the carbon canister and into the air cleaner, to be drawn into the engine.
14 A check valve, located in the evaporative line between the fuel tank and the carbon canister, allows fuel vapours and ventilation to flow under normal circumstances. In theory, if this line became blocked or frozen, ventilation would not occur and the engine fuel supply would be cut off. However, in practice, the valve is of the three-way type, and is opened by the partial vacuum caused in the fuel tank by such a blockage, thus allowing a ventilation passage to the atmosphere. Conversely, when fuel vapour in the fuel tank is expanded due to intense heat, a pressure rise occurs, and once again the valve opens to vent the pressure to the atmosphere. The carbon canister, the purpose of which is to store fuel vapours until they are burnt, is installed in the engine compartment.

15 Emission control system – tests and adjustments

Air pump drivebelt (California)
1   To check the belt tension, press down midway between the air pump pulley and the water pump pulley. At a downwards pressure of 22 lbf ft (10 kgf m) there should be a deflection of 0.43 to 0.51 in (11 to 13 mm) for a used belt, and 0.31 to 0.39 in (8 to 10 mm) for a new

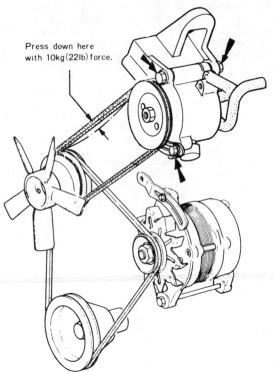

Press down here with 10kg(22lb) force.

Fig. 3.33 The drivebelt adjustment points (Sec 15)

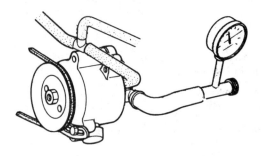

Fig. 3.34 The pressure gauge set-up for testing the air pump (Sec 15)

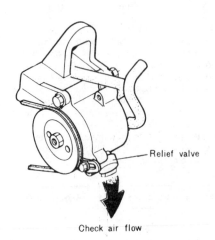

Relief valve

Check air flow

Fig. 3.35 Checking the air pump relief valve (Sec 15)

belt.

2   To adjust the belt, loosen the air pump adjusting bolt and the attaching bolt. Pry the pump outwards until the correct belt tension is obtained, then tighten the bolts.

Air pump (California)

3   Disconnect the air pump outlet hose and connect a pressure gauge and T-piece as shown in Fig. 3.34.

4   Ensure that the belt tension is correct and run the engine at 1500 rpm. Provided that a pump reading of 1 lbf/in^2 (0.07 Kgf/cm^2) or more is obtained, the pump is satisfactory.

Air pump relief valve (California)

5   Operate the engine at the specified idle speed and check for airflow at the relief valve; any signs of airflow from the relief valve indicate a faulty relief valve. If the test previously mentioned is satisfactory, proceed to paragraph 6.

6   Disconnect the vacuum tube between the air control valve and the inlet manifold at the air control valve. Increase the engine speed to 4000 rpm and ensure that air does flow from the relief valve. Any faults in the air pump can only be rectified by purchasing a new, or reconditioned, assembly.

Air manifold check valve (California)

7   Run the engine until it reaches normal operating temperatures. Disconnect the air hose at the check valve. Slowly increase the engine speed to 1500 rpm and check for any exhaust gas leakage at the air inlet fitting on the check valve. If the valve is faulty, purchase a replacement.

Air control valve (California)

8   Disconnect the air pump outlet hose from the air control valve, start the engine and run it at the specified idle speed. Place a finger over the outlet pipe to ensure that air is passing through the valve. If this test proves satisfactory, proceed to the next paragraph (Fig. 3.37).

9   Disconnect the vacuum tube between the air control valve and the inlet manifold at the air control valve and plug the open end of the tube. Now check to see that no air is passing through the valve at the outlet pipe (Fig. 3.38). If all is well, carry out the test described in paragraph 10.

10   Reconnect the vacuum tube between the air control valve and the inlet manifold (disconnected in paragraph 9). Disconnect the vacuum tube between the air control valve and the check valve on the inlet manifold, and check that there is a good flow of air at the outlet pipe (Fig. 3.39). Failure of any of the tests mentioned in paragraphs 8, 9 or 10 will necessitate the fitting of a replacement air control valve.

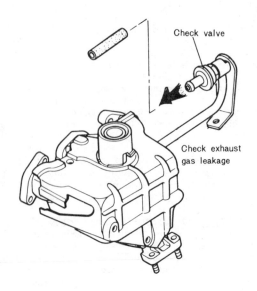

Check valve

Check exhaust gas leakage

Fig. 3.36 The air manifold check valve (Sec 15)

Check valve for the air control valve (California)

11 From the air control valve, disconnect the vacuum tube between the air control valve and the check valve, which is located on the inlet manifold. With the engine switched off, ensure that air only passes through the valve when blowing through the pipe. Sucking through the pipe should immediately close the valve allowing no air to pass. If there is any doubt as to the efficiency of the check valve it should be replaced (Fig. 3.40).

Reed valve (except California and Canada)

12 Run the engine until it reaches its normal operating temperature, then disconnect the air hose from the reed valve. With the engine at idle speed, place a finger over the reed valve inlet and ensure that air is being sucked into the valve. Increase the engine speed to 1500 rpm and check for any exhaust gas leakage by placing a finger over the valve inlet (Fig. 3.41).

Anti-afterburn (Canada and USA)

13 Disconnect the air hose between the anti-afterburn valve and the air cleaner, from the air cleaner. Run the engine at the specified idle speed and close the hose opening with a finger and make sure the engine speed does not change. Increase the engine speed and quickly release the accelerator. Air should be sucked into the air hose for a few seconds (no more than three seconds) after releasing the accelerator.

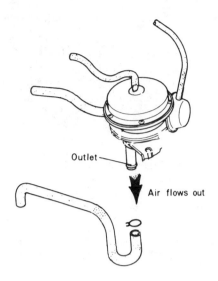

Fig. 3.37 Testing the air control valve for airflow (Sec 15)

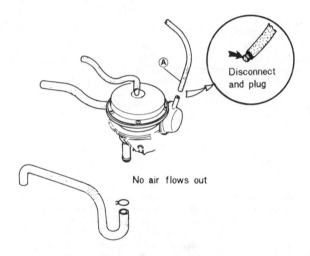

Fig. 3.38 Checking that the air control valve passes no air with the inlet manifold vacuum tube 'A' disconnected (Sec 15)

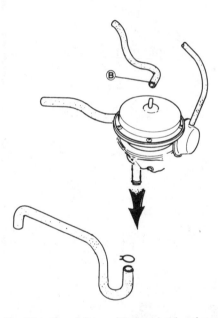

Fig. 3.39 Checking for airflow with the check valve vacuum tube 'B' disconnected (Sec 15)

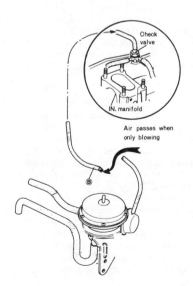

Fig. 3.40 Disconnect the vacuum tube 'B' from the air control valve and blow through it (Sec 15)

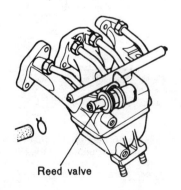

Fig. 3.41 Testing the reed valve (Sec 15)

3

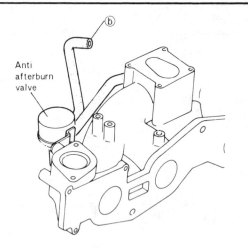

Fig. 3.42 Disconnect the air hose 'B' from the air cleaner to check the anti-afterburn valve (Sec 15)

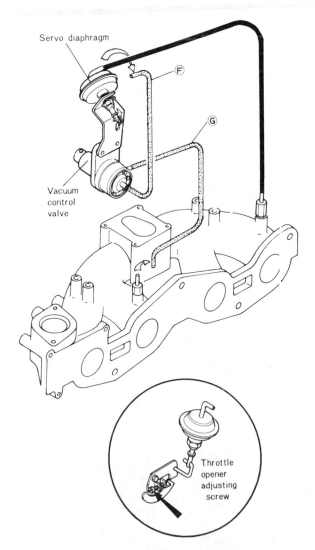

Fig. 3.43 Testing the servo diaphragm. Inset shows the throttle opener adjusting screw (Sec 15)

F *Vacuum tube between the servo diaphragm and the vacuum control valve*
G *A substituted length of tube between the inlet manifold and the vacuum control valve*

If a fault is evident in the valve, it will have to be renewed as a complete assembly (Fig. 3.42).

Throttle gear system; vacuum control valve and servo diaphragm (Canada and USA)

14 To check the servo diaphragm, ensure that the engine is at its correct operating temperature and remove the air cleaner, as described in Section 3. Disconnect the vacuum tube between the servo diaphragm and the vacuum control valve, at the servo diaphragm. Disconnect the vacuum tube between the vacuum control valve and the inlet manifold, at the inlet manifold, and connect the inlet manifold to the servo diaphragm with a suitable tube so that the inlet manifold vacuum can be fed directly to the servo diaphragm (Fig. 3.43).

15 Disconnect the vacuum tube between the carburettor and the distributor, at the distributor, and plug the tube. Start the engine and check to see that the engine speed increases from the specified idle rpm to 1400 ± 100 rpm. If the engine speed does not come within the figures specified, turn the throttle opener adjusting screw in or out until the specified engine speed is obtained.

16 To check the servo diaphragm's vacuum control valve, warm the engine up to its normal operating temperature. Ensure that the engine operates at its specified idle speed. Stop the engine and remove the air cleaner, as described in Section 3.

17 Disconnect the vacuum tube from the anti-afterburn valve and plug the open end of the tube.

18 Disconnect the vacuum tube between the vacuum control valve and the inlet manifold, at the inlet manifold, and connect a vacuum gauge (Fig. 3.44).

19 Start the engine and increase the speed to about 3000 rpm and quickly release the accelerator. The vacuum gauge reading should reach a high reading soon after releasing the accelerator and then descend to 22.4 ± 0.4 in Hg (570 ± 10 mm Hg). The vacuum reading should come to a rest for a few seconds at this point, then gradually fall until it reaches the idle speed vacuum. For vehicles operating in Canada the vacuum reading should read 22 ± 0.4 in Hg (560 ± 10 mm Hg).

EGR Control valve (USA only)

20 To check the EGR control valve, remove the air cleaner, as described in Section 3.

21 Start the engine and allow it to run at its specified idling speed.

22 Disconnect the vacuum tube between the three-way solenoid and the EGR valve, at the EGR control valve. Disconnect the vacuum tube between the vacuum control valve and the inlet manifold, at the vacuum control valve; connect this tube to the EGR control valve so that the inlet manifold vacuum can be fed directly to the EGR valve. When this connection is made, the engine should stop or run roughly (Fig. 3.45).

Water temperature switch (USA only)

23 To check the water temperature switch, disconnect the two bullet connectors, and unscrew it from its location in the inlet manifold.

24 Place the switch in water, and whilst testing the temperature of the water with a thermometer, heat the water to a temperature of 55° ± 5° C (131° ± 9° F). Connect an ohmmeter to the water temperature switch. There should be no continuity between the switch terminals at this temperature.

Three-way solenoid valve (USA only)

25 To check the three-way solenoid, disconnect the bullet connectors from the water temperature switch, and using a suitable piece of wire, connect the two bullet connectors together.

26 Disconnect the vacuum tube from the EGR control valve to the three-way solenoid, at the EGR control valve. Disconnect the vacuum tube between the three-way solenoid and the exhaust manifold, at the three-way solenoid (Fig. 3.47).

27 Turn the ignition switch to the on position and blow through the tube, which was disconnected from the EGR control valve. Air should flow from the air filter in the three-way solenoid.

28 Now disconnect the wire which was used to join the water temperature switch bullet connectors together. Blow through the tube again, as mentioned in paragraph 27. Air should now flow through the three-way solenoid instead of venting at the filter.

Positive crankcase ventilation valve (Canada and USA)

29 To check the PCV valve, run the engine until it reaches its normal

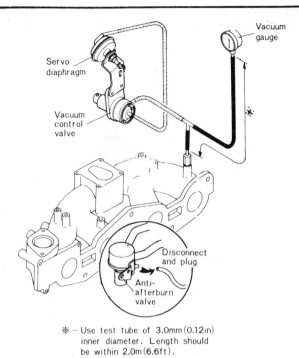

※ — Use test tube of 3.0mm (0.12in) inner diameter. Length should be within 2.0m (6.6ft).

Fig. 3.44 Testing the vacuum control valve (Sec 15)

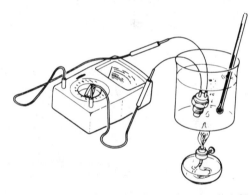

Fig. 3.46 Checking the water temperature switch (Sec 15)

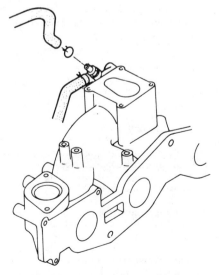

Fig. 3.48 The PCV valve test (Sec 15)

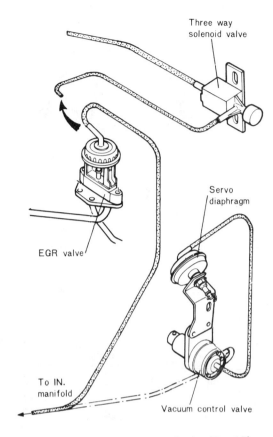

Fig. 3.45 Testing the EGR control valve (Sec 15)

3

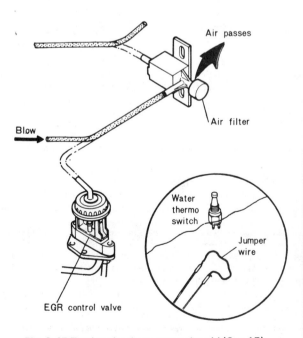

Fig. 3.47 Testing the three-way solenoid (Sec 15)

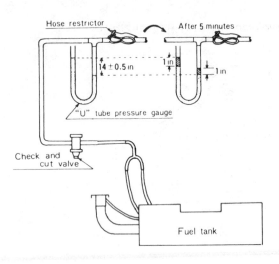

Fig. 3.49 The U-tube set-up for testing the evaporative fuel line
(Sec 15)

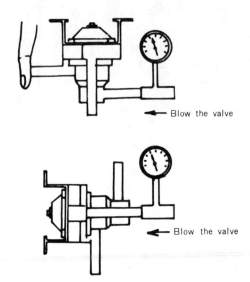

Fig. 3.50 Testing the check and cut valve (Sec 15)

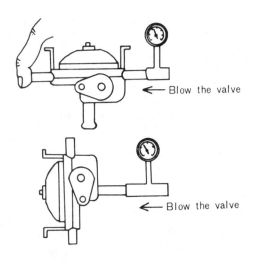

Fig. 3.51 Checking the check valve (Sec 15)

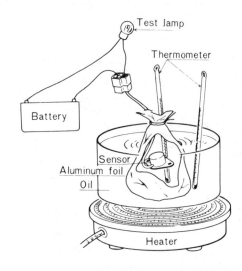

Fig. 3.52 The set-up for testing the heat hazard sensor (Sec 15)

operating temperature. Disconnect the hose from the PCV valve, and with the engine idling, place a finger over the valve inlet pipe. If the engine speed slows down slightly, the ventilation valve is working properly.

Evaporative fuel line (Canada and USA)

30 Disconnect the evaporative hose from the carbon canister inlet. Connect the disconnected hose to a U-tube pressure gauge as shown in Fig. 3.49.

31 Gradually apply low pressure compressed air in the U-tube, so that the difference between the two water levels is 14 ± 0.5 in (356 ± 12 mm). Now tightly blank off the inlet side of the U-tube and leave it for five minutes. If the water level drops within the shaded lines, the evaporative line is in good order.

Check and cut valve (USA only)

32 The check and cut valve is located behind the side panel in the rear of the vehicle, and it will have to be removed for checking.

33 With the check and cut valve removed, connect a pressure gauge to the side of the valve that connects to the fuel tank (Fig. 3.50). Then, using a finger, blank off the other side of the valve. Blow through the T-piece into the valve. With an applied pressure of 0.57 lbf/in^2 (0.04 Kgf/cm^2) or more, the valve should be open.

34 Now connect the pressure gauge to the passage to atmosphere. Blow through the T-piece and into the valve. With an applied pressure of 0.14 lbf/in^2 (0.01 Kgf/cm^2) or more, the valve should open.

Check valve (Canada only)

35 The check valve is located in the rear of the vehicle, behind the side panel, and it will to be removed for checking.

36 With the check valve removed, connect a pressure gauge to the passage to the fuel tank (Fig. 3.51), and using a finger, blank off the other end of the valve. Blow through the valve. The valve should open with a pressure of 0.57 lbf/in^2 (0.04 kgf/cm^2) or more.

37 Now connect the pressure gauge to the passage to atmosphere, blow through the valve as before, and it should open at a pressure of 0.14 lbf/in^2 (0.01 kgf/cm^2) or more.

Heat hazard warning system (USA only)

38 Disconnect the wire coupler of the heat hazard sensor, which is located under the floor mat near the parking brake lever. Using a suitable piece of wire, connect the two terminals in the coupler together;

the warning light should now come on.

Heat hazard sensor (USA only)

39 Remove the heat hazard sensor, which is located under the floor mat near the parking brake lever, by removing its two attaching screws.

40 Wrap the sensor and a suitable thermometer in aluminium foil to prevent oil penetration. Place the sensor and thermometer into a suitable oil bath.

41 Connect a test lamp and battery to the sensors coupler as shown in Fig. 3.52. Gradually heat up the oil, using another thermometer to test the temperature of the oil. The test lamp should light up when the temperature in the aluminium foil reaches 150 ± 10° C (302 ± 18° F). **Note**: *Never allow the temperature of the oil to exceed 200° C (392° F).*

---

16 Emission control system – removal and refitting of component parts

---

Air pump

1   Remove the alternator and alternator V-belt. Refer to Chapter 10, if necessary.

2   Disconnect the air pump inlet and outlet hose.

3   Remove the air pump adjustment bolt and the drivebelt.

4   Remove the air pump mounting bolt and lift away the pump.

5   Refitting is the reverse of the removal procedure. Adjust the alternator drivebelt, as described in Chapter 2, and the air pump drivebelt, as described in paragraph 2 of the previous Section.

Air manifold check valve (California)

6   Disconnect the air hose and unscrew the check valve from the manifold.

7   Refitting is the reverse of the removal procedure.

Air manifold (USA and Canada)

8   Remove the air manifold check valve and the heat insulator.

9   Apply penetrating oil to the nozzle unions. Unscrew the unions and remove the air manifold.

10 Refitting is the reverse of the removal procedure.

Air injection nozzles (USA and Canada)

11 Disconnect the hose from the air manifold check valve. Remove the heat insulator.

12 Remove the manifold assembly with check valve (refer to paragraph 9).

13 Apply penetrating oil to the nozzle screw threads then remove them from the exhaust ports.

14 Refitting is the reverse of the removal procedure.

Air control valve (California)

15 Disconnect the two vacuum tubes and the two air hoses at the valve.

16 Remove the bolts and washers and lift the valve, together with its mounting bracket, away from the engine compartment.

17 Refitting is the reverse of the removal procedure.

Check valve for the air control valve (California)

18 Pull off the three vacuum tubes at the check valve and remove the check valve.

Reed valve (except California and Canada)

19 Disconnect the air hose from the reed valve and unscrew it from the air manifold.

Anti-afterburn valve (USA and Canada)

20 Remove the air cleaner as described in Section 3.

21 Disconnect the two air hoses and the vacuum tube from the valve. Remove the bolt(s) securing the valve to the inlet manifold..

22 Refitting is a reversal of the removal procedure.

Servo diaphragm (USA and Canada)

23 Remove the air cleaner as described in Section 3.

24 Disconnect the vacuum tube to the servo diaphragm.

25 Loosen the nut securing the diaphragm to its mounting bracket.

26 Disconnect the linkage at the carburettor and lift the servo diaphragm from its bracket.

27 To refit the servo diaphragm, reverse the removal procedure.

Vacuum control valve (USA and Canada)

28 Remove the air cleaner as described in Section 3.

29 Disconnect the two vacuum tubes at the valve.

30 Remove the two bolts securing the control valve mounting bracket to the inlet manifold. Lift away the control valve.

31 Refitting is the reverse of the removal procedure.

EGR control valve (USA only)

32 Remove the air cleaner as described in Section 3.

33 Disconnect the vacuum tube at the top of the valve.

34 Remove the bolts securing the EGR valve to its mounting flange. Lift away the EGR valve, taking care not to lose the gasket between the valve and its mounting flange.

35 Refitting is the reverse of the removal procedure.

Three-way solenoid valve (USA only)

36 Disconnect the two bullet connectors.

37 Pull off the two vacuum tubes at the solenoid valve.

38 Remove the two bolts securing the solenoid valve to the engine bulkhead and lift the solenoid valve from the engine compartment.

39 Refitting is a reversal of the removal procedure.

Positive crankcase ventilation valve (USA and Canada)

40 Remove the air cleaner as described in Section 3.

41 Pull off the air hose at the PCV valve and unscrew it from its location on the inlet manifold.

42 To refit the PCV valve, reverse the removal procedure.

Carbon canister (USA and Canada)

43 Disconnect the two hoses at the canister. Remove the bolt securing the canister to its mounting place.

44 Refitting is the reverse of the removal procedure.

Check and cut valve (USA)

45 In the rear of the vehicle, remove the appropriate trim panel.

46 Disconnect the three hoses at the valve.

47 Remove the two bolts securing the valve to its location point and lift out the valve.

48 Refitting is the reverse of the removal procedure, but ensure that the orifice is assembled to the outlet side of the valve.

Check valve (Canada)

49 The removal procedure is identical to that of the check and cut valve (USA), so carry out the operations described in paragraphs 45, 46, 47 and 48.

Catalytic converter

50 Raise the wheel on suitable stands or drive it over an inspection pit.

51 Remove the bolts securing the heat insulator to the bottom of the converter. Remove the heat insulator.

52 Remove the two nuts and washers (from each end of the converter) that attach the converter to the exhaust pipe. Remove the converter.

53 Refitting is the reverse of the removal procedure.

---

17 Fault diagnosis – fuel, exhaust and emission control systems

---

Unsatisfactory engine performance and excessive fuel consumption are not necessarily the fault of the fuel system or carburettor. In fact they more commonly occur as a result of faults in the emission control system or ignition system. Because of the complexities and interdependence of the components in the emission control system, it is difficult to give a satisfactory diagnostic procedure; however, the checks and adjustments given in this Section are fairly straightforward and should always be carried out when problems occur. Fault finding and servicing of the ignition system should always be carried out before attending to problems which are leading to excessive fuel consumption, erratic or unsatisfactory performance, etc.

The table below assumes that the associated systems are correct.

Symptom	Reason/s	Remedy
Smell of petrol when engine is stopped	Leaking fuel lines or unions	Repair or renew as necessary.
	Leaking fuel tank	Fill fuel tank to capacity and examine carefully at seams, unions and filler pipe connections. Repair as necessary.
Smell of petrol when engine is idling	Leaking fuel line unions between pump and carburettor	Check line and unions and tighten or repair.
	Overflow of fuel from float bowl due to wrong level setting or ineffective needle valve or punctured float	Check fuel level setting and condition of float and needle valve, and renew if necessary.
Excessive fuel consumption for reasons not covered by leaks or float bowl faults	Worn jets	Renew jets.
	Sticking choke mechanism	Check correct movement and operation of choke plate.
	Accelerator pump incorrectly set	Check and adjust as necessary.
Difficult starting, uneven running, lack of power, cutting out	One or more jets blocked or restricted	Dismantle and clean carburettor.
	Float bowl fuel level too low or needle sticking	Dismantle and check fuel level and needle.
	Fuel pump not delivering sufficient fuel	Check pump delivery and clean or repair as required.
	Intake manifold gaskets leaking, or manifold fractured	Check tightness of mounting nuts and inspect manifold.
	Check valve sticking	Check and renew as necessary.
	PCV valve stuck open	Check and clean as necessary.
Engine will not idle	PCV valve sticking	Check and clean as necessary.
	Incorrect idle settings	Check and adjust as necessary.
Engine will not run	Fuel pump inoperative	Check and repair as necessary.

Chapter 4 Ignition system

See Chapter 13 for specifications and information applicable to 1979 thru 1983 North American models

Contents

Specifications

Spark plugs

Type:

Europe	NGK BPR–6ES or Nippon Denso W20 EXR–U
Except Europe and North America	NGK BPR–6ES, BP–6ES, BPR–7ES, BP–7ES or Nippon Denso W20 EP
North America	NGK BPR–6ES, BP–6ES, BPR–7ES, BP–7ES or Nippon Denso W20 EX–U, W22 Ex–U
Thread	14 mm
Electrode gap	0·031 in (0·8 mm)

Engine firing order .... 1 – 3 – 4 – 2 (No. 1 nearest radiator)

Distributor

Type	Single points or transistorized, with vacuum and centrifugal advance and retard; gear driven from camshaft
Direction of rotation	Clockwise
Electronic ignition (USA except California):	
Gap between reluctor and pick-up coil	0·010 to 0·012 in (0·25 to 0·35 mm)
Pick-up coil resistance	670 to 790 ohms at 20°C (68°F)
Contact breaker points (Other markets):	
Gap	0·020 in (0·5 mm)
Point pressure	0·99 to 1·32 lb (450 to 600 gm)
Dwell angle	49 to 55°
Centrifugal advance (1000 cc):	
Starts	0° at 550 rpm
Maximum	14° at 2310 rpm
Centrifugal advance (1300 cc):	
Starts	0° at 600 rpm
Maximum	12·5° at 2750 rpm
Vacuum advance – 1000 cc and 1300 cc (Europe), and California:	
Starts	0° at 4·72 in Hg (120 mm Hg)
Maximum	11° at 14·17 in Hg (360 mm Hg)
Vacuum advance – 1000 cc and 1300 cc (Except Europe and North America:	
Starts	0° at 7·09 in Hg (180 mm Hg)
Maximum	6° at 13·78 in Hg (350 mm Hg)
Vacuum advance – Canada:	
Starts	0° at 8·66 in Hg (220 mm Hg)
Maximum	6° at 15·35 in Hg (390 mm Hg)
Vacuum advance – USA (Except California)	
Starts	0° at 4·72 in Hg (120 mm Hg)
Maximum	11° at 13·8 in Hg (350 mm Hg)
Condenser capacity	0·20 to 0·24 mfd

Ignition timing

1000 cc (Europe) and 1300 cc	11° BTDC
1000 cc (except Europe)	7° BTDC

Timing mark location .... Crankshaft pulley

Torque wrench settings

	lbf ft	kgf m
Spark plugs	11 to 15	1·5 to 2·1

1   General description

In order that the engine can run correctly it is necessary for an electrical spark to ignite the fuel/air mixture in the combustion chamber at exactly the right moment in relation to engine speed and load. The ignition system is based on feeding low tension (LT) voltage from the battery to the coil where it is converted to high tension (HT) voltage. The high tension voltage is powerful enough to jump the spark plug gap in the cylinders many times a second under high compression pressures, provided that the system is in good condition and that all adjustments are correct.

The ignition system is divided into two circuits; the low tension circuit and the high tension circuit.

The low tension (sometimes known as the primary) circuit consists of the battery, lead to the ignition switch, lead from the ignition switch to the low tension or primary coil windings (terminal +), and the lead from the low tension coil windings (coil terminal -) to the contact breaker points and condenser in the distributor.

The high tension circuit consists of the high tension or secondary coil windings, the heavy ignition lead from the centre of the coil to the centre of the distributor cap, the rotor arm, and the spark plug leads and spark plugs.

The system functions in the following manner. Low tension voltage is changed in the coil into high tension voltage by the opening and closing of the contact breaker points in the low tension circuit. High tension voltage is then fed via the carbon brush in the centre of the distributor cap to the rotor arm of the distributor, and each time it comes in line with one of the four metal segments in the gap, which are connected to the spark plug leads, the opening and closing of the contact breaker points causes the high tension voltage to build up, jump the gap from the rotor arm to the appropriate metal segment and so via the spark plug lead to the spark plug, where it finally jumps the spark plug gap before going to earth.

The ignition advance is controlled both mechanically and by a vacuum operated system. The mechanical governor mechanism comprises two lead weights, which move out from the distributor shaft as the engine speed rises due to centrifugal force. As they move outwards they rotate the cam relative to the distributor shaft, and so advance the spark. The weights are held in position by two light springs and it is the tension of the springs which is largely responsible for correct spark advancement.

The vacuum control consists of a diaphragm, one side of which is connected via a small bore tube to the carburettor, and the other side to the contact breaker plate. Depression in the inlet manifold and carburettor, which varies with engine speed and throttle opening, causes the diaphragm to move, so moving the contact breaker plate, and advancing or retarding the spark. A fine degree of control is achieved by a spring in the vacuum assembly.

Vehicles built for use in the USA (but not California) use a transistorized ignition system. The essential difference between this and the mechanical type is that the mechanical type 'make-and-break' contact points are replaced by a reluctor and pick-up coil which carry out the function electronically. As each of the projections of the reluctor passes the pick-up coil, the coil flux density changes and the resultant electrical signal is passed to a transistorized circuit. This circuit cuts off the ignition coil primary feed which generates a high voltage in the ignition coil secondary windings. After a pre-determined time the ignition coil primary circuit is restored until the next reluctor projection passes the pick-up coil, when the cycle is repeated. The centrifugal vacuum advance and retard assemblies are of the same type as used in the mechanical-type distributor.

2   Contact breaker points – adjustment

1   To adjust the contact breaker points to the correct gap, first pull off the two clips securing the distributor cap to the distributor body, and lift away the cap. Clean the cap inside and out with a dry cloth, It is unlikely that the four segments will be badly burned or scored, but if

they are the cap will have to be renewed.

2   Inspect the carbon brush contact located in the top of the cap; see that it is unbroken and stands proud of the plastic surface.

3   Check the rotor arm for cracks or excessive burning at the contact face. Remove the rotor arm for inspection (photo).

4   Gently prise the contact breaker points open to examine the condition of their faces. If they are rough or pitted it will be necessary to remove them for resurfacing, or for replacement points to be fitted (see Section 3).

5   Presuming the points are satisfactory, or they have been cleaned or renewed, measure the gap between the points by turning the engine over until the heel of the breaker arm is on the highest point of the cam.

6   A 0.020 in (0.5 mm) feeler gauge should now just fit between the points.

7   If the gap varies from this amount slacken the contact breaker plate securing screws.

8   Adjust the control gap by moving the stationary contact point until the correct gap is obtained. When the gap is correct, tighten the securing screws and check the gap again

9   Examine the rubber seal, between the distributor body and the cap, if it is damaged or perished, renew it.

10   Making sure that the rotor arm is in position, replace the distributor cap and clip the spring blade retainers into position.

3   Contact breaker points – removal and refitting

1   Slip back the spring clips which secure the distributor cap in position. Remove the distributor cap and lay it to one side.

2   Pull the rotor arm from the distributor shaft.

3   Remove the two screws and washers securing the contact set to the base plate. Take great care not to drop the screws and washers into the distributor.

4   Slacken the screw retaining the condenser lead and the LT lead to the contact set. Slide out the two leads from the contact set.

5   The contact set can now be removed from the distributor.

6   Inspect the faces of the contact points. If they are only lightly bound or pitted then they may be ground square on an oilstone or by rubbing a carborundum strip between them. Where the points are found to be severely burned or pitted, then they must be renewed and, at the same time, the cause of the erosion of the points established. This is most likely to be due to poor earth connections from the battery negative lead to the body, or the engine to earth strap. Remove the connecting bolts at these points, scrape the surfaces free from rust and corrosion and tighten the bolts using a star type lock washer. Other screws to check for security are: the baseplate to distributor body securing screws, the condenser securing screw and the distributor body to lockplate bolt. Looseness in any of these could contribute to a poor earth connection. Check the condenser (Section 4), as a fault in this case can also cause pitting.

7   Refitting the contact breaker assembly is a reversal of removal. When fitted, adjust the points gap as described in the preceding Section and sparingly apply a slight smear of grease to the high points of the cam.

4   Condenser (capacitor) – removal, testing and refitting

1   The condenser ensures that with the contact breaker points open, the sparking between them is not so excessive as to cause severe pitting. The condenser is fitted in parallel and its failure will automatically cause failure of the ignition system as the points will be prevented from interrupting the low tension circuit.

2   Testing for an unserviceable condenser may be effected by switching on the ignition and separating the contact points by hand. If this action is accompanied by a blue flash the condenser failure is indicated. Difficult starting, misfiring of the engine after several miles running, or badly pitted points are other indications of a faulty condenser.

3   The surest test is by substitution of a new unit.

4   Removal of the condenser is achieved by loosening the screw securing the lead to the contact breaker arm and sliding out the connection, then removing the screw securing the condenser to distributor body. Where applicable, remove the condenser from the distributor base plate by removing the securing screw. Replacement of the condenser is a reversal of the removal procedure.

5   Distributor - removal and refitting

1   Before any attempt is made to remove the distributor from the engine, it is wise to position the engine's number one piston in the top dead centre (TDC) position (Figs. 4.2 and 4.3) (see Section 6), then mark the position of the distributor in relation to the engine crankcase; remove the distributor cap and mark the exact position of the rotor arm to the distributor body.

2   Pull off the vacuum tube to the vacuum advance and retard take off pipe.

3   Disconnect the low tension wire from the ignition coil.

4   Undo and remove the bolt which retains the distributor clamp

2.3 Remove the rotor arm ...

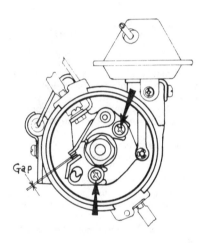

Fig. 4.1 The contact breaker points adjusting screws (Sec 2)

2.4 ... and inspect the condition of the points

4

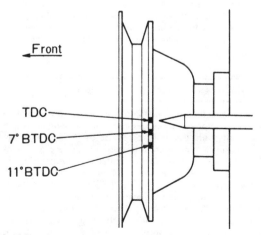

Fig. 4.2 Ignition timing marks on three notch crank pulley

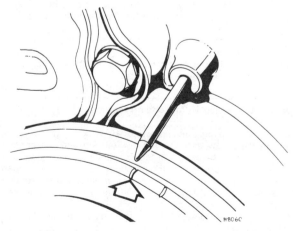

Fig. 4.3 The crankshaft pulley set at TDC (two notch pulley). The other notch represents the relevant timing mark (7° BTDC or 11° BTDC)

5.4A Undo and remove the distributor clamp plate bolt …

5.4B … and lift out the distributor

plate to the crankcase and lift out the distributor (photos).
5   Refitting is a reversal of the removal procedure, but ensure that the marks made previously are aligned.

6   Ignition timing (contact breaker type distributor)

1   Rotate the engine until number one piston is rising on its compression stroke. This may be checked by removing number one spark plug and placing a finger over the plug hole to feel the compression being generated or alternatively, removing the distributor cap and observing that the rotor arm is coming up to align with the position of number one contact segment in the distributor cap.
2   There are notches on the crankshaft pulley and a pointer on the front of the crankcase. Continue turning the engine until the appropriate notch on the crankshaft pulley is opposite the pointer on the engine crankcase (refer to the Specifications for the setting according to engine type). See Figs. 4.2 and 4.3.
3   Slacken the distributor clamp plate bolt.
4   Connect a test bulb between the LT terminal of the distributor and a good earth and switch on the ignition.
5   Turn the distributor body until the position is obtained where any movement in either direction will either illuminate or extinguish the test bulb.
6   Tighten the distributor clamp plate, remove the test bulb and switch off the ignition.
7   Timing the ignition using a stroboscope may be used as an alternative method, as described in Section 9.

7   Distributors – dismantling and reassembly

1   Remove the cap and pull off the rotor arm.
2   Loosen the screw securing the low tension and the condenser wires to the contact breaker arm. Slide out the connector and remove the rubber block from the distributor body, remove the condenser retaining screw and remove the wires, rubber block and the condenser from the distributor. Where applicable, remove the condenser by removing the screw retaining it to the base plate.
3   Remove the two contact breaker retaining screws and lift out the contact breaker points.
4   Prise off the clip retaining the vacuum diaphragm link to the breaker base plate.
5   On vehicles fitted with transistor type ignition, remove the screw retaining the pick-up coil to the breaker base plate. Prise off the clip

retaining the vacuum diaphragm link to the breaker base plate.
6   Remove the vacuum control unit from the distributor body, which is retained by one screw and washer.
7   Remove the screw and washer securing the breaker base plate to the distributor body. Lift out the breaker base plate.
8   Knock out the pin from the shoulder of the driven gear. Remove the driven gear and washer from the shaft.
9   The shaft and counterweight assembly may now be withdrawn through the upper end of the distributor body.
10 Where it is necessary to remove the cam from the top of the distributor shaft, first mark the relative position of the cam to the shaft and then unscrew and remove the screw and washer from the cam recess.
11 Where the counterweights and their springs are to be dismantled, take care not to stretch the springs.
12 Check all the components for wear and renew as appropriate.
13 It is advisable to fit new springs in view of the modest cost. Check the diaphragm unit by sucking through the inlet pipe and observing the retraction of the link, with the vacuum held in place by the tongue; the link should remain retracted; if it fails to do so, the vacuum unit must be renewed. Ensure that the driven gear is securely pinned to the shaft, and that the gear is undamaged.
14 Grease the counterweight pivots and reassemble by reversing the dismantling procedure. Refit the distributor as described in Section 5, then check the ignition timing as described in Section 6 or Section 9 as appropriate.

8   Distributor (transistor type) – air gap adjustment

1   This type of distributor employs a reluctor and a pick-up coil. The pick-up coil comprises a magnet and coil, and whenever the reluctor is in alignment with the pole piece of the coil, the magnetic plate between them is instrumental in generating the signal in the pick-up coil. This signal is relayed to the igniter unit, which in turn breaks the primary circuit and so generates high voltage in the ignition coil secondary windings.
2   To adjust the air gap between the reluctor and the pick-up coil, remove the distributor cap and pull off the rotor arm.
3   Turn the engine until the highest point of the cam, on the distributor shaft, comes into alignment with the reluctor. Using a 0.010 to 0.014 in (0.25 to 0.35 mm) feeler gauge, check the gap between the reluctor and the pick-up coil. If the gap is incorrect, loosen the two set screws and move the pick-up coil base until the correct air gap is obtained; tighten the screws and recheck the air gap.

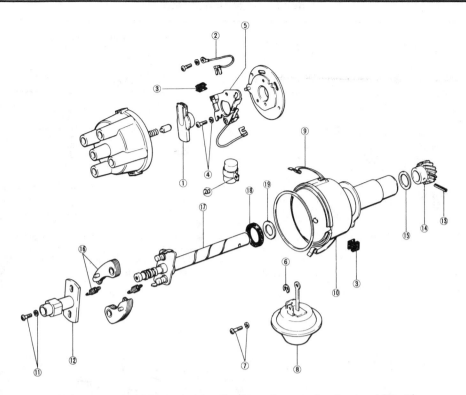

Fig. 4.4 An exploded view of the distributor (contact breaker type) (Sec 7)

1 Rotor arm
2 Wire and connector
3 Rubber block
4 Screw and washer
5 Breaker base plate assembly
6 Clip
7 Screw and washer
8 Vacuum control unit
9 Cap retaining clip
10 Cap retaining clip
11 Cam retaining screw
12 Cam assembly
13 Retaining pin
14 Driven gear
15 Washer
16 Governor weight and spring
17 Distributor shaft
18 Oil seal
19 Washer
20 Condenser

4

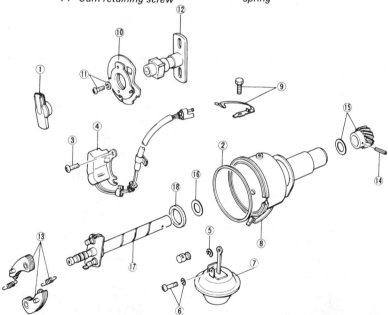

Fig. 4.5 An exploded view of a transistor type distributor (Sec 7)

1 Rotor arm
2 Rubber seal
3 Retaining screw
4 Pick-up coil
5 Clip
6 Vacuum unit retaining screw
7 Vacuum control unit
8 Cap retaining clip
9 Cap retaining clip and screw
10 Pick-up coil plate
11 Reluctor retaining screw
12 Reluctor assembly
13 Governor weights and spring
14 Retaining pin
15 Driven gear and washer
16 Thrust washer
17 Distributor shaft
18 Oil seal

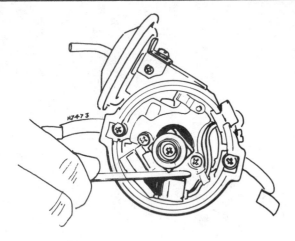

Fig. 4.6 Checking the air gap (Sec 8)

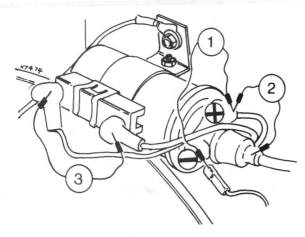

Fig. 4.7 Checking the coil resistances (contact breaker type ignition) (Sec 10)

1   Primary coil resistance ; 1.5 ohms
2   Secondary coil resistance ; 9000 ohms
3   External resistor resistance ; 1.6 ohms

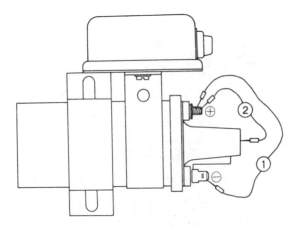

Fig. 4.8 Checking the coil resistances (transistor type ignition) (Sec 10)

1   Primary coil resistance ; 1.28 ohms
2   Secondary coil resistance ; 13 500 ohms

4   Refit the rotor arm and the distributor cap.

9   Ignition timing (transistor type distributor)

1   Mark the appropriate notch on the crankshaft pulley with white chalk or paint (see Specifications for correct setting).
2   Mark the pointer on the engine crankcase with white paint or chalk.
3   Disconnect the vacuum pipe (which runs from the vacuum control unit on the distributor) from its connection at the carburettor, and plug the pipe.
4   Connect a stroboscope in accordance with the maker's instructions (usually between number one spark plug and its HT lead).
5   Start the engine (which should have previously been run to normal operating temperature) and let it idle slowly (see recommended idle speeds in Chapter 3 Specifications) otherwise the mechanical advance mechanism will operate and give a false ignition timing.
6   Point the stroboscope at the ignition timing marks when they will appear stationary and, if the ignition timing is correct, in alignment. If the marks are not in alignment, loosen the distributor clamp plate bolt and gently rotate the distributor in the appropriate direction until the marks align.
7   Switch off the engine, tighten the distributor clamp plate bolt and remove the stroboscope.

10   Ignition coil - removal, testing and refitting

1   Pull out the HT lead from the centre of the coil. Disconnect the wires to the positive(+) and negative (-) terminals and the external resistor. (Vehicles with transistorized ignition do not have an external resistor).
2   Unscrew the bolt retaining the coil clamp to the side of the wing. Remove the coil from the vehicle.
3   Refer to Fig. 4.7 and, using an ohmmeter, check the resistance.
4   On vehicles fitted with transistorised ignition, refer to Fig. 4.8 and check the resistance.
5   Before testing coil resistances, the coil should be at its normal operating temperature.
6   Refitting of the coil is a reverse of the removal sequence.

11   Spark plugs and HT leads

1   The correct functioning of the spark plugs is vital for the proper running and efficient operation of the engine.
2   At the intervals specified, the plugs should be removed, examined, cleaned, and if worn excessively, renewed. The condition of the spark plug can also tell much about the general condition of the engine.
3   If the insulator nose of the spark plug is clean and white, with no deposits, this is indicative of a weak mixture, or too hot a plug (a hot plug transfers heat away from the electrode slowly - a cold plug transfers heat away quickly).
4   If the insulator nose is covered with hard black-looking deposits, then this is indicative that the mixture is too rich. Should the plug be black and oily then it is likely that the engine is fairly worn, as well as the mixture being too rich.
5   If the insulator nose is covered with light tan to greyish brown deposits, then the mixture is correct, and it is likely that the engine is in good condition.
6   If there are any traces of long brown tapering stains on the outside of the white portion of the plug, then the plug will have to be renewed, as this shows that there is a faulty joint between the plug body and the insulator, and compression is being allowed to leak away.
7   Plugs should be cleaned by sand blasting machine, which will free them from carbon more than cleaning by hand. The machine will also test the condition of the plugs under compression. Any plug that fails to spark at the recommended pressure should be renewed.
8   The spark plug gap is of considerable importance, as if it is too large or too small the size of the spark and its efficiency will be seriously impaired. The spark plug gap is given in the Specifications Section.
9   To set it, measure the gap with a feeler gauge, and then bend open, or closed, the outer plug electrode until the correct gap is

CARBON DEPOSITS

Symptoms: Dry sooty deposits indicate a rich mixture or weak ignition. Causes misfiring, hard starting and hesitation.

Recommendation: Check for a clogged air cleaner, high float level, sticky choke and worn ignition points. Use a spark plug with a longer core nose for greater anti-fouling protection.

OIL DEPOSITS

Symptoms: Oily coating caused by poor oil control. Oil is leaking past worn valve guides or piston rings into the combustion chamber. Causes hard starting, misfiring and hesition.

Recommendation: Correct the mechanical condition with necessary repairs and install new plugs.

TOO HOT

Symptoms: Blistered, white insulator, eroded electrode and absence of deposits. Results in shortened plug life.

Recommendation: Check for the correct plug heat range, over-advanced ignition timing, lean fuel mixture, intake manifold vacuum leaks and sticking valves. Check the coolant level and make sure the radiator is not clogged.

PREIGNITION

Symptoms: Melted electrodes. Insulators are white, but may be dirty due to misfiring or flying debris in the combustion chamber. Can lead to engine damage.

Recommendation: Check for the correct plug heat range, over-advanced ignition timing, lean fuel mixture, clogged cooling system and lack of lubrication.

HIGH SPEED GLAZING

Symptoms: Insulator has yellowish, glazed appearance. Indicates that combustion chamber temperatures have risen suddenly during hard acceleration. Normal deposits melt to form a conductive coating. Causes misfiring at high speeds.

Recommendation: Install new plugs. Consider using a colder plug if driving habits warrant.

GAP BRIDGING

Symptoms: Combustion deposits lodge between the electrodes. Heavy deposits accumulate and bridge the electrode gap. The plug ceases to fire, resulting in a dead cylinder.

Recommendation: Locate the faulty plug and remove the deposits from between the electrodes.

NORMAL

Symptoms: Brown to grayish-tan color and slight electrode wear. Correct heat range for engine and operating conditions.

Recommendation: When new spark plugs are installed, replace with plugs of the same heat range.

ASH DEPOSITS

Symptoms: Light brown deposits encrusted on the side or center electrodes or both. Derived from oil and/or fuel additives. Excessive amounts may mask the spark, causing misfiring and hesitation during acceleration.

Recommendation: If excessive deposits accumulate over a short time or low mileage, install new valve guide seals to prevent seepage of oil into the combustion chambers. Also try changing gasoline brands.

WORN

Symptoms: Rounded electrodes with a small amount of deposits on the firing end. Normal color. Causes hard starting in damp or cold weather and poor fuel economy.

Recommendation: Replace with new plugs of the same heat range.

DETONATION

Symptoms: Insulators may be cracked or chipped. Improper gap setting techniques can also result in a fractured insulator tip. Can lead to piston damage.

Recommendation: Make sure the fuel anti-knock values meet engine requirements. Use care when setting the gaps on new plugs. Avoid lugging the engine.

4

SPLASHED DEPOSITS

Symptoms: After long periods of misfiring, deposits can loosen when normal combustion temperature is restored by an overdue tune-up. At high speeds, deposits flake off the piston and are thrown against the hot insulator, causing misfiring.

Recommendation: Replace the plugs with new ones or clean and reinstall the originals.

MECHANICAL DAMAGE

Symptoms: May be caused by a foreign object in the combustion chamber or the piston striking an incorrect reach (too long) plug. Causes a dead cylinder and could result in piston damage.

Recommendation: Remove the foreign object from the engine and/or install the correct reach plug.

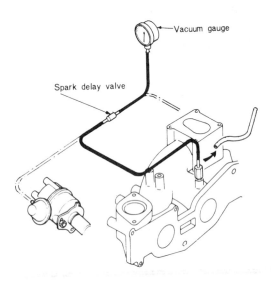

Fig. 4.9 Testing the spark delay valve (Sec 12)

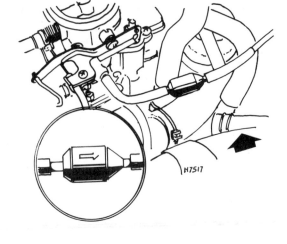

Fig. 4.10 The correct fitment of the spark delay valve (Sec 12)

achieved. The centre electrode should never be bent as this may crack the insulation and cause plug failure, if nothing worse.

10 When refitting the plugs, remember to connect the leads from the distributor cap in the correct firing order which is, 1, 3, 4, 2, No 1 cylinder being the one nearest the radiator.

11 The plug leads require no maintenance other than being kept clean and wiped over regularly. The leads used are of the carbon cored type which are used to suppress high frequency radio interference from the ignition system. Although these leads can give trouble-free performance for many years, they can sometimes cause starting problems after a considerable period of usage. If a fault is suspected, check the resistance of each wire, which should be 16 000 ohms/metre (4877 ohms/ft).

12 Spark delay valve – testing and renewal

1 Certain vehicles with emission control systems have a spark delay valve fitted between the distributor vacuum control unit and the carburettor. In operation, this valve controls the vacuum advance and retard mechanism in accordance with the prevailing engine operating conditions to reduce the emission of noxious exhaust fumes, particularly during periods of deceleration.

2 To check the valve's operation proceed as follows: Remove the plug from the inlet manifold. Disconnect the vacuum tube between the valve and the distributor, at the distributor, and connect it to the inlet manifold. Disconnect the vacuum tube between the valve and the carburettor, at the carburettor, and connect this tube to a vacuum gauge.

3 Start the engine and record the vacuum reading with the engine idling.

4 Disconnect the vacuum tube from the inlet manifold and check the time taken for the reading at the gauge to fall 11.8 in Hg (300 mm Hg) from the vacuum reading at engine idle. The correct time for USA models (except Californian automatic transmission models) and Canadian manual transmission models is 2 to 10 seconds. For Californian and Canadian automatic transmission models, this is 9 to 21 seconds. If the spark delay valve is suspect, renew it.

5 To renew the valve, merely disconnect it from its vacuum tube by pulling.

6 When refitting a new unit, ensure that the arrow mark on the valve is directed to the distributor vacuum control unit.

13 Ignition system (contact breaker type) – fault diagnosis

Engine fails to start

1 If the engine fails to start and the car was running normally when it was last used, first check there is fuel in the fuel tank. If the engine turns over normally on the starter motor and the battery is evidently well charged, then the fault may be in either the high or low tension circuits. First check the HT circuit. **Note:** If the battery is known to be fully charged, the ignition light comes on, and the starter motor fails to turn the engine **check the tightness of the leads on the battery terminals** and also the secureness of the earth lead to its **connection to the body.** It is quite common for the leads to have worked loose, even if they look and feel secure. If one of the battery terminal posts gets very hot when trying to work the starter motor this is a sure indication of a faulty connection to that terminal.

2 One of the most common reasons for bad starting is wet or damp spark plug leads and distributor. Remove the distributor cap. If condensation is visible internally, dry the cap with a rag and also wipe over the leads. Refit the cap.

3 If the engine still fails to start, check that current is reaching the plugs, by disconnecting each plug lead in turn at the spark plug end, and hold the end of the cable about $\frac{3}{16}$ inch (5 mm) away from the cylinder block. Spin the engine on the starter motor.

4 Sparking between the end of the cable and the block will be fairly strong with a regular blue spark. (Hold the lead with rubber to avoid electric shocks). If the current is reaching the plugs, then remove them and clean and regap them. The engine should now start.

5 If there is no spark at the plug leads take off the HT lead from the centre of the distributor gap and hold it to the block as before. Spin the engine on the starter once more. A rapid succession of blue sparks between the end of the lead and the block indicate that the coil is in order and that the distributor cap is cracked, the rotor arm faulty, or the carbon brush in the top of the distributor cap is not making good contact with the rotor arm. Possibly the points are in bad condition. Clean and reset them as described in this Chapter.

6 If there are no sparks from the end of the lead from the coil, check the connection at the coil end of the lead. If it is in order start checking the low tension circuit.

7 Use a 12 v voltmeter or a 12 v bulb and two lengths of wire. With the ignition switch on and the points open test between the low tension wire to the coil (it is marked SW or+) and earth. No reading indicates a break in the supply from the ignition switch. Check the connections at the switch to see if any are loose. Refit them and the engine should run. A reading shows a faulty coil or condenser, or broken lead between the coil and the distributor.

8 Take the condenser wire off the points and earth. If there is now a reading, then the fault is in the condenser. Fit a new one and the fault is cleared.

9 With no reading from the moving point to earth, take a reading between earth and the CB or - terminal of the coil. A reading here shows a broken wire, which will need to be renewed, between the coil

and distributor. No reading indicates that the coil has failed and must be renewed after which the engine will run once more. Remember to refit the condenser wire to the points assembly. For these tests it is sufficient to separate the points with a piece of dry paper while testing with the points open.

Engine misfires

10 If the engine misfires regularly run it at a fast idling speed. Pull off each of the plug caps in turn and listen to the note of the engine. Hold the plug cap in a dry cloth or with a rubber glove as additional protection against a shock from the HT supply.

11 No difference in engine running will be noticed when the lead from the defective circuit is removed. Removing the lead from one of the good cylinders will accentuate the misfire.

12 Remove the plug lead from the end of the defective plug and hold it about $\frac{3}{16}$ inch away from the block. Restart the engine. If the sparking is fairly strong and regular the fault must lie in the spark plug.

13 The plug may be loose, the insulation may be cracked, or the points may have burnt away giving too wide a gap for the spark to jump. Worse still, one of the points may have broken off. Either renew the plug, or clean it, reset the gap, and then test it.

14 If there is no spark at the end of the plug end, or if it is weak and intermittent, check the ignition lead from the distributor to the plug. If the insulation is cracked or perished, renew the lead. Check the connections at the distributor cap.

15 If there is still no spark, examine the distributor cap carefully for lines running between two or more electrodes, or between an electrode and some other part of the cap. These lines are paths which now conduct electricity across the cap, thus letting it run to earth. The only answer is to fit a new distributor cap.

16 Apart from the ignition timing being incorrect, other causes of misfiring have already been dealt with under the section dealing with the failure of the engine to start. To recap - these are that:

a) *The coil may be faulty giving an intermittent misfire.*
b) *There may be a damaged wire or loose connection in the low tension circuit.*
c) *The condenser may be short circuiting.*
d) *There may be a mechanical fault in the distributor (broken driving spindle or contact breaker spring).*

17 If the ignition timing is too far retarded, it should be noted that the engine will tend to overheat, and there will be a quite noticeable drop in power. If the engine is overheating and the power is down, and the ignition timing is correct, then the carburettor should be checked, as it is likely that this is where the fault lies.

14 Ignition system (transistor type) – fault diagnosis

1 Expensive and special equipment is required to test the ignitor unit. It is therefore recommended that the unit, which is attached to the ignition coil, should be removed and tested by a competent automobile electrician.

2 Apart from this, check the security of all HT and LT leads and examine the distributor cap for cracks.

3 The testing procedure described in the preceding Section will apply, except in respect of the contact breaker, which of course should be ignored. The air gap between the reluctor and the pick-up coil should, however, be checked as described earlier in this Chapter.

4

Chapter 5 Clutch

Contents

Specifications

Type	Single dry plate, diaphragm spring
Pressure plate permissible lateral run-out	0·002 in (0·05 mm)
Clutch disc permissible lateral run-out	0·039 in (1·0 mm)
Clutch release mechanism	Mechanical (cable)
Release cable free play	0·06 to 0·09 in (1·5 to 2·25 mm)
Clutch pedal free play at pedal pad	0·39 to 0·59 in (10 to 15 mm)

Torque wrench settings	lbf ft	kgf m
Clutch cover-to-flywheel bolts	13 to 20	1·8 to 2·7
Flywheel bolts	60 to 65	8·0 to 9·0

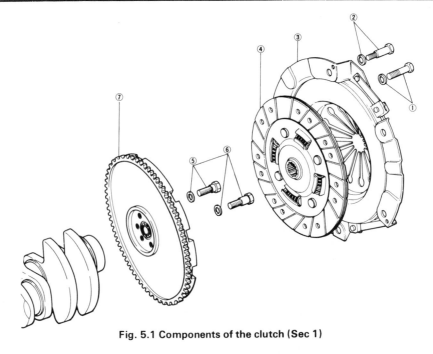

Fig. 5.1 Components of the clutch (Sec 1)

1 Standard bolt and washer (cover assembly)
2 Pilot bolt and washer (cover assembly)
3 Pressure plate and cover assembly

4 Clutch driven plate (friction disc)
5 Bolt and washer (flywheel)

6 Pilot bolt and washer (flywheel)
7 Flywheel

1 General description

The clutch is of single dry plate type with a diaphragm spring. Actuation is by means of a cable through a pendant mounted pedal. The unit comprises a pressure plate and cover assembly, and the release mechanism. In operation the clutch driven plate is free to slide along the splined first motion shaft, and is held in position between the flywheel and the pressure plate by the pressure of the pressure plate spring. When the clutch pedal is depressed, the cable moves the release arm, to which is attached the release bearing, and so moves the centre of the diaphragm inwards. The diaphragm spring is sandwiched between two annular rings which act as fulcrum points. As the centre of the spring is pushed in, the outside of the spring is pushed out, so moving the pressure plate backwards and disengaging the pressure plate from the clutch driven plate. When the clutch pedal is released, the diaphragm spring forces the pressure plate into contact with the friction linings on the clutch driven plate, and at the same time pushes the clutch driven plate a fraction of an inch forwards on its splines, so engaging the clutch driven plate with the flywheel. The clutch driven plate is now firmly sandwiched between the pressure plate and the flywheel and the drive is taken up.

2 Clutch – adjustment

1 Every 4000 miles (6500 km) check, and if necessary, adjust the clutch free movement. Dependent upon market destination, there are two types of cable mechanism fitted to actuate the clutch. The cable may be of the conventional pull type, or of a heavy duty flexible push operation type.

2 To adjust the push operated cable, proceed as described in the following paragraphs.

3 Raise the vehicle and support it with suitable stands, or alternatively drive the vehicle over an inspection pit.

4 Unhook the spring between the clutch release arm and the cable mounting bracket (photo).

5 Now grip the end of the clutch release arm and, moving it fore and aft, check the amount of movement needed before it touches the dome of the cable end. The movement should be between 0·060 to 0·090 in (1·5 to 2·25 mm). Adjustment is carried out by loosening the locknut under the rubber boot, and turning the dome on the end of the cable in the required direction (photo).

6 When a satisfactory adjustment has been obtained, tighten the locknut, pull the rubber boot into position, and refit the return spring. Lower the vehicle to the ground.

7 To adjust the pull type cable, proceed as described in the following paragraphs.

8 Working beneath the bonnet, loosen the locknut at the adjustment point on the engine rear bulkhead (Fig. 5.2).

9 Pull the outer cable towards the bulkhead, and at the same time, turn the adjustment nut until the distance shown in Fig. 5.2 is obtained.

10 When adjustment is correct, tighten the locknut.

3 Clutch cable – renewal

1 To renew the push operated cable, proceed as described in the following paragraphs.

2 Raise the vehicle and support it with suitable stands, or alternatively, drive the vehicle over an inspection pit.

3 Unhook the spring between the clutch release arm and the cable mounting bracket.

4 Undo and remove the two nuts, bolts and washers that attach the cable mounting bracket to the clutch bellhousing (photo).

5 Working inside the engine compartment, remove the two nuts attaching the cable flange to the rear bulkhead. Pull the flange, together with the cable, away from the bulkhead. Remove the rubber mounting sleeve (if applicable), and pull the cable up into the engine compartment, and out of the vehicle.

6 To renew the pull type cable, proceed as described in the following paragraphs.

7 Loosen the locknut, and unscrew it, together with the adjusting nut 'A' (Fig. 5.3).

8 Pull the inner cable 'B' towards the clutch pedal, and disconnect

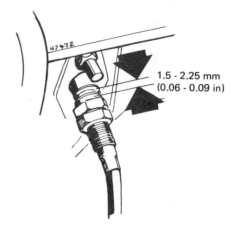

Fig. 5.2 Adjusting the clutch to the correct dimension (Sec 2)

1.5 - 2.25 mm
(0.06 - 0.09 in)

2.4 Unhook the return spring

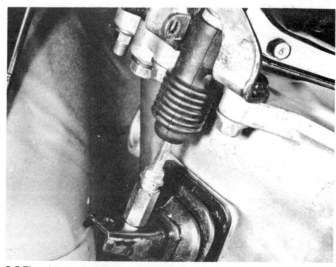

2.5 The clutch adjustment point

5

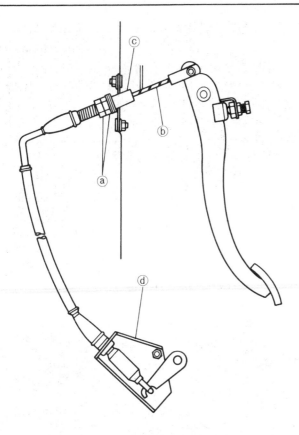

Fig. 5.3 Removing the clutch cable (Sec 3)

a   *Adjusting nut and locknut*
b   *Inner cable*
c   *Stop ring*
d   *Mounting bracket*

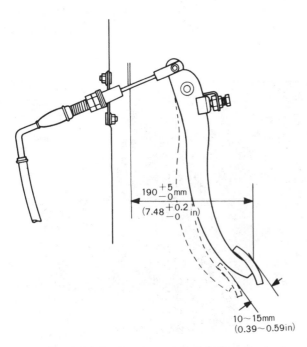

Fig. 5.4 Adjusting the pedal height (Sec 4)

the inner cable from the clutch pedal.
9   Remove the clutch cable by pulling it through the stop ring 'C'.
10 Working underneath the vehicle, disconnect the inner cable at the clutch release lever.
11 Disconnect the outer cable from the mounting bracket 'D', by removing the retaining clip.
12 Refitting, in both cases, is a reversal of the removal procedure, but check, and if necessary, adjust the clutch free play as described in Section 2.

4   Clutch pedal – height adjustment

Clutch pedal free travel is determined by the correct adjustment of the cable which has been described in Section 2. However, clutch pedal height can be adjusted separately. The clutch pedal return spring holds the pedal shaft against an adjusting bolt and lock nut. Adjustment is carried out merely by loosening the lock nut and screwing the adjustment bolt in the direction required to obtain the correct pedal height (Fig. 5.4). After pedal height has been adjusted, check and, if necessary, adjust the clutch cable.

5   Clutch – removal

1   Access to the clutch may be obtained in one of two ways. Either remove the engine (Chapter 1) or remove the gearbox (Chapter 6). Unless the engine requires major overhaul, removal of the gearbox is much the easier and quicker method.
2   Mark the relationship of the clutch cover to the flywheel, and unscrew the bolts which secure the clutch pressure plate cover to the flywheel. Unscrew the bolts only a turn at a time, until the pressure of the diaphragm spring is relieved before completely withdrawing them.
3   Lift away the pressure plate/diaphragm spring assembly and the driven plate (friction disc) from the face of the flywheel.

6   Clutch – inspection and renovation

1   The clutch will normally need renewal when all the free movement adjustment on the cable has been taken up, or it can be felt to be slipping under conditions of hard acceleration or when climbing a hill. Sometimes squealing noises can be heard when the clutch is engaged. This may be due to the friction linings having worn down to the rivets and/or a badly worn release bearing.
2   Examine the surfaces of the pressure plate and flywheel for signs of scoring. If this is only slight it may be left, but if very deep the pressure plate assembly will have to be renewed. If the flywheel is deeply scored it should be taken off and advice sought from an engineering firm. Flywheels can be repaired by remachining the scored face; provided it is machined completely across the face, balance of the engine and flywheel should not be too severely upset. If the only solution is a new flywheel, this will have to be balanced to match the original.
3   Renew the driven plate (friction disc) if the linings are worn down to the rivets (or nearly so) or if the linings appear to be contaminated with oil. Always purchase a new driven plate or one that has been professionally reconditioned. It is unwise to waste time and effort trying to rivet new linings yourself.
4   If the driven plate and the interior of the clutch bellhousing are saturated with oil, check the gearbox front bearing oil seal and the crankshaft rear oil seal. Renew whichever has failed.
5   The diaphragm spring and pressure plate should be examined for cracks or scoring and, if evident, a reconditioned (or new) unit should be purchased. Do not attempt to dismantle it yourself.
6   Finally, inspect the condition of the spigot bush in the centre of the flywheel mounting flange. If it is worn, renew it as described in Chapter 1, Section 8. Inspect the condition of the release bearing; if it is worn, or badly scored, replace it as described in Section 7.

7   Release bearing – removal and refitting

1   Dependent upon market destination, there are two types of release mechanism. The release bearing may be mounted on a shaft passing through the bellhousing, or it could be mounted into a release lever arm which pivots from the gearbox front bearing cover (photo).

2 To remove the pivot type bearing, proceed as described in the following paragraphs.

3 Grip the release bearing, and whilst pulling it as far forward as possible, give it a sharp tug. This will release the spring that attaches the release lever arm to the front bearing cover.

4 Remove the bearing and lever assembly from the input shaft, and the dust boot in the side of the bellhousing (photo).

5 The bearing can be removed from the release lever arm by tilting it until its location lips clear the cut-outs in the lever.

6 To remove the bearing which is mounted on the shaft passing through the bellhousing, proceed as described in the following paragraphs.

7 Undo and remove the location bolt securing the bearing to the cross-shaft.

8 Disconnect the cross-shaft from the bellhousing and slide it out of the bellhousing, releasing the bearing which can now be removed.

9 Refitting, in both cases, is the reverse of the removal procedure.

---

8 Clutch – refitting

1 Refitting the clutch assembly to the flywheel will necessitate centralising the driven plate, and for this either an old input shaft or a

3.4 Removing the cable mounting bracket

7.1 The clutch release bearing and lever assembly

7.4 Removing the clutch release bearing and lever assembly

5

8.3 Fitting the clutch assembly

8.4 Centralizing the clutch using the transmission input shaft

suitable piece of dowel (with a step at one end to engage in the input spigot bush in the centre of the flywheel) must be obtained.

2   Locate the driven plate against the face of the flywheel so that the appropriately marked side is towards the flywheel.

3   Offer up the pressure plate assembly to the flywheel and locate it on the positioning dowels. Check that the alignment marks made before removal are in line (photo).

4   Insert the guide tool through the splined hub of the driven plate so that the end of the tool locates in the flywheel spigot bush (photo).

5   Assemble the four standard and two pilot bolts to the pressure plate assembly and, ensuring that the guide tool is firmly located, tighten the bolts a turn at a time and in a diametrically opposite sequence to the specified torque.

6   Refit the gearbox or engine (whichever was removed) as described in Chapter 6 or Chapter 1, respectively. The purpose of centralising the driven plate is to permit the input shaft of the gearbox to pass through the driven plate and to engage in the spigot bush in the centre of the flywheel. Even so, when refitting either the gearbox or the engine, it may be necessary to rotate the crankshaft a little to align the splines on the driven plate and the input shaft.

7   When refitting is completed, the clutch free movement must be adjusted as described in Section 2.

9   Fault diagnosis – clutch

Symptom	Reason/s	Remedy
Judder when taking up drive	Loose engine/transmission mountings	Check and tighten all mounting bolts.
	Badly worn friction surfaces or friction plate contaminated with oil	Renew clutch parts as required, rectify oil leakage source.
	Worn splines on the transmission input shaft or driven plate hub	Renew friction plate and/or input shaft.
	Worn spigot bush in flywheel	Renew spigot bush.
Clutch drag (failure to disengage) so that gears cannot be meshed	Clutch clearance too great	Adjust cable.
	Damaged or misaligned pressure plate assembly	Renew pressure plate assembly.
	Driven plate sticking on input shaft splines due to rust. (After vehicle has been standing any length of time)	As a temporary measure engage top gear, apply handbrake, depress clutch and start engine. If badly stuck engine will not turn; renew friction plate.
Clutch slip (increase in engine speed does not result in increase in vehicle road speed especially on hills)	Friction surfaces worn out or oil contaminated	Renew friction plate, rectify oil leaks.
	Incorrect release bearing to diaphragm spring clearance	Adjust cable.
Noise evident on depressing clutch pedal	Dry, worn or damaged release bearing	Renew release bearing.
	Excessive play between friction plate hub splines and input shaft	Renew friction plate and/or input shaft.
Noise evident as clutch pedal released	Distorted friction plate	Renew friction plate.
	Insufficient pedal free travel	Adjust cable.
	Distorted or worn input shaft	Renew input shaft.

Chapter 6 Part A Manual transmission

See Chapter 13 for specifications and information applicable to 1979 thru 1983 North American models

Contents

Specifications

Transmission types .... 4 forward and 1 reverse gear, with synchromesh on all forward gears; optional 5 forward and 1 reverse gear, with synchromesh on all forward gears

Gear ratios

Except North America:

	4-speed (1300 cc)	5-speed (1300 cc)	4-speed (1000 cc)
1st	3·337 : 1	3·337 : 1	3·655 : 1
2nd	1·995 : 1	1·95 : 1	2·185 : 1
3rd	1·301 : 1	1·301 : 1	1·425 : 1
4th	1·000 : 1	1·000 : 1	1·000 : 1
5th	—	0··831 : 1	—
Reverse	3·337 : 1	3·337 : 1	3·655 : 1

North America:

	4-speed		5-speed
1st	3·655 : 1		3·655 : 1
2nd	2·185 : 1		2·185 : 1
3rd	1·425 : 1		1·425 : 1
4th	1·000 : 1		1·000 : 1
5th	—		0·827 : 1
Reverse	3·655 : 1		3·655 : 1

Lubricant capacity

4-speed	1·1 Imp quarts (1·4 US quarts) (1·3 litres)
5-speed	1·5 Imp quarts (1·8 US quarts) (1·7 litres)

Lubricant type:

Above −18°C (0°F)	SAE 90EP
Below −18°C (0°F)	SAE 80EP

Mainshaft

Maximum permissible run-out	0·0012 in (0·03 mm)
Maximum clearance between mainshaft and gears	0·006 in (0·15 mm)

Reverse idler gear

Maximum clearance between reverse idler gear bush and shaft	0·006 in (0·15 mm)

Shift fork

Maximum clearance between shift fork and clutch sleeve	0·020 in (0·5 mm)
Maximum clearance between shift fork and reverse idler gear	0·020 in (0·5 mm)
Maximum clearance between shift rod gate and control lever	0·031 in (0·8 mm)

Synchronizer ring

Clearance between synchronizer ring and side of gear when fitted:

New	0·059 in (1·5 mm)
Wear limit	0·031 in (0·8 mm)

Torque wrench settings

	lbf ft	kgf m
Shift lock spring caps	7 to 11	1·0 to 1·5
Plug for interlock pin hole	7 to 11	1·0 to 1·5
Reverse lock spring cap	33 to 40	4·5 to 5·5
Control lever-to-control rod end	20 to 25	2·8 to 3·4

6A

Shift fork set bolts	6 to 9	0·8 to 1·2
Mainshaft locknut:		
4-speed	116 to 174	16·0 to 24·0
5-speed	94 to 152	13·0 to 21·0
Reverse lamp switch	20 to 33	2·8 to 4·5

1 General description

A four-speed manual transmission is fitted to these vehicles as standard equipment. Dependent on the market destination, a five-speed manual transmission is optional. All forward gears are helically cut, which ensures quiet running; the reverse gear and reverse idler gear are spur (straight) cut.

The transmission case is of an aluminium alloy construction, with a detachable clutch housing and rear extension housing. The rear extension carries the change lever mechanism.

Note: *On some early models with a four-speed transmission, difficulty may be experienced with engaging second gear. A modified third/fourth synchronizer clutch hub has been introduced to overcome this. Further information should be available from Mazda dealers.*

2 Transmission – removal and refitting (4-speed and 5-speed)

1   In order to gain access to the underside of the vehicle, raise the front end and support it with wooden blocks. Alternatively, drive the vehicle onto ramps.
2   Disconnect the battery earth lead, and select neutral.
3   Working inside the vehicle, unscrew and remove the gear change lever knob. Lift off the gear lever cover, which is held to the floor aperture surround by elastic sewn into the cover. Remove the four bolts that retain the gear lever housing; lift out the gear lever and housing assembly (photo).
4   Working beneath the vehicle, drain the transmission oil into a container of suitable capacity (photo).
5   Undo and remove the nut and bolt that secure the exhaust pipe clip to the mounting plate, then remove the upper nut and bolt and the lower bolt, that retain the mounting plate to the clutch bellhousing. Remove the mounting plate, taking notice of the earth strap which is located to the upper nut and bolt.
6   Unhook and remove the clutch release bearing return spring. Unbolt and remove the clutch cable mounting bracket; tie the cable and mounting bracket to a convenient anchorage point, well out of the way.

7   Unbolt and remove the starter motor (refer to Chapter 10 if necessary).
8   Unscrew and remove the speedometer drive cable.
9   Disconnect the wires to the reversing light switch.
10 Undo and remove the four bolts that secure the lower cover to the clutch bellhousing. Remove the lower cover.
11 Mark the fitted position of the propeller shaft to ensure correct refitting. Remove the propeller shaft (refer to Chapter 7 if necessary).
12 Position a suitable jack beneath the engine oil sump; place a wooden block on the jack head for protection.
13 Remove the remaining nuts and bolt which retain the clutch housing to the rear of the cylinder block.
14 Remove the four nuts and washers that retain the crossmember to the transmission and the vehicle's floor pan. Remove the crossmember.
15 Carefully lower the jack and slide the transmission rearwards until the input shaft clears the clutch assembly. It is wise at this stage to seek the assistance of a second person to prevent the weight of the transmission hanging on the input shaft; if the transmission is allowed to hang on the input shaft, serious damage to the shaft may occur.
16 Remove the transmission away from the vehicle, to a suitable workbench or storage area.
17 When refitting, first ensure that the mating surfaces and locating dowels are free from dirt, paint, machining burrs etc, which would prevent easy assembly.
18 Mount the transmission on a suitable jack or supports beneath the vehicles.
19 Start the input shaft into the clutch disc, aligning the splines and move the transmission forward to locate it on the engine rear plate dowels.
20 Fit the bolts, flat washers and lockwashers and tighten.
21 Remove the transmission jack and refit the rear crossmember. Tighten the nuts.
22 Position the engine as necessary then fit the bolts, nuts and washers attaching the transmission mount to the rear crossmember. Tighten the nuts and remove the engine jack.
23 Refit the starter motor.
24 The remainder of the refitting procedure is the reverse of the removal sequence.

2.3 Removing the gear lever and housing

2.4 The transmission oil drain plug

3   Transmission – dismantling (4-speed)

1   Having removed the transmission from the vehicle, remove the clutch release bearing and lever from the clutch bellhousing. This is best removed by holding the lever as far back into the bellhousing as it will go, and pulling the release bearing forward off its spring mountings on the lever. Remove the lever by pulling it from its dome mounting. If required, ease out the rubber dust boot.

2   Where applicable, remove the release bearing by undoing the location bolt that secures the release bearing to the cross-shaft, disconnecting the cross-shaft from the bellhousing, and sliding the

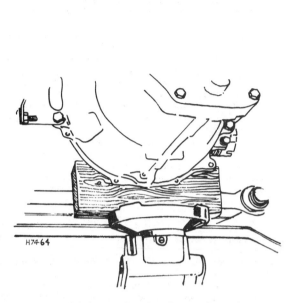

Fig. 6.1 Supporting the oil sump (Sec 2)

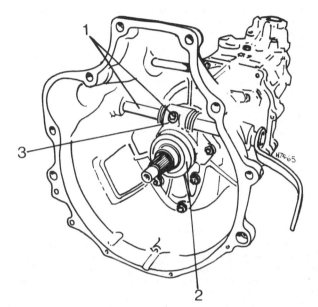

Fig. 6.2 An alternative type of clutch release bearing mechanism (Sec 3)

1   Bolt and shaft   3   Release lever
2   Release bearing

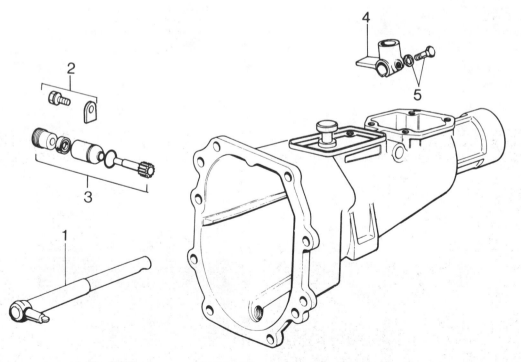

Fig. 6.3 The component parts of the rear extension housing (Sec 3)

1   Control lever shaft   3   Speedometer drive gear   4   Control lever end
2   Bolt and clamp   assembly   5   Bolt and washer

6A

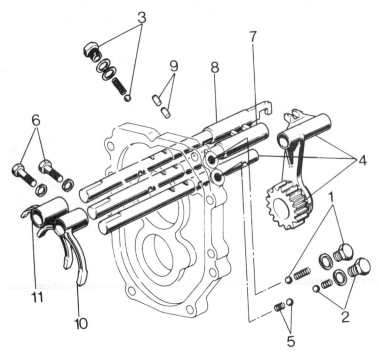

Fig. 6.4 The bearing housing and selector mechanisms (Sec 3)

1 Spring cap bolt, washer, spring and lock ball (3rd and 4th)
2 Spring cap bolt, washer, spring and lock ball
3 Spring cap bolt, washer, spring and lock ball (reverse)
4 Reverse idler gear, shift fork and rod
5 Spring and lock ball
6 Bolts and washers
7 Shift rod (3rd and 4th)
8 Shift rod (1st and 2nd)
9 Interlock pins
10 Shift fork (3rd and 4th)
11 Shift fork (1st and 2nd)

3.4 Removing the bearing front cover

3.8 The reverse light switch

3.9A Remove the bolt and clamp ...

3.9B ... and pull out the speedometer drive gear assembly

3.13 The speedometer drive gear circlip

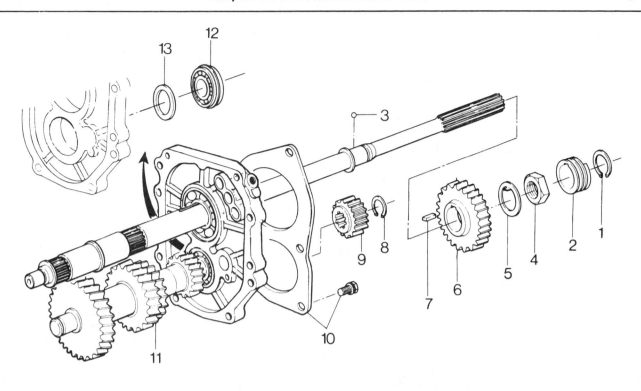

Fig. 6.5 The bearing housing assembly, showing the positions of the reverse gears and the speedometer drive gear (Sec 3)

1 Circlip	5 Lockwasher	8 Circlip	11 Countershaft
2 Speedometer drive gear	6 Reverse gear (mainshaft)	9 Countershaft reverse gear	12 Countershaft rear bearing
3 Ball	7 Woodruff key	10 Bearing cover bolt	13 Shim
4 Locknut			

shaft out; the release bearing can then be removed.

3   Remove the nuts attaching the front bearing cover to the transmission case.

4   Remove the front cover, shim and gasket (photo).

5   Prise out the circlip surrounding the mainshaft bearing.

6   Invert the transmission onto the bellhousing face and undo and remove the nuts that retain the rear extension housing to the bearing housing.

7   Rotate the gear change lever socket fully anti-clockwise to free the selector forks from the selector shaft. Lift off the rear extension housing.

8   Unscrew the reverse light switch from the extension housing (photo).

9   Remove the bolt and clamp plate and pull the speedometer drive gear assembly out of the rear extension (photos).

10 Unscrew and remove the bolt and washer that secure the control lever end to the control lever shaft. Draw the control lever shaft out of the rear extension housing, catching the control lever end as the shaft disengages from it.

11 Using a suitable extractor, remove the oil seal and bush from the rear end of the extension housing.

12 If considered necessary, remove the small top cover and gasket.

13 Remove the circlip which secures the speedometer gear to the mainshaft (photo). Slide the gear off the mainshaft and dislodge the ball from its recess.

14 Using a soft-faced hammer, tap the front end of the mainshaft and countershaft in turn and remove the bearing housing, as an assembly, from the case.

15 Unscrew the three spring caps, then remove the springs and shift locking balls.

16 Remove the shift lever attaching nut then push out the reverse shift rod, together with the reverse idler gear and shift lever, from the bearing housing.

17 Remove the bolts attaching the shift forks to the shift rods.

18 Push each shift rod rearwards through the fork and bearing housing to remove them.

19 Remove the reverse shift rod locking ball and spring, followed by the interlock pins.

20 Fold back the tab on the mainshaft reverse gear retaining nut. Whilst preventing the mainshaft from turning by clamping it in a vise with protective jaw clamps, unscrew the nut and remove the reverse gear and key.

21 Remove the circlip from the rear end of the countershaft and slide off the reverse countergear.

22 Remove the bearing cover from the bearing housing (five bolts and washers).

23 Using a soft-faced hammer, tap the rear end of the mainshaft and countershaft in turn to remove them from the housing. Take care that the shafts are not damaged. Separate the input shaft and bearing from the front end of the mainshaft. When removing the input shaft, watch out for the needle bearing, which may come off with the input shaft, or it could remain on the mainshaft; in either event it is best to remove the bearing and stow it somewhere safe. Place the countershaft and bearing in a safe place.

24 Press the bearings out of the bearing housing, retaining any shims. Ensure that load is only applied to the bearing outer race, or damage may occur. Note that the countershaft bearing is retained in the housing by a circlip.

25 Remove the circlip from the front end of the mainshaft and slide off the 3rd/4th clutch hub and sleeve assembly, synchronizer ring and 3rd gear.

26 Remove the thrust washer, 1st gear and sleeve, 1st/2nd clutch hub and sleeve assembly, synchronizer ring and 2nd gear from the mainshaft.

27 Store all the parts until they can be cleaned and inspected, as described in Section 5.

6A

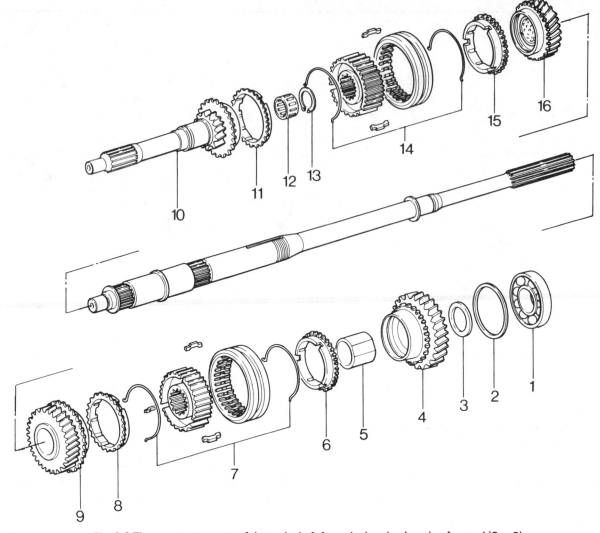

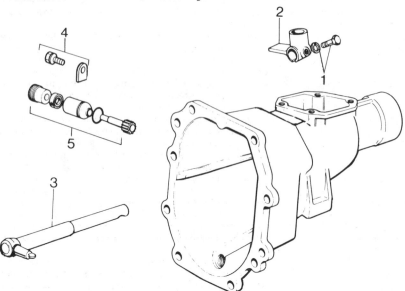

Fig. 6.6 The component parts of the mainshaft from the bearing housing forward (Sec 3)

1	Mainshaft bearing	6	Synchronizer ring
2	Shim	7	Clutch hub assembly (1st and 2nd)
3	Thrust washer		
4	First gear	8	Synchronizer ring
5	Gear sleeve	9	Second gear

10 Input shaft
11 Synchronizer ring
12 Needle bearing
13 Circlip

14 Clutch hub assembly (3rd and 4th)
15 Synchronizer ring
16 Third gear

Fig. 6.7 The component parts of the rear extension housing (5-speed) (Sec 4)

1 Bolt and washer
2 Control lever end
3 Control lever shaft
4 Bolt and clamp
5 The speedometer drive gear assembly

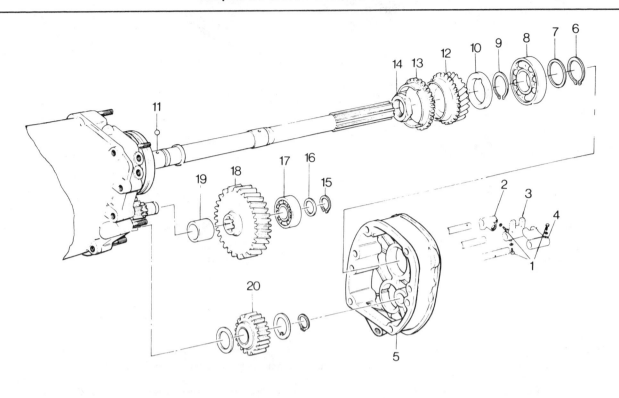

Fig. 6.8 An exploded view of the component parts after the intermediate housing has been removed (Sec 4)

1	Bolt and washers		reverse)
2	Shift rod end (1st and 2nd)	5	Intermediate housing
3	Shift rod end (3rd and 4th)	6	Circlip
		7	Thrust washer
4	Shift rod end (5th and	8	Mainshaft rear bearing
		9	Circlip

10	Thrust washer	16	Thrust washer
11	Ball	17	Countershaft rear bearing
12	Fifth gear (mainshaft)	18	Countershaft fifth gear
13	Synchronizer ring	19	Spacer
14	Locknut	20	Reverse idler gear
15	Circlip		

4 Transmission – dismantling (5-speed)

1   The four-speed and five-speed transmissions are very similar in design, so initially proceed as described in Section 3, paragraphs 1 to 11.
2   Remove the two circlips that retain the speedometer drive gear to the mainshaft. Slide off the gear and dislodge the location ball from its recess in the mainshaft. Stow the ball in a safe place.
3   Remove the bolts and washers securing the three shift rod ends to their shift rods. Pull off the shift rod ends.
4   Undo and remove the nuts and washers securing the intermediate housing to the bearing housing. Pull off the intermediate housing; if it is stubborn to remove, a few taps with a soft faced hammer around the mating faces of the flanges should break the joint. On no account try to prise it with sharp instruments, or damage to the sealing faces may result.
5   Using a soft faced hammer and a suitable piece of wood, tap on the end of the input shaft and the countershaft bearing in turn, until the bearing housing and gear assembly separate from the clutch bellhousing.
6   Remove the three spring cap bolts, washers, springs and lock balls from the bearing housing.
7   Undo and remove the bolts and washers securing the 3rd and 4th, and the 1st and 2nd, shift forks to their shift rods. Slide the shift rods out from the shift forks and the bearing housing. When the shift rods disengage from the bearing housing, watch out for the interlock pins and the lock ball and spring, which could be lost.
8   Undo and remove the bolt and washer securing the 5th and reverse gear shift fork, and slide the shift rod out of the shift fork and the bearing housing.
9   From the rear end of the mainshaft, remove the circlip and thrust

washer. Using a suitable puller, draw off the mainshaft rear bearing. Remove the circlip retaining the 5th gear thrust washer; pull off the thrust washer, taking care not to lose the location ball, followed by the 5th gear and 5th gear synchronizer ring.
10   Remove the circlip and thrust washer from the countershaft, and using a suitable puller, draw off the countershaft rear bearing, then remove the countershaft 5th gear and spacer, followed by the countershaft reverse gear.
11   Remove the circlip that retains the thrust washer, reverse idler gear and the spacer. Remove these items from their bearing housing shaft.
12   Undo and remove the five bolts securing the bearing cover to the bearing housing. Remove the cover.
13   Clamp the mainshaft between protected vice jaws and undo and remove the mainshaft lock nut.
14   Withdraw the 5th and reverse clutch hub assembly, and the mainshaft reverse gear and needle bearing from the mainshaft; pull off the inner race and thrust washer.
15   Using a soft faced hammer, tap the rear end of the mainshaft and the countershaft, until they disengage from their respective bearings in the housing.
16   When the mainshaft and the countershaft have been separated from the bearing housing, place the counter gear and bearing somewhere safe to await cleaning and inspection. Separate the input shaft and bearing from the front end of the mainshaft; watch out for the needle bearing, which may stay attached to the mainshaft, or remain in the input shaft.
17   With the bearing housing now separated from the shafts, prise out the circlip surrounding the mainshaft bearing and carefully press out the two bearings, retaining any shims that may be under them.
18   At the front end of the mainshaft, remove the circlip and draw off the 3rd and 4th clutch hub assembly, and the synchronizer ring,

6A

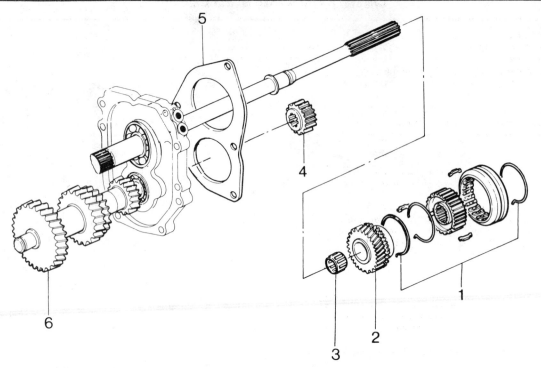

Fig. 6.9 The reverse gear components removed from the mainshaft and countershaft (5-speed) (Sec 4)

1 Clutch and hub assembly (5th and reverse)	2 Reverse gear	4 Countershaft reverse gear	6 Countershaft
	3 Needle bearing	5 Bearing cover	

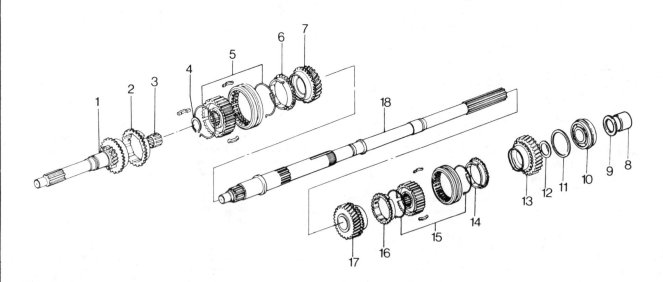

Fig. 6.10 The components parts of the mainshaft from the bearing housing forwards (Sec 4)

1	Input shaft		(3rd and 4th)	10 Mainshaft front bearing	15 Clutch and hub assembly
2	Synchronizer ring	6	Synchronizer ring	11 Shim	(1st and 2nd)
3	Needle bearing	7	Third gear	12 Thrust washer	16 Synchronizer ring
4	Circlip	8	Bearing inner race	13 First gear	17 Second gear
5	Clutch and hub assembly	9	Thrust washer	14 Synchronizer ring	18 Mainshaft

followed by the third gear.

19 From the rear end of the mainshaft, remove the thrust washer, first gear, the synchronizer ring, 1st and 2nd clutch hub assembly, synchronizer ring and the second gear. Stow all the components in a safe place, until it is time to clean and inspect them.

5 Transmission – cleaning and inspection

Cleaning

1 All parts, except seals and ball bearing assemblies, should be soaked in petrol or paraffin, brushed or scraped as necessary, and dried with compressed air.

2 Ball bearing assemblies may be carefully dipped into clean petrol or paraffin, and spun with the fingers. Whilst being prevented from turning, they should be dried with compressed air. After inspection (paragraphs 12 to 14) they should be lubricated with SAE 90EP transmission oil and stored carefully in clean conditions until ready for use.

Inspection – general

3 Inspect the transmission case and extension housing for cracks, worn or damaged bearings and bores, and damaged threads and machined surfaces. Small nicks and burrs can be locally dressed out using a fine file.

4 Examine the shift levers, forks, shift rods and associated parts for wear and damage.

5 Renew roller bearings which are chipped, corroded or rough running.

6 Examine the countershaft for damage and wear; renew it if this is evident.

7 Examine the reverse idler and sliding gears for damage and wear. Renew them if this is evident.

8 Renew the input shaft and gear if damaged or worn. If the roller bearing surface in the counterbore is damaged or worn, or the cone surface is damaged, renew the gear and gear rollers.

9 Examine all the gears for wear and damage, renewing as necessary.

10 Renew the mainshaft if bent, or if the splines are damaged.

11 Renew the seal in the transmission front bearing cover.

Ball bearing assemblies – inspection

12 Examine the inner and outer raceways for pitting and corrosion, renewing any bearings where this is found.

13 Examine the ball cage and races for signs of cracking, renewing any bearings where this is found.

14 Lubricate the raceways with a small quantity of SAE 90EP transmission oil then rotate the outer race slowly until the balls are lubricated. Spin the bearings by hand in various attitudes, checking for roughness. If any is found after the bearings have been cleaned

Fig. 6.11 Checking for wear between the shift control rods (Sec 5)

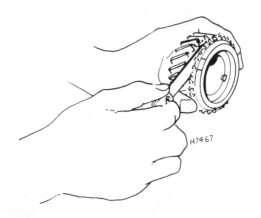

Fig. 6.12 Checking clearance on a synchronizing ring – typical (Sec 5)

6A

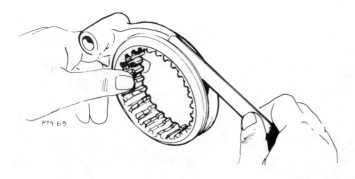

Fig. 6.13 Checking clearance between a shift fork and hub – typical (Sec 5)

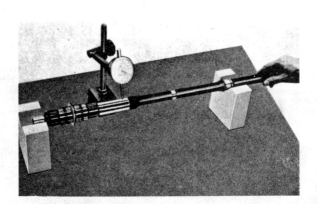

Fig. 6.14 Checking the mainshaft for run-out (Sec 5)

6.1A The components of the 3rd/4th synchronizer mechanism

6.1B Place the synchronizer keys into their slots ...

6.1C ... and fit the key springs

6.4 Fitting the circlip

6.7 Fitting the 1st/2nd clutch hub and sleeve assembly

6.12A The synchronizer ring on the input shaft

6.12B Fitting the input shaft to the mainshaft

6.16 The mainshaft and countershaft assembled to the bearing housing

6.17 The bearing cover in position

6.18A Fit the reverse gear, ...

6.18B ... the tab washer and locknut to the mainshaft

6.18C Bend the tab washer

(paragraph 2) they should be renewed. If they are satisfactory they should be stored carefully whilst awaiting assembly into the transmission.

Synchronizer mechanism

15 Inspect the synchronizer ring for chipped and worn teeth, renewing parts as necessary.

16 Fit the synchronizer ring to the gear cone and measure the clearance between the side faces of the synchronizer ring and the gear as shown in Fig. 6.12. If outside the specified limits, the parts must be renewed.

17 To check the contact between the cone surface and the inner surface of the synchronizer ring, apply a little engineer's blue to the cone surface and fit it to the ring. If the contact pattern is poor, it can be corrected by lapping with a fine lapping paste.

18 Check that the clutch sleeve slides freely on the clutch hub.

19 Check the synchronizer insert (key), the insert groove in the clutch hub and the inner surface of the clutch sleeve for wear.

20 Check that the synchronizer insert springs are undamaged.

Mainshaft run out

21 Mount the mainshaft on V-blocks and check for run-out using a dial test indicator. If the run-out exceeds the specified limit, the shaft must be renewed.

Reverse idler gear bushings and shaft

22 Check the fit of the reverse idler gear bush and shaft. If outside the specified limit, the gear must be renewed.

6 Transmission – reassembly (4-speed)

1 Assemble the 3rd/4th synchronizer mechanism by fitting the clutch hub onto the sleeve, placing the three synchronizer keys into the clutch hub key slots and fitting the key springs onto the clutch hub. To ensure uniform tension of the key springs, fit them as shown in Fig. 6.15 (photos).

2 Fit the 3rd gear and synchronizer ring onto the front section of the mainshaft.

3 Fit the 3rd/4th clutch hub and sleeve assembly onto the mainshaft (see Fig. 6.16).

4 Fit the circlip to the front end of the mainshaft (photo).

5 Assemble the 1st/2nd synchronizer mechanism in the same manner, as described for the 3rd/4th mechanism (paragraph 1).

6 Position the synchronizer ring on the 2nd gear and slide the gear onto the mainshaft with the synchronizer ring towards the rear.

7 Slide the 1st/2nd clutch hub and sleeve assembly onto the mainshaft, with the oil grooves of the clutch hub towards the front of the shaft. Ensure that the three synchronizer keys in the synchronizer mechanism engage the notches in the second synchronizer ring (see Fig. 6.16 for direction of fitting the clutch hub assembly) (photo).

8 Slide the 1st gear sleeve onto the mainshaft.

9 Place the synchronizer ring on the first gear and slide it onto the mainshaft with the synchronizer ring towards the front. Rotate the 1st gear as necessary to engage the three notches in the synchronizer ring with the keys in the clutch hub assembly.

10 Fit the thrust washer to the rear end of the mainshaft.

11 Position the needle roller bearing onto the front end of the mainshaft.

12 Position the synchronizer ring onto the input shaft. Fit this shaft to the front end of the mainshaft (photos).

13 It is now time to press the countershaft bearing into the bearing housing, but first the bearing end play must be established. Using a depth micrometer, measure the depth of the bearing bores. Record these figures. Now measure across the width of each bearing and subtract each of these figures from the respective bore depths. The difference should be no more than 0.0039 in (0.1 mm). If the figures are found to be in excess of this, there are two shim thicknesses available, namely 0.0039 in (0.1 mm) and 0.0118 in (0.3 mm), to correct the endplay. Having arrived at the correct shim size, lightly grease each shim and fit them into their bores in the bearing housing.

14 Press the countershaft rear bearing into the bearing housing. Take care that press loads are applied to the outer race only. Fit the circlip.

15 Fit the countershaft to its rear bearing. This is best carried out with a piece of tube of suitable length that will allow the countershaft spline

to enter it. Stand the tube on a firm surface, place the bearing housing onto the tube, with the bearing inner race firmly supported by the tube. Locate the countershaft into the bearing and, using a soft-faced hammer, tap the shaft into its bearing.

16 Position the mainshaft assembly into the bearing housing, checking that the thrust washer and 1st gear do not fall off, and that the mainshaft and countershaft gears line up properly. Use a soft-faced hammer to drive the bearing into the housing (see Fig. 6.19) (photo).

17 Fit the bearing cover to the bearing housing (photo).

18 Fit the reverse gear, tab washer and key onto the mainshaft and tighten the locknut whilst preventing the mainshaft from turning. Fold over the tab on the washer. **Note**: *Ensure that the mainshaft and countershaft reverse gears are fitted with the tooth chamfer to the rear* (photos).

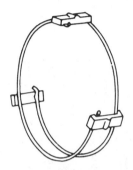

Fig. 6.15 The synchronizer key springs correctly positioned (Sec 6)

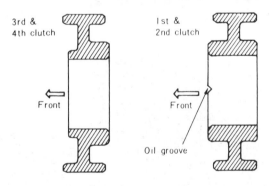

Fig. 6.16 Direction of fitting of the clutch hub assemblies (Sec 6)

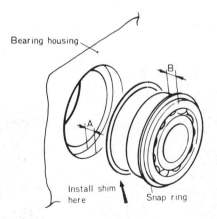

Fig. 6.17 Measuring the countershaft rear bearing endplay (Sec 6)

A – Bore depth　　　*B – Bearing width*

6A

6.19A Fit the countershaft reverse gear ...

6.19B ... and retain it with the circlip

6.20A The spring should be assembled first ...

6.20B ... followed by the locking ball

6.22 Fitting the shift rod together with the reverse idler gear

6.24 The first interlock pin in position

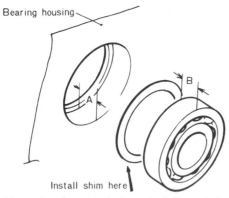

Fig. 6.18 Measuring the mainshaft rear bearing endplay (Sec 6)

A – Bore depth   B – Bearing width

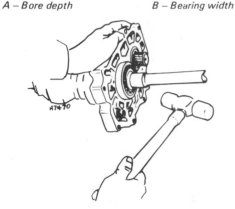

Fig. 6.19 Fitting the mainshaft bearing (Sec 6)

19 Fit the reverse gear to the countershaft and retain it with the circlip (photos).
20 Insert the locking ball and spring into the bearing housing, as shown in position 'A' of Fig. 6.20 (photos).
21 Push the ball down with a suitable screwdriver.
22 Fit the reverse shift rod and lever, together with the reverse gear idler (photo).
23 Place the reverse shift rod into the neutral position then fit the 1st/2nd shift fork and 3rd/4th shift fork to their respective clutch sleeves.
24 Insert the first interlock pin (see Fig. 6.21). The simplest way of doing this is to 'stick' it to the end of a screwdriver using a blob of grease and feed it through the passageway in the direction of the arrow (photo).
25 Fit the 3rd/4th shift rod through the holes in the bearing housing and fork (photo).
26 Place the shift rod in neutral then fit the second interlock pin using the same procedure as previously (paragraph 24) (See Fig. 6.22).
27 Fit the 1st/2nd shift rod (photo).
28 Align the lockbolt holes of the shift fork and rod, and fit the lockbolts (photo).
29 Fit the shift locking balls and springs into their respective locations and fit the retaining caps.
30 Fit the speedometer drive gear and ball, and secure it with the circlip (photo).
31 Apply a thin film of non-setting gasket sealant to the contact faces of the bearing housing. Fit the housing to the transmission case (photo).
32 Fit the input shaft front bearing and countershaft front bearing to the transmission case, applying load to the outer race only.
33 Secure the input shaft front bearing with the circlip (photo).
34 Fit the speedometer driven gear to the extension housing. Retain it with the locking plate and bolt.
35 Insert the control lever shaft through the hole from the front of the extension housing. Fit the control lever end to the shaft and secure them with the bolt and washer.
36 Fit the extension housing to the bearing housing, with the control

6.25 Fitting the 3rd/4th shift rod

6.27 Fitting the 1st/2nd shift rod

6.28 A shift fork lockbolt

6.30 Fitting the speedometer drive gear

6.31 Fitting the bearing housing assembly

6.33 The front bearing circlip in position

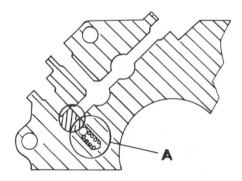

Fig. 6.20 The position of the shift locking ball (Sec 6)

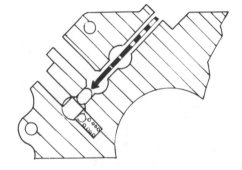

Fig. 6.21 Inserting the first interlock pin (Sec 6)

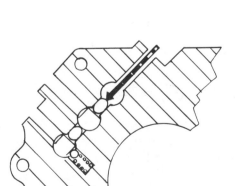

Fig. 6.22 Inserting the second interlock pin (Sec 6)

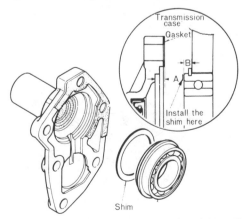

Fig. 6.23 Input shaft bearing endplay (A-B) (Sec 6)

rod end turned fully in the anti-clockwise position. Tighten the attaching nuts, then check that the control rod operates properly.

37 Fit the reverse light switch.

38 Fit the oil seal and bush at the tail end of the extension housing, after first lubricating the seal lips with transmission oil.

39 Lubricate the lips at the front bearing cover oil seal; fit the gasket, shim and cover (photo).

40 Check the clearance between the front bearing cover and the bearing outer race. Alter the shim thickness (paragraph 39) to obtain a clearance of less than 0.0039 in (0.1 mm) (see Fig. 6.23). Two shim thicknesses are available, namely 0.0039 in (0.1 mm) and 0.0118 in (0.3 mm).

41 Fit the release bearing and release fork. This is best carried out by fitting the bearing to the fork, then sliding the lever through the bellhousing aperture, locating the bearing over the input shaft. A sharp tap will locate the spring behind the lever to its mating dome on the front cover.

42 Where applicable, assemble the clutch release bearing by locating the cross-shaft into the bellhousing, fitting the bearing over the input shaft, and sliding the cross-shaft through the bearing and into its oposite location. Secure the bearing and shaft with the bolt.

7 Transmission – reassembly (5-speed)

1 Initially proceed as described in Section 6, paragraphs 1 to 13.

2 With the correct shims fitted in the bearing housing, press the countershaft centre bearing into the bearing housing. Take care that press loads are applied to the outer race only.

3 Fit the countershaft to its centre bearing. This is best carried out with a piece of tube of suitable length, that will allow the countershaft spline to enter it. Standing the tube on a firm surface, place the bearing housing onto the tube with the bearing inner race firmly supported by the tube. Locate the countershaft into the bearing and, using a soft-faced hammer, tap the shaft into its bearing.

4 Position the mainshaft assembly into the bearing housing, checking that the thrust washer and 1st gear do not fall off, and that the mainshaft and countershaft gears line up properly. Use a soft-faced hammer to drive the bearing into the housing (see Fig. 6.19). Don't forget the circlip.

5 Fit the bearing cover to the bearing housing.

6 To the countershaft rear end, press on the countershaft reverse gear. Then, to the mainshaft, assemble the thrust washer and bearing inner race, followed by the needle bearing, reverse gear and the 5th and reverse clutch hub assembly. Fit the lock tab and lock nut; tighten the nut to the specified torque and bend over the lock tab.

7 Assemble the synchronizer ring to the 5th gear and fit them to the mainshaft. Insert the location ball and slide the thrust washer over it. Fit the circlip, then check the 5th gear endplay using feeler gauges (see Fig. 6.24). This should be 0.0039 to 0.0118 in (0.1 to 0.3 mm). The following oversize thrust washers are available to correct the endplay; 0.2520 in (6.4 mm), 0.2559 in (6.5 mm), 0.2598 in (6.6 mm) or 0.2638 in (6.7 mm).

8 Using a soft-faced hammer and a suitable piece of hard wood, drive on the mainshaft rear bearing. Ensure that the load is applied to the bearing inner race. Fit the thrust washer.

9 Fit the circlip and, using feeler gauges, measure the bearing endplay (see Fig. 6.25), which should be less than 0.0039 in (0.1 mm). If the endplay is beyond this figure, select a thrust washer from the following available oversizes; 0.0748 in (1.9 mm), 0.0787 in (2.0 mm), 0.0827 in (2.1 mm) or 0.0866 in (2.2 mm).

10 To the countershaft, assemble the spacer and the countershaft 5th gear. Fit the countershaft rear bearing, and using a suitable size of tube, drive the bearing onto the shaft. Ensure that the load is applied to the bearing inner race.

11 Fit the thrust washer and circlip and measure the endplay as previously described (Fig. 6.26). The endplay should be less than 0.0039 in (0.1 mm). To correct the endplay, oversize thrust washers are available as follows; 0.0708 in (1.8 mm). 0.0748 in (1.9 mm), 0.0787 in (2.0 mm) or 0.0827 in (2.1 mm).

12 Assemble the spacer and the reverse idler gear to their shaft. Retain them with the spacer and circlip.

13 Insert the locking ball and spring into the bearing housing, as shown in position 'A' of Fig. 6.20. Push the ball down with a suitable screwdriver and slide in the 5th and reverse shift rod. Locate the 5th and reverse shift fork over its clutch and hub assembly, and slide the rod into the shift fork; secure with the bolt and washer.

14 Place the shift rod into the neutral position then fit the 1st/2nd shift fork and 3rd/4th shift fork to their respective clutch hub assemblies.

6.39 The front bearing cover and shims

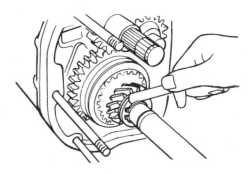

Fig. 6.24 Measuring the fifth gear endplay (Sec 7)

Fig. 6.25 Measuring the mainshaft rear bearing endplay (Sec 7)

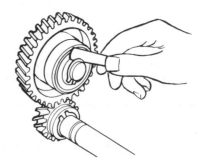

Fig. 6.26 Measuring the countershaft rear bearing endplay (Sec 7)

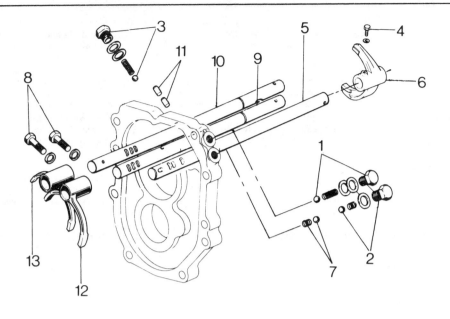

Fig. 6.27 The selector mechanism (5-speed) (Sec 7)

1   *Spring cap bolt, washer,*
    spring and lock ball
2   *Spring cap bolt, washer,*
    spring and lock ball
3   *Spring cap bolt, washer,*

    spring and lock ball
4   *Bolt and washer*
5   *Shift rod (5th and*
    reverse)
6   *Shift fork (5th and*

    reverse)
7   *Spring and lock ball*
8   *Bolt and washer*
9   *Shift rod (3rd and 4th)*

10   *Shift rod (1st and 2nd)*
11   *Interlock pins*
12   *Shift fork (3rd and 4th)*
13   *Shift fork (1st and 2nd)*

15 Insert the first interlock pin (see Fig. 6.21). The simplest way of doing this is to 'stick' it to the end of a screwdriver using a blob of grease and feed it through the passageway in the direction of the arrow.

16 Fit the 3rd/4th shift rod through the hole in the bearing housing and fork.

17 Place the shift rod in neutral then fit the second interlock pin using the same procedure as previously (paragraph 15) (see Fig. 6.22).

18 Fit the 1st/2nd shift rod.

19 Align the lockbolt holes of the shift fork and rod, and fit the lockbolts.

20 Fit the shift locking balls and springs into their respective locations and fit the retaining caps.

21 Stand the bellhousing on end, and carefully lower the gear assembly onto the studs; use a little non-setting gasket sealant on the mating faces.

22 Smear a little non-setting gasket sealant to the mating faces of the intermediate housing and the bearing housing. Fit the intermediate housing, and tap it into position with a soft-faced hammer.

23 Fit the input shaft front bearing and countershaft front bearing to the transmission case, applying load to the outer races only. Secure the input shaft bearing with the circlips.

24 Fit the 1st/2nd shift rod end to its shift rod and secure it in position

with the bolt and washer.

25 Fit the 3rd/4th and the 5th/reverse shift rod ends to their respective shift rods. Secure them in their positions with the bolts and washers.

26 Insert the speedometer drive gear locating ball into its recess, fit the first circlip, slide on the drive gear and fit the second circlip.

27 Fit the speedometer driven gear assembly to the rear extension housing. Secure it in position with the clamp plate and bolt.

28 Insert the control lever shaft through the hole from the front of the extension housing. Fit the control lever end to the shaft and secure with the bolt and washer.

29 Fit the extension housing to the bearing housing, with the control rod end turned fully in the anti-clockwise position. Tighten the attaching nuts then check that the control rod operates correctly.

30 Fit the reverse light switch.

31 Fit the oil seal and bush at the tail end of the extension housing, after first lubricating the seal lips with transmission oil.

32 Lubricate the lips at the front bearing cover oil seal; fit the gasket, shim and cover.

33 Check the clearance between the front bearing cover and the bearing outer race, as described in Section 6, paragraph 40.

34 Fit the clutch release bearing and release lever, as described in Section 6, paragraph 41, (or where applicable, paragraph 42).

6A

8 Fault diagnosis – manual transmission

Symptom	Reason/s	Remedy
Weak or ineffective synchromesh		
General wear	Synchronizing cones worn, split or damaged	Dismantle and overhaul transmission Fit new synchromesh unit
	Synchromesh teeth worn or damaged	Dismantle and overhaul transmission Fit new synchromesh unit

Jumps out of gear

General wear or damage	Shift rods worn or damaged	Dismantle and fit new shift rods
	Detent springs broken	Fit new detent springs

Excessive noise

Lack of maintenance	Incorrect grade of transmission oil, or oil level too low	Drain, refill and top up with correct grade of oil
	Bush or needle roller bearings worn or damaged	Dismantle and overhaul transmission Renew bearings
	Gear teeth excessively worn or damaged	Dismantle and overhaul transmission Renew gears

Excessive difficulty in engaging gear

Clutch not fully disengaging	Clutch pedal adjustment incorrect	Adjust clutch pedal correctly
Shift rod interlock ineffective	Shift rods or interlock pins damaged or worn	Dismantle and fit new shift rails and interlock pins
General wear	Worn shift linkage	Dismantle shift linkage and renew parts as necessary

Chapter 6 Part B Automatic transmission

See Chapter 13 for specifications and information applicable to 1979 thru 1983 North American models

Contents

Specifications

Type .... JATCO

Gear ratios

1st	2·458 : 1
2nd	1·458 : 1
3rd	1·000 : 1
Reverse	2·181 : 1

Fluid capacity .... 5·0 Imp quarts (6·0 US quarts) (5·7 litres)

Fluid type .... M2C33F (Type F, eg Castrol TQF)

Stall speed .... 1750 to 2000 rpm

Shift speeds

Throttle condition (manifold vacuum)	Gear shift	Shift speed; mph (km/h)
Wide open throttle (0–3·94 in Hg) (0–100 mm Hg)	D_1–D_2	29–40 (45–63)
	D_2–D_3	50–66 (79–107)
	D_3–D_2	44–57 (70–92)
	D_2–D_1	20–29 (31–47)
Half throttle (7·87 ± 0·39 in Hg) (200 ± 10 mm Hg)	D_1–D_2	8–16 (11–26)
	D_2–D_3	15–35 (23–55)
Fully closed throttle	D_3–D_1	6–12 (9–19)
Manual 1	1_2–1_1	23–30 (36–48)

Torque wrench settings

	lbf ft	kgf m
Driveplate to crankshaft	60 to 69	8·3 to 9·5
Driveplate to torque converter	25 to 36	3·5 to 5·0
Converter housing to engine	23 to 34	3·2 to 4·7
Oil pan	3·6 to 5·1	0·5 to 0·7
Inhibitor switch	3·6 to 5·1	0·5 to 0·7
Manual shaft locknut	22 to 29	3·0 to 4·0
Oil cooler pipe set bolt	17 to 26	2·4 to 3·6
Actuator for parking rod to extension housing	5·8 to 8·0	0·8 to 1·1

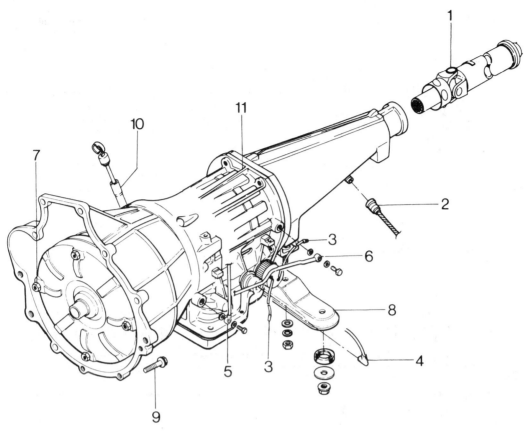

Fig. 6.28 The automatic transmission (Sec 10)

1	Propeller shaft	4 Vacuum tube	7 Bolt	10 Oil filler tube
2	Speedometer cable	5 Oil pipe	8 Crossmember	11 Transmission
3	Downshift solenoid leads	6 Oil pipe	9 Bolt	

9 General description

The automatic transmission is conventional in design, having a torque converter and one reverse and three forward gears. The gears are epicyclic and are engaged by hydraulically actuated clutches and brake bands.

The transmission, manufactured by the Japanese Automatic Transmission Company, has a floor-mounted gear selector lever by which six positions can be selected, these being 'P', 'N', 'R', 'D', '2' and '1'. These operate as follows: When 'P' (Park) is selected, the drive to the rear wheels is locked. This range should never be selected while the vehicle is in motion, but should be used when parking. This position can also be used for starting the engine. When 'N' (Neutral) is selected, there is no drive to the rear wheels. This position can be used for engine starting, and should always be used when the engine is idling for any period of time. 'R' (Reverse) corresponds to a normal reverse gear on manual transmission. Before selecting 'R', ensure that the vehicle is stationary. The 'D' (Drive) selection is the normal driving position where gear changing is automatic. '2' (second gear manual) should be used for slippery surfaces, traffic braking and steep descents. The transmission remains in second gear; it should not be selected at road speeds above 60 mph (95 km/h). '1' (low gear manual) is for sustained pulling power or braking on hilly roads. To avoid skidding, this range should not be selected above 20 mph (31 km/h) on slippery surfaces. When in first gear, but with 'D' selected, the car will free-wheel if the engine is on overrun, as the drive goes through the one way clutch. If in '1' then the low and reverse brake is selected, and there is engine braking.

There is no oil pump on the output shaft, so tow (bump) starting is not possible with the automatic transmission. If the car needs a tow in the event of a breakdown, the propeller shaft should be removed.

In view of the complex nature of the automatic transmission, it is not recommended that any attempt is made to carry out repair operations other than those given in this Chapter. In the event of a transmission fault occurring, the transmission may be removed (see Section 9) and should then be forwarded to a vehicle main dealer or automatic transmission specialist for repair.

10 Automatic transmission – removal and fitting

1 Disconnect the battery earth lead.
2 Raise the vehicle on a hoist or suitable axle stands, or alternatively place it over an inspection pit, for access beneath.
3 Place a large drain pan beneath the transmission. Starting at the rear of the transmission oil pan, and working towards the front, loosen the attaching bolts and allow the fluid to drain.
4 Remove the oil pan bolts except for two at the front and allow the fluid to drain further. Now fit two bolts at the rear to hold the oil pan in place.
5 Remove the propeller shaft (refer to Chapter 7 for further information).
6 Disconnect the speedometer drive cable from the extension housing.
7 Disconnect the shift rod from the manual control lever at the transmission.
8 Remove the hose connection from the vacuum diaphragm. Disconnect the wires from the downshift solenoid and inhibitor switch; remove the wires from the clip.
9 Disconnect the transmission oil pipes.
10 Remove the access cover from the converter housing lower end.
11 Index mark the driveplate and torque converter relationship, then remove the four bolts attaching the converter to the drive plate.

6B

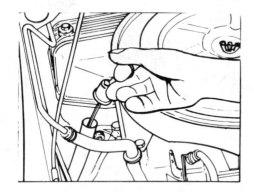

Fig. 6.29 The dipstick/filler tube (Sec 11)

Fig. 6.30 The dipstick level markings (Sec 11)

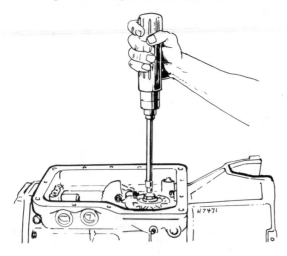

Fig. 6.31 Adjusting the brake band (Sec 12)

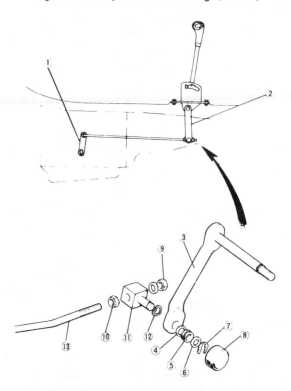

Fig. 6.32 The manual linkage (Sec 14)

1	Manual lever	7	Retaining clip
2	Selector lever operating arm	8	Dust cover
3	Selector lever operating arm	9	Nut
		10	Nut
4	Washer	11	T-joint
5	Bush	12	Wave washer
6	Washer	13	Shift rod

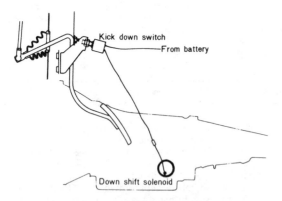

Fig. 6.33 The kickdown switch (Sec 15)

12 Remove the transmission rear support to crossmember nuts and bolts.

13 Support the transmission with a suitable jack then remove the crossmember to frame attaching bolts and remove the crossmember.

14 Carefully lower the transmission, whilst at the same time supporting it.

15 Remove the starter motor, then the converter housing to engine attaching bolts. Remove the transmission dipstick/filler tube.

16 Using a suitable lever, exert pressure between the driveplate (flexplate) and the converter to prevent the converter from disengaging from the transmission.

17 Lower the transmission and move it rearwards as an assembly,

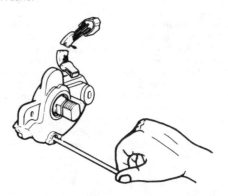

Fig. 6.34 Adjusting the inhibitor switch (Sec 16)

from beneath the vehicle.

18 When refitting ensure that the converter is properly fitted and raise the transmission on a suitable jack beneath the vehicle.

19 Raise the transmission into position and fit the converter housing to engine attaching bolts. Tighten the bolts to the specified torque.

20 Lower the transmission and fit the starter motor.

21 Fit the dipstick/filler tube and O-ring on the transmission case and secure it to the cylinder block with the bolt.

22 Raise the transmission. Fit the crossmember to the frame and the rear support to the crossmember.

23 Align the index marks on the torque converter and driveplate. Fit and tighten the bolts to the specified torque.

24 Fit the converter housing access cover and remove the jack.

25 Connect the oil pipes, electrical leads and the vacuum diaphragm hose.

26 Connect the shift rod to the transmission manual lever.

27 Connect the speedometer drive cable.

28 Fit the driveshaft (refer to Chapter 7, if necessary).

29 If the oil pan has not been refitted properly, remove it completely, fit a new gasket using a non-setting gasket sealant, position the oil pan and fit the attaching bolts. Tighten the bolts in a crosswise order to the specified torque.

30 Lower the vehicle and connect the battery. Fill the transmission with the specified fluid, run the engine and check for fluid leaks.

11 Automatic transmission – checking the fluid level

1 The automatic transmission fluid level should be checked at intervals of 4000 miles (6500 km) using the procedure given in the following paragraphs.

2 Park the vehicle on level ground and apply the handbrake.

3 Run the engine at idle speed. (If the engine is cold, run at a fast idle for about five minutes then slow down to normal idle speed).

4 With engine idling, select each position in turn then leave it at 'N' or 'P'.

5 With the engine still idling, pull out the dipstick, wipe it clean with a lint-free cloth, push it back fully into the tube then pull it out again.

6 Observe the level on the dipstick. After idling for about two minutes the level should be read on the cold side of the dipstick; after idling for about five minutes the level should be read on the hot side of the dipstick. The fluid level should be between the 'L' and 'F' marks **Note**: A cold fluid temperature is approximately 104°F(40°C) and a hot fluid temperature is approximately 176°F (80°C).

7 When any topping-up is done, fluid should be poured into the filler tube to bring it up to the correct level. Under no circumstances must the transmission be overfilled, or foaming of oil will occur once it gets hot and expands; this will cause loss of fluid from the vent which may lead to a transmission malfunction.

12 Brake band – adjustment

1 Although not part of the routine maintenance procedure, brake band adjustment may become necessary. Typical faults which can be attributed to an incorrectly adjusted brake band are:

   a) *Poor acceleration, with a low maximum speed*
   b) *The vehicle is braked when 'R' is selected*
   c) *No 'D$_1$' – 'D$_2$' change, or excessive shock when changing*
   d) *Unsatisfactory change to 'N' from 'D$_3$'*
   e) *Transmission overheating and/or fluid discharge*

2 To adjust the brake band, drain the transmission fluid, as described in Section 10.

3 Remove the oil pan completely, thoroughly clean it and discard the gasket.

4 Loosen the locknut on the brake band adjusting screw then tighten the screw to a torque of 9 to 11 lbf ft (1.24 to 1.52 kgf m). Back off the screw two full turns then hold the screw while the locknut is tightened to a torque of 22 to 29 lbf ft (3 to 4 kgf m).

5 Place a new gasket on the oil pan using a non-setting gasket sealant. Fit the oil pan, torque-tightening the bolts in a crosswise order.

6 Lower the vehicle to the ground and fill the transmission with the recommended fluid.

13 Vacuum diaphragm – removal and refitting

1 Disconnect the vacuum diaphragm hose.

2 Unscrew the diaphragm unit by hand. Remove the O-ring and diaphragm rod from the case.

3 When refitting, first position the diaphragm rod in the transmission case.

4 Fit a new O-ring on the vacuum unit. Screw in the vacuum unit firmly by hand and connect the hose.

14 Selector lever – removal, refitting and linkage adjustments

1 Raise the vehicle, then remove the nut and lockwasher securing the selector lever operating arm to the selector lever.

2 Lower the vehicle, then remove the bolts which secure the selector lever assembly to the floorpan; the selector lever assembly can then be removed.

3 When refitting, follow the reverse of the removal procedure.

4 To adjust the linkage, select 'N' then disconnect the clevis from the lower end of the selector lever operating arm.

5 Move the transmission manual lever to the third detent from the rear of the transmission.

6 Loosen the clevis retaining nuts and adjust the clevis so that it freely enters the hole in the lever. Tighten the retaining nuts.

7 Connect the clevis to the lever and secure it with the spring washer, flat washer and the retainer.

8 Lower the vehicle and test the transmission operation during a test drive.

15 Kickdown switch – removal, fitting and adjustment

1 Disconnect the wire from the kickdown switch.

2 Remove the switch retaining nut and remove the switch from the bracket.

3 When refitting the switch, follow the reverse procedure to removal. Adjust the switch to engage when the accelerator pedal is between $\frac{7}{8}$ and $\frac{15}{16}$ of full pedal travel.

4 On completion, test drive the vehicle to ensure that the switch is functioning correctly.

16 Inhibitor switch – removal, fitting and adjustment

1 Place the transmission lever in the 'N' position, then raise the vehicle.

2 Disconnect the multiple connector, then disengage the shift rod from the manual lever.

3 Remove the transmission manual lever retaining nut and washer; remove the lever.

4 Undo and remove the switch retaining bolts. Lift away the switch.

5 When refitting, leave the switch retaining bolts loose, then remove the screw from the alignment pin hole at the bottom of the switch.

6 Rotate the switch and insert an alignment pin of 0.078 in (2 mm) (the unmarked shank of a No. 47 drill should be suitable) through the alignment hole and into the hole of the internal rotor.

7 Tighten the switch attaching bolts and remove the alignment pin.

8 Fit the alignment pin screw in the switch body and connect the multiple connector.

9 Position the transmission manual lever on the manual lever shaft; secure with the washer and nut. Connect the shift rod to the manual lever.

10 Check that the engine will only start when 'N' or 'P' is selected.

6B

Chapter 7 Propeller shaft

See Chapter 13 for specifications and information applicable to 1979 thru 1983 North American models

Contents

Specifications

Maximum permissible run-out .... 0.016 in (0.4 mm)

Universal joint
Spider diameter wear limit .... 0.4996 in (12.689 mm)
Selective circlip availability .... 0.0480 to 0.543 in (1.22 to 1.38 mm) in 0.0008 in (0.02 mm) increments

Torque wrench settings

	lbf ft	kgf m
Yoke to rear axle companion flange	25 to 27	3.5 to 3.8

1 General description

Drive is transmitted from the transmission to the rear axle via a propeller shaft of single piece tubular steel construction, with a universal joint at either end. At the front end of the shaft, an internally splined sliding sleeve connects with the gearbox mainshaft. The sliding sleeve is so designed to cater for fore-and-aft movement of the propeller shaft which occurs due to the deflection of the rear axle on the coil type road springs. The rear end of the propeller shaft is connected to the companion flange of the rear axle by the yoke.

2 Universal joints – testing for wear

1 Wear in the needle roller bearings is characterized by vibration in the transmission, 'clonks' on taking up drive, and, in extreme cases, metallic squeaking, ultimately ending in grating and shrieking sounds as the bearings break up.
2 To check if the needle roller bearings are worn with the propeller-shaft in position, try to turn the shaft with one hand, the other hand holding the rear axle companion flange, when checking the rear universal joint. When checking the front universal joint, hold the splined coupling. Any movement between the propeller shaft and the rear axle companion flange, or the shaft and the splined coupling, is indicative of considerable wear.
3 Now lift the shaft and watch for any movement between the universal joints and their respective adjoining components.
4 If wear is evident as described in the previous paragraphs, the propeller shaft will have to be removed and overhauled.

3 Propeller shaft – removal and refitting

1 Jack up the rear of the vehicle or position the vehicle over a pit.
2 If the rear of the vehicle is jacked up, supplement the jack with support blocks or axle stands under the bodyframe members so that any danger of the jack collapsing is minimized.
3 If the rear wheels are off the ground, place the car in gear and also apply the handbrake to prevent the propeller shaft turning when the rear flange securing bolts are loosened.
4 The propeller shaft is carefully balanced to fine limits and it is

extremely important that it is replaced in exactly the same position it was in prior to removal. Scratch marks on the driveshaft and rear axle flange to ensure accurate mating when the time comes for refitting.
5 Undo and remove the four bolts which secure the rear yoke to the companion flange (photo).
6 Carefully withdraw the propeller shaft from the transmission, taking care not to damage the rear oil seal. Suitably plug the end of the transmission to prevent loss of the transmission oil (photo).
7 Refitting is the reverse of the removal procedure but ensure that the transmission end of the propeller shaft is well lubricated with transmission oil and that care is taken not to damage the oil seal as the shaft is inserted. Also ensure that the alignment marks on the rear yoke are correctly mated with those on the companion flange. Top up the transmission if any lubricant has been lost.

4 Propeller shaft – dismantling and reassembly

1 Before work commences clean away all traces of dirt from around the universal joint. If the joints are of the type where a steel plug is pressed in on top of the circlips, these will have to be removed using a small chisel or steelshaft screwdriver, which can be driven through to prise them out.
2 Using suitable pliers or a small screwdriver, remove the circlips. It does not matter if the circlips are distorted because new ones must be fitted on reassembly. If they are difficult to remove, a hammer blow on the end of the joint spider may relieve the pressure on them.
3 To remove the bearing cups select two sockets, one large enough to fit completely over the bearing cup and the other to fit on the end of the cup. By compressing in the jaws of a vice, or by hammering carefully on the smaller socket, the spider and cups can be driven out of one side of the yoke. Penetrating oil will usually assist if the job proves really stubborn. Having removed one cup, the opposite one can be removed by reversing the direction of applied pressure. The remaining ones are then treated in a similar manner.
4 Having dismantled the universal joint, carefully clean the parts and inspect them for wear and other damage. If a joint spider is worn below the permissible limit, it must be renewed regardless of the bearing condition.
5 Examine the propeller shaft splines for wear. If worn it will be necessary to purchase a new front half coupling, or if the yokes are badly worn, an exchange shaft will probably prove to be the best course of action, both time-wise and financially.

3.5 The rear yoke-to-companion flange bolts

3.6 Withdraw the propeller shaft from the transmission

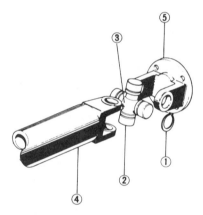

Fig. 7.1 Components of the propeller shaft rear end (Sec 4)

1 Circlip   4 Propeller shaft
2 Needle bearing   5 Yoke
3 Spider

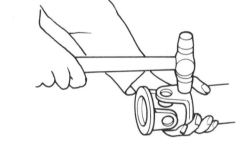

Fig. 7.2 Removal of universal joint bearings

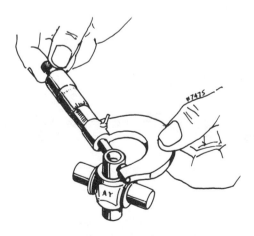

Fig. 7.3 Measuring a bearing spider

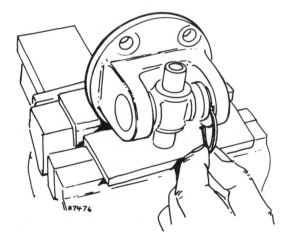

Fig. 7.4 Assembling a universal joint

7

6   When reassembling the universal joints, ensure that all the parts are clean, then lightly grease the holes in the yokes. Ensure that the bearing cups are approximately $\frac{1}{3}$ full of general purpose grease.

7   Insert the spider into the yoke and carefully press in two opposing cups and seals. The vice and sockets can again be used for this job as they were when dismantling. Circlips are available in selective sizes so that the universal joint can be positioned centrally and then all endplay can be taken up. Ensure that the opposing circlips are of equal size to obtain optimum balance.

8   Repeat paragraph 7 for the remaining pair of bearings on the universal joint.

9   Where applicable, apply a little grease to the end of the bearing cups and fit the steel plugs. Apply soft hammer blows to the centre of the plugs, if necessary, to enable them to spread and so obtain a tight fit.

5   Fault diagnosis – propeller shaft

Symptom	Reason
Vibration when vehicle running on road	Out of balance or distorted propeller shaft
	Wear in splined shaft coupling
	Loose flange securing bolts
	Worn universal joint bearings
Knock or 'clunk' when taking up drive or changing gear	Loose rear flange bolts
	Worn universal joint bearings
	Worn rear axle pinion splines
	Excessive wear in differential gears
	Loose roadwheel nuts

Chapter 8 Rear axle

See Chapter 13 for specifications and information applicable to 1979 thru 1983 North American models

Contents

Specifications

Type	Semi-floating, hypoid	
	1000 cc	**1300 cc**
Ratio (Except North America)		
Manual transmission	4.10 : 1	3.909 : 1
Automatic transmission	–	4.10 : 1
Ratio (North America)		
Manual transmission	3.727 : 1	
Automatic transmission	4.10 : 1	

Lubricant
Type:
    Above −18°C (0°F) . . . .   SAE 90
    Below −18°C (0°F) . . . .   SAE 80
Capacity . . . .   0.7 Imp quarts (0.8 US quarts) (0.8 litres)

Crownwheel to pinion clearance . . . .   0.0059 to 0.0067 in (0.15 to 0.17 mm)

Maximum allowable variation of clearance . . . .   0.0028 in (0.07 mm)

Side gear to pinion gear clearance . . . .   0 to 0.004 in (0 to 0.1 mm)

Pinion bearing preload (without pinion oil seal) . . . . . .   2.6 to 6.1 lbf in (3 to 7 kgf cm)

Differential side bearing preload (without pinion) . . . . .   4.3 to 8.7 lbf in (5 to 10 kgf cm)

Torque wrench settings	**lbf ft**	**kgf m**
Crownwheel	40 to 47	5.5 to 6.5
Differential side bearing caps	23 to 34	3.2 to 4.7
Companion flange to pinion	87 to 130	12 to 18

1 General description

The rear axle is conventional in design. In the centre of the axle casing is the final drive that turns the drive from the propeller shaft through 90° to the wheels and gives it a final gear reduction. Also in the centre is the differential which allows the two wheels to rotate at different speeds (when turning corners), yet share the torque equally between them. Halfshafts carry the drive from the differential out to both hubs.

The construction of the hubs and halfshafts is of the type known as 'semi-floating'. This makes servicing of the shafts or bearings difficult, especially so if the special halfshaft withdrawing tool is not available. It is considered beyond the scope of the average home mechanic to go into the complexities of setting up the differential, since a number of special tools are required. The procedure is also rather complicated, and for those people without a good knowledge of this type of gear train there is a possibility of incorrect clearance, which in turn will lead to early failure. In this Section you will find details of the removal procedures for the halfshafts and differential carrier. In the

event of any failure of the differential assembly, it is recommended that a complete exchange unit is purchased or the existing faulty assembly is repaired by a vehicle main dealer.

2 Lubrication

1 As the final drive is by a hypoid bevel the use of the correct type of extreme pressure lubricant is important.
2 Some manufacturers do not call for any oil changes for their final drives, and the oil is reputed to last the life of the axle. Indeed, some cars have no drain plug.
3 The additives that give the oil its extreme pressure qualities do get used up. Also, with time the oil absorbs damp. Bearings corrode, pitting their running surface. It is probably fairer to put the phrase the other way, and say the axle lasts the life of the oil. As the oil is easy to change, but the axle very difficult, it is suggested that Toyo Kogyo are quite right to recommend changes every 25 000 miles (40 000 Km), it being especially timely to do so after the winter, when damp will have been absorbed in the oil.

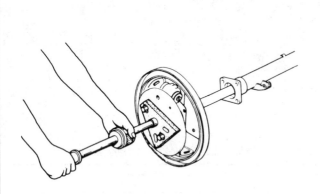

Fig. 8.1 Extracting a halfshaft (Sec 4)

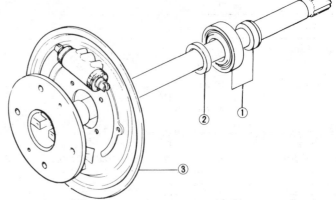

Fig. 8.2 The halfshaft withdrawn (Sec 4)

1 The retaining collar and bearing
2 Spacer
3 Brake backplate

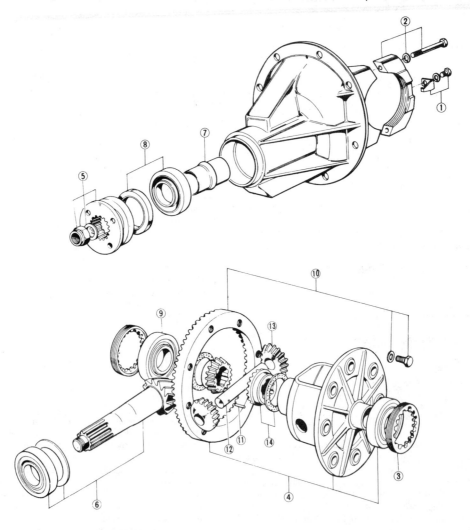

Fig. 8.3 The component parts of the differential (Sec 6)

1 Adjuster lockplate, bolt and washer	bearing	7 Collapsible spacer	11 Lock pin
2 Bearing cap, bolt and washer	5 Companion flange, nut and washer	8 Front bearing and oil seal	12 Pinion shaft
3 Adjuster screw	6 Drive pinion, adjusting washer and bearing	9 Bearing	13 Pinion gear
4 Differential assembly and		10 Crownwheel, bolt and washer	14 Side gear and thrust washer

3 Hub bearing construction

1 The single ball race bearing is mounted on the halfshaft, with its outer race inside the axle casing. The wheel and brake drums are hung outboard on the end of the shaft.
2 The bearing outer race is held in the casing by a plate outside the brake backing plate.
3 The halfshaft is held in the inner race by a collar that is an interference fit on the shaft.
4 To remove the bearing, the halfshaft is removed as described in the next Section.

4 Halfshaft and hub bearing – removal and refitting

Note: *Although removal of the halfshaft should cause no problems, special tools will be required to remove the bearing retaining collar and the bearing. If the bearing is to be removed, this part of the job should be entrusted to your Mazda dealer, together with fitting replacement parts.*
1 Jack up the rear end of the vehicle and support it with axle stands.
2 Remove the appropriate roadwheel.
3 If necessary, refer to Chapter 9, back off the brake linings from the brake drum and remove the brake drum.
4 Thoroughly clean the area around the brakes and the axle casing; in particular, there must be no dirt left that could work its way between the brake backplate and the end of the axle casing when these are undone.
5 Remove the brake shoes as described in Chapter 9.
6 Disconnect the handbrake lever, and the hydraulic supply pipe. Suitably plug, or tape, the end of the hydraulic pipe to prevent the ingress of dirt.
7 Remove the four nuts, bolts and washers holding the brake backplate and bearing retainer to the axle casing. Break the joint between the retaining plate and the backplate.

8 The halfshaft, complete with bearing, must now be drawn out of the axle casing. It may be possible to pull this out, but it will probably be tightly held. Temporarily refit the brake drum so that the inside is towards you, then use a heavy soft-faced hammer on the shoulder of the drum to drive out the shaft while at the same time turning the drum.
9 With the halfshaft removed, detach the brake drum.
10 Prise the oil seal out of the end of the axle casing.
11 Make sure that the seal recess in the axle casing is clean, then lubricate a new seal with clean axle oil and press it in carefully with the lip towards the differential.
12 The halfshaft can now be refitted by reversing the removal procedure, but first work plenty of general purpose grease into the bearing with your fingers. After refitting, don't forget to bleed the brakes as described in Chapter 9, and top up the axle oil level if necessary.

5 Differential carrier assembly – removal and refitting

1 Remove the halfshafts as previously described in Section 4. Additionally remove the axle drain plug and drain the oil into a container of appropriate capacity. Clean the drain plug (it is magnetic and will probably be very dirty) and refit it once the oil is drained.
2 Mark the relative fitted position of the propeller shaft and the differential companion flange. Disconnect the shaft, as described in Chapter 7.
3 Remove the nuts which retain the differential carrier to the axle housing. Remove the carrier.
4 When refitting, clean the sealing surfaces of the housing and carrier, then apply a non-setting gasket sealant to these surfaces.
5 Position the carrier to the case then fit and tighten the nuts.
6 Refit the propeller shaft, referring to Chapter 7 for torque figures and other information.
7 Refit the halfshafts as previously described then top up the axle with lubricant of the specified grade.

6 Fault diagnosis – rear axle

Symptom	Reason/s
Vibration	Worn halfshaft bearing
	Loose bolts (propeller shaft to companion flange)
	Tyres require balancing
	Propeller shaft out of balance
Noise on turns	Worn differential gear
Noise on drive or coasting*	Worn or incorrectly adjusted ring and pinion gear
'Clunk' on acceleration or deceleration	Worn differential gear cross-shaft
	Worn propeller shaft universal joints
	Loose bolts (propeller shaft to companion flange)

*It must be appreciated that tyre noise, wear in the rear suspension bushes and worn or loose shock absorber mountings can all mislead the mechanic into thinking that components of the rear axle are the source of trouble.

8

Chapter 9 Braking system

See Chapter 13 for specifications and information applicable to 1979 thru 1983 North American models

Contents

Specifications

Type
Front . . . . Disc, or drum (twin leading shoes)
Rear . . . . Drum (leading and trailing shoes)
Handbrake . . . . Mechanical, hand operated through cable and linkage to rear wheels

Master cylinder
Type . . . . Tandem
Bore diameter . . . . $\frac{13}{16}$ in (20.64 mm)
Clearance between piston and bore:
    New . . . . 0.0016 to 0.0049 in (0.040 to 0.125 mm)
    Wear limit . . . . 0.006 in (0.15 mm)

Front disc brake
Brake disc outer diameter . . . . 8.150 in (207 mm)
Brake disc thickness:
    New . . . . 0.5118 in (13.0 mm)
    Wear limit . . . . 0.4724 in (12.0 mm)
Maximum allowable lateral run-out of brake disc . . . . 0.0024 in (0.06 mm)
Thickness of pad and backing plate:
    New . . . . 0.551 in (14.0 mm)
    Wear limit . . . . 0.276 in (7.0 mm)
Caliper piston bore diameter . . . . 2.0 in (50.8 mm)

Front drum brake
Drum diameter:
    New . . . . 7.874 in (200 mm)
    Wear limit . . . . 7.913 in (201 mm)
Thickness of lining:
    New . . . . 0.157 in (4.0 mm)
    Wear limit . . . . 0.039 in (1.0 mm)
Wheel cylinder bore diameter . . . . $\frac{15}{16}$ in (23.81 mm)
Clearance between piston and bore:
    New . . . . 0.0016 to 0.0049 in (0.040 to 0.125 mm)
    Wear limit . . . . 0.006 in (0.15 mm)

Rear drum brake
Drum diameter:
    New . . . . 7.874 in (200 mm)
    Wear limit . . . . 7.913 in (201 mm)
Thickness of lining:
    New . . . . 0.157 in (4.0 mm)
    Wear limit . . . . 0.039 in (1.0 mm)
Rear wheel cylinder bore diameter:
    Europe and North America . . . . $\frac{3}{4}$ in (19.05 mm)

Models fitted with front wheel disc brakes (except Europe and North America)	$\frac{5}{8}$ in (15.87 mm)
Models fitted with front wheel drum brakes (except Europe and North America)	$\frac{11}{16}$ in (17.46 mm)

Clearance between piston and bore:

New:

Europe and North America	0.0016 to 0.0049 in (0.040 to 0.125 mm)
Models fitted with front wheel disc brakes (except Europe and North America)	0.0013 to 0.0040 in (0.032 to 0.102 mm)
Models fitted with front wheel drum brakes (except Europe and North America)	0.0016 to 0.0049 in (0.040 to 0.125 mm)

Wear limit:

Europe and North America	0.006 in (0.15 mm)
Models fitted with front wheel disc brakes (except Europe and North America)	0.0048 in (0.122 mm)
Models fitted with front wheel disc brakes (except Europe and North America)	0.006 in (0.15 mm)

Brake pedal free travel

With servo assistance	0.28 to 0.35 in (7 to 9 mm)
Without servo assistance	0.02 to 0.09 in (0.5 to 2.4 mm)

Hydraulic brake fluid . . . . FMVSS 116, DOT 3, DOT 4 or SAE J1703

1 General description

The braking system is of the dual circuit hydraulic type, with either drum or disc brakes on the front wheels, and drum brakes on the rear wheels; a vacuum servo unit is fitted to most models. A differential proportioning valve is fitted in the hydraulic circuit to prevent the rear wheels locking before the front wheels during heavy applications of the brakes. The tandem master cylinder incorporates a brake fluid level sensor, which, when the brake fluid level is low, operates a warning light in the instrument cluster. The warning light also operates when the handbrake, which is mechanically operated on the rear wheels, is in operation.

2 Disc pads – inspection and renewal

1   At intervals of 4000 miles (6500 Km) the front disc pads should be inspected for wear.
2   Raise the front end of the vehicle and support it under the bodyframe with suitable stands.
3   Remove the appropriate roadwheel.
4   Shine a torch beam through the aperture in the caliper, or alternatively, look down from above the caliper; the pads can be seen beside the disc. If the pads appear to be well worn, they will have to be removed for a closer inspection.
5   To remove the pads, proceed as described in the following paragraphs.
6   Using suitable long-nosed pliers, pull out the retaining clip at each end of the stopper plate. Insert a screwdriver between the caliper and the caliper mounting bracket, and whilst levering the caliper away from the bracket, pull out the stopper plate. Remove the lower stopper plate in the same manner (photos).
7   The caliper can now be removed from the bracket by sliding it inwards until it clears the bracket cut-outs. Suspend the caliper from a convenient anchorage point with string or wire. Never allow the caliper to hang by the flexible brake hose (photo).
8   Now remove the anti-rattle spring, which is located in a hole at the edge of the inboard disc pad (photo).
9   The disc pads can now be removed from their recess in the mounting bracket. Take care to keep any shims that may be against the back face of the disc pads.
10   If the pads are worn beyond the specified limit, they must be renewed. It is advisable to renew them anyway, if they are nearing the limit, to save work at a later date. Never renew pads on one wheel alone; always renew them in complete sets.
11   To commence reassembly, place the disc pads into their recess in the mounting bracket. This is best carried out by first pressing them up against their stop spring, then locating them into their lower mounting bracket recess. Remember to refit any shims that were evident when removing the pads (photo).

2.6A Remove the retaining clips ...

2.6B ... and pull out the stopper plates

2.7 Remove the caliper from the mounting bracket

2.8 The anti-rattle spring

2.11 Fitting a disc brake pad

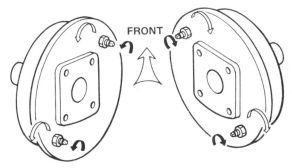

FRONT

Turning Direction

⤴ — Anchor pin (to expand brake shoe)

⤵ — Lock nut (to tighten)

Fig. 9.1 Adjusting a front drum brake (Sec 3)

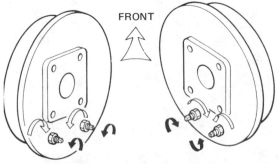

FRONT

Turning Direction

⤴ — Anchor pin (to expand brake shoe)

⤵ — Lock nut (to tighten)

Fig. 9.2 Adjusting a rear drum brake (Sec 3)

H7477

Fig. 9.3 Adjusting the handbrake lever (Sec 4)

12 Locate the anti-rattle spring into the hole in the inboard disc pad.

13 Before refitting the brake caliper the piston must be fully retracted into the piston bore. This is best carried out by first removing a small quantity of brake fluid from the fluid reservoir; then using a suitable wooden block, levering the piston back into the piston bore. Keep an eye on the brake fluid reservoir whilst doing this, as the retraction of the piston will increase the brake fluid level.

14 Now locate the caliper over the anti-rattle spring and into position on the mounting bracket. The loose end of the anti-rattle spring is now firmly located against the inside wall of the caliper.

15 With the caliper now in position, slide in the lower stopper plate and secure it with the retaining clips.

16 Place a screwdriver between the mounting bracket and the top of the caliper, and whilst levering downwards, slide in the upper stopper plate. Fit the retaining pins as previously mentioned.

17 Refit the roadwheel. Lower the vehicle to the ground. Road test it in a safe, convenient place to check for satisfactory brake operation.

3 Brake adjustment

1 In order to maintain maximum braking efficiency, the front and rear drum brakes should be adjusted at the intervals given in the Routine Maintenance Section at the beginning of this manual.

Front brakes

2 Chock the rear wheels to prevent movement of the vehicle and raise the front end until the roadwheels are clear of the ground. Ensure that a secondary means of support is used to maintain maximum safety standards.

3 Each brake shoe is adjusted by means of an anchor pin, which, when correctly positioned, is secured by a locknut.

4 Undo the locknuts on both adjustment anchor pins in an anti-clockwise direction (Fig. 9.1).

5 Turn the adjustment anchor pin clockwise until the roadwheel is unable to rotate, then turn the adjustment anchor pin anti-clockwise until the roadwheel rotates freely, and with the anchor pin held in this position, tighten the locknut. After the locknut has been tightened check that the roadwheel is still free to rotate. Adjust the other brake shoe in a similar manner. Repeat the procedure for the other front wheel.

6 Lower the vehicle to the ground.

Rear brakes

7 Firmly check the front wheels of the vehicle and release the hand-brake.

8 Raise the rear of the vehicle until the roadwheels are clear of the ground and support the vehicle with axle stands.

9 The adjustment procedure is identical to that described for the front brakes earlier in this Section, except that both adjustment anchor pins are located at the bottom of the brake drum (Fig. 9.2).

10 On satisfactory completion of drum brake adjustment, lower the vehicle to the ground and test drive the car in a safe, convenient place to check for satisfactory brake operation.

4 Handbrake – adjustment

1 After some time the handbrake cable may stretch and require adjustment.

2 Firmly chock the front road wheels and raise the rear of the vehicle until the rear roadwheels are clear of the ground. Support the vehicle on axle stands.

3 Adjust the handbrake lever adjusting nut so that the rear wheels are locked when the handbrake lever is pulled up 3 to 7 notches of the ratchet.

4 Release the handbrake lever and check that the wheels are free to rotate.

5 Apply the handbrake several times, then recheck that the wheels lock between 3 and 7 notches of the handbrake, and the wheels rotate freely when the handbrake lever is released.

6 When satisfactory adjustment has been obtained lower the vehicle to the ground.

5 Brake pedal – adjustment

1 At the intervals given in the Routine Maintenance Section, the brake pedal height and pedal free travel should be checked and, if necessary, adjusted using the following procedure.

2 Pedal height should be adjusted to 7.87 to 8.07 in (200 to 205 mm) from the centre of the pedal pad to the bulkhead (Fig. 9.4).

3 Adjustment is made merely by loosening the locknut (2) and turning the stop lamp switch (1) in the required direction.

4 After the pedal height has been adjusted, the pedal free travel will have to be adjusted to ensure the correct clearance exists between the brake pedal pushrod and the primary piston cup in the master cylinder; or in the case of vehicles fitted with a vacuum servo unit, the operating piston.

5 To adjust brake pedal free travel, loosen the locknut (4) (Fig. 9.4) and turn the pushrod (3) until the specified free travel is obtained.

6 When satisfactory adjustments have been carried out always make sure that the locknuts have been retightened.

6 Hydraulic brake system – bleeding

1 The system should need bleeding only when some part of it has

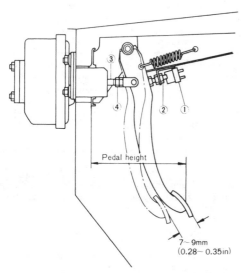

Fig. 9.4 Brake pedal adjustment components; pedal free travel illustrated is for vehicles fitted with vacuum servo units

1 *Stop lamp switch*	3 *Pushrod*
2 *Locknut*	4 *Locknut*

Pedal height

7~9mm
(0.28~0.35in)

9

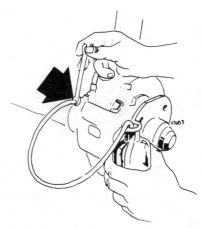

Fig. 9.5 Bleeding a brake caliper unit (Sec 6)

Fig. 9.6 Bleeding a brake drum wheel cylinder (Sec 6)

Fig. 9.7 Checking a brake disc for run-out (Sec 10)

9.4 The hose retaining clip on the suspension strut

been dismantled which would allow air into the fluid circuit, or if the level in the reservoir has been allowed to drop so far that air has entered the master cylinder.

2   Before commencing bleeding, check all brake pipe unions and connections for possible leakage, and at the same time check the condition of the rubber hoses, which may be perished.

3   Remove the brake drums and inspect the wheel cylinder for possible signs of leakage.

4   Check the brake calipers for any signs of leakage.

5   If there is any possibility of incorrect fluid having been put into the system, drain all the fluid out and flush through with methylated spirit. Renew all piston seals and cups since these will be affected and could possibly fail under pressure.

6   Gather together a clean jar, a length of tubing which fits tightly over the bleed nipples, and a tin of the specified brake fluid.

7   On vehicles with vacuum servo assistance depress the brake pedal several times to destroy the vacuum in the servo system.

8   As the front and rear circuits are independent, it will be obvious that only the circuit that has air in it need be bled.

9   Regardless of the system type, bleeding must commence at the longest run from the master cylinder (whichever side is being bled), and the last nipple to be bled should be the nearest to the master cylinder.

10   With all the points previously mentioned in mind, it is now time to seek the assistance of a second person, as the bleed nipples have to be opened while the brake pedal is pumped, and the reservoir level has to be continuously watched and replenished.

11   Pour a little hydraulic fluid into the clean jar, sufficient to immerse the tube. Push the tube over the bleed nipple while keeping the open end of the tube completely immersed. Keep it immersed throughout the entire bleeding operation.

12   Open the bleed nipple a half of a turn, and slowly push the pedal down through its full travel. While the pedal is travelling downwards a mixture of air and hydraulic fluid will be entering the jar via the bleed nipple and tube. With the pedal at the end of its travel the bleed nipple must be tightened up, then the pedal can be allowed to return to its original position. Failure to tighten the bleed nipple before the pedal returns, will suck all the air and possible dirt back into the hydraulic circuits. Repeat the procedure until all air is expelled from that nipple, and clean fluid is being pumped through.

13   Repeat the procedure described in paragraph 12, for the remaining bleed nipples.

14   It is vital that the fluid in the hydraulic master cylinder is kept at its correct level throughout the bleeding operation. **Never re-use brake fluid that has been pumped through the system; this should be discarded.**

15   Finally, when bleeding is complete, refit the dust excluding caps to the bleed nipples and road test the vehicle in a safe, convenient place to test for satisfactory operation of the brakes.

7   Flexible hoses – inspection, removal and refitting

1   Regularly inspect the condition of the flexible hydraulic hoses. If they are perished, chafed or swollen they must be renewed.

2   To remove a flexible hose, extract the clip from the support bracket.

3   Hold the flats of the flexible hose end fitting with a spanner and unscrew the union nut which couples it to the rigid brake pipe.

4   Disconnect the flexible hose from the rigid pipe and from the support bracket and then unscrew the hose from the component at its opposite end.

5   Refitting is a reverse of removal. Screw the hose into the caliper, wheel cylinder, or T-piece, using the sealing washer supplied and then connect it to the rigid pipe, having first passed it through the opening in the support bracket. Fit the securing clip.

6   It is important to note that all pipe unions and connections are manufactured to metric standards **and hoses or unions with other threads should on no account be used.**

7   Always bleed the hydraulic system as described in Section 6 if a flexible hose has been disconnected or renewed.

8   Rigid brake pipes – inspection– removal and refitting

1   At regular intervals wipe the steel brake pipes clean and examine

them for signs of rust or denting caused by flying stones.

2   Examine the securing clips; bend the tongues of the clips if necessary to ensure that they hold the brake pipes securely without letting them rattle or vibrate.

3   Check that the pipes are not touching any adjacent components or rubbing against any part of the vehicle. Where this is observed, bend the pipe gently away to clear.

4   Although the pipes are plated, any section of pipe may become rusty through chafing and, if so, should be renewed. Brake pipes are available to the correct length and fitted with end unions from most Mazda dealers, and can if necessary be made to pattern by many accessory suppliers. When fitting the new pipes use the old pipes as a guide to bending and do not make any bends sharper than necessary.

5   The system will have to be bled after any disconnections of pipes; see Section 6 for details of bleeding.

9   Disc caliper and mounting bracket – removal and refitting

1   Raise the front end of the vehicle and support it under the bodyframe side rails with suitable stands.

2   Remove the appropriate roadwheel.

3   The disc caliper and mounting bracket can be removed as an assembly by removing the two bolts and washers securing the mounting bracket to the backplate, which are located behind the backplate.

4   Never allow the assembly to hang suspended by the brake flexible hose after removing the hose retaining clip (photo).

5   When refitting the caliper and mounting bracket assembly, locate the assembly over the disc carefully to avoid any damage to the disc pads.

6   Refit the bolts and tighten them to a torque of 33 to 40 lbf ft (4.5 to 5.5 kgf m).

7   Refit the roadwheel and lower the vehicle to the ground.

10   Brake disc – examination, removal and refitting

1   The brake disc should be inspected for deep scoring ((light scoring is normal). If severe, the disc should be removed and either renewed or ground provided the thickness will not be reduced below the minimum specified.

2   To check for brake disc run-out, ideally a dial gauge should be used, but feeler blades can be used against a fixed block as the hub/disc is slowly rotated.

3   If the disc has to be removed, first remove the caliper and mounting bracket as described in Section 9. Tie the caliper and mounting bracket block out of the way; never allow the caliper to hang by the flexible hose.

4   Prise off the wheel bearing grease cap, remove the split pin and lock nut, undo and remove the hub retaining nut, and draw off the brake disc as described in Chapter 11.

5   Refitting is a reversal of removal, but be sure to adjust the wheel bearing preload as described in Chapter 11.

11   Disc caliper – dismantling, servicing and reassembly

1   Remove the disc caliper as described in Section 9.

2   Before attempting to dismantle the disc caliper, brush off any loose dirt and clean the exterior surfaces thoroughly, and prepare a clean surface where dismantling can be carried out without the fear of possible dirt ingress to the piston and seal.

3   If necessary, remove the flexible brake hose from the caliper. This is best done by carefully holding the caliper in a vice and unscrewing the flexible hose.

4   Referring to Fig. 9.8, prise out the retainer (2) and remove the dust boot (3).

5   Remove the piston (4) by applying compressed air from a foot pump or similar source to the hydraulic fluid inlet. If the piston is seized or stubborn to remove, try **gently** tapping the caliper around the piston area with a hammer whilst maintaining the air pressure.

6   Remove the piston seal (5).

7   Clean the disassembled parts in clean brake fluid or alcohol. **Never use petrol or paraffin.** Blow the parts dry with compressed air.

8   Inspect the caliper bore and the piston for signs of scoring, scratches or rust. If any of these conditions are found, renew the

caliper and/or piston. Minor damage can be eliminated by means of polishing with a very fine grade of crocus paper.

9   Discard the old piston seal and dust boot, and obtain replacement items, available in the form of a repair kit.

10   To reassemble, ensure that all parts are clean, and locate the piston seal into its groove in the caliper; ensure that it does not become twisted and is seated fully in the groove. To aid assembly, smear the seal with clean hydraulic fluid.

11   Lubricate the piston with clean hydraulic fluid and assemble it into the caliper bore, ensuring that it is located square to the bore, and will not jam in the bore.

12   Fit the dust boot and the dust boot retainer.

13   If applicable, connect the brake flexible hose to the caliper.

14   Fit the caliper unit to the mounting bracket, connect the flexible hose to the rigid pipe, and bleed the hydraulic system as described in Section 6.

12   Front brake shoes – inspection, removal and refitting

1   Apply the handbrake and raise the front of the vehicle. Remove the roadwheel.

2   Refer to Section 3 and loosen the locknuts that secure the adjusting anchor pins. Turn the anchor pins anti-clockwise to ensure that the brake linings are completely clear of the brake drums.

3   Fig. 9.9 shows an exploded view of the front brake assembly; refer to this and prise off the wheel bearing grease cap (1), remove the split pin (2) and nut lock (3).

4   Undo and remove the wheel bearing retaining nut (4) and the washer (5).

5   Using a suitable extractor, draw off the drum (6).

6   At this stage brush all the dust from the backing plate and the interior of the drum. Check for any signs of brake fluid leakage at the wheel cylinders and signs of lubricant loss from the hub; refer to Section 13 of this Chapter, and Chapter 11 for corrective action for the wheel cylinders and hub, respectively.

7   Inspect the linmings for excessive wear or shoe damage, and contamination from brake fluid or hub grease. Linings that are worn to the specified limits must be renewed with exchange shoes; never attempt to fit new linings to the brake shoes yourself as this rarely proves successful.

8   If the linings have worn to the specified limits, or are contaminated with oil or grease they must be renewed.

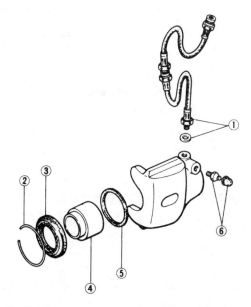

Fig. 9.8 Components of the front disc caliper (Sec 11)

1 Brake hose and washer	4 Piston
2 Retainer	5 Piston seal
3 Dust boot	6 Bleed screw and cap

9

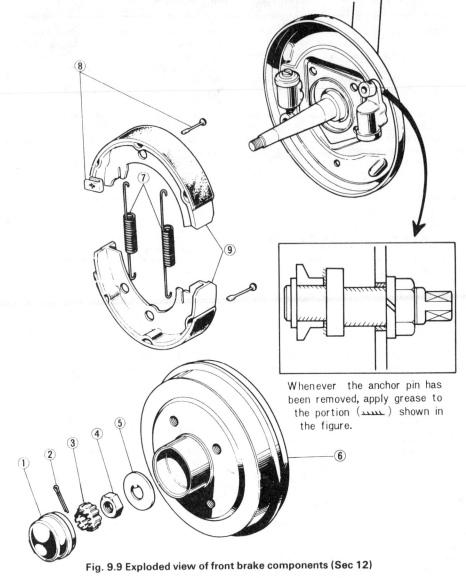

Whenever the anchor pin has been removed, apply grease to the portion (⟲⟲⟲) shown in the figure.

Fig. 9.9 Exploded view of front brake components (Sec 12)

1 Grease cap
2 Split pin
3 Nut lock
4 Nut
5 Washer
6 Brake drum
7 Shoe return spring
8 Retaining spring and guide
pin
9 Brake shoes

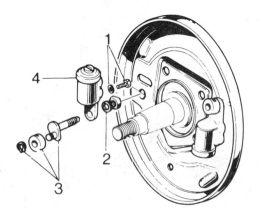

Fig. 9.10 Removing a front brake wheel cylinder (Sec 13)

1 Wheel cylinder retaining bolt and washer
2 Anchor pin nut and washer
3 Anchor pin, adjusting spacer and circlip
4 Wheel cylinder

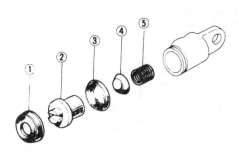

Fig. 9.11 Components of the front brake wheel cylinder (Sec 13)

1 Dust cover
2 Piston
3 Piston cup
4 Spacer block
5 Spring

9   Linings that have worn badly may have scored the brake drum; light scoring can be removed professionally by machining, providing that the amount of material removed while machining does not increase the diameter of the brake drum beyond the specified figure. Heavily scored brake drums will have to be renewed.

10   To remove the brake shoes, remove the shoe return springs (7).

11   Remove the retaining spring and guide pin (8) from each brake shoe, by holding the guide pin against the backing plate and pressing down the spring and turning it through 90° to release the guide pin.

12   The brake shoes can now be removed from the wheel cylinders, but to prevent the wheel cylinder pistons from coming out, rubber bands can be used to hold them in.

13   Now that the brake shoes have been removed, the backing plate area can be thoroughly cleaned with petrol or proprietary solvents.

14   It is wise, at this stage, to examine the adjustment anchor pins for signs of wear and/or damage to the eccentric cam and peg. To do this, remove the locknut and washer from the backing plate and pull the adjustment anchor pin from its location, through the lower wheel cylinder mounting hole and the backing plate (Fig. 9.10). Prise off the circlip and lift off the adjusting spacer. When reassembling the adjustment anchor pin, smear a little high melting point grease to the eccentric cam and peg and the adjusting spacer.

15   If the brake shoe return springs have been overheated - caused by incorrect adjustment, abuse or extremely heavy braking - it is a good policy to renew them.

16   Commence reassembly by refitting the adjustment anchor pins (if applicable).

17   Locate each brake shoe onto the wheel cylinder; if rubber bands are used to retain the wheel cylinder pistons, these should now be removed. Fit the retaining spring and guide pin to each brake shoe.

18   Hook the shoe return springs into their locating holes in the brake shoes and, using a pair of pliers, stretch them until they can be located into their respective holes in the opposite brake shoe. Ensure that the brake shoe tongues locate into the wheel cylinder slots at one end of the brake shoes, and at the other end they locate onto the adjusting spacer.

19   Refit the brake drum and secure in position with the washer and nut. Adjust the wheel bearing preload as described in Chapter 11. Refit the roadwheel.

20   Refer to Section 3, and adjust the brake shoes. Lower the vehicle to the ground, and repeat the entire procedure for the other front wheel drum brake. After completion, road test the vehicle in a safe, convenient place to check for satisfactory brake operation.

13   Front brake wheel cylinder – removal, servicing and refitting

1   Apply the handbrake and raise the front end of the vehicle. Remove the roadwheel.

2   Remove the brake shoes, as described in the previous Section.

3   Disconnect the hydraulic brake pipe at the rear of the wheel cylinder(s).

4   Fig. 9.10 shows an exploded view of the wheel cylinder and back-plate; refer to this and remove the bolt and washer (2), then remove the nut and washer (3) that retain the adjustment anchor pin. The wheel cylinder can then be lifted out. Repeat this procedure if the other wheel cylinder requires attention.

5   Working on a clean surface, prise off the dust cover (1) and pull out the piston (2). If the piston is seized in its bore, or stubborn to remove, apply compressed air from a foot pump to the fluid inlet orifice. **Gentle** tapping with a hammer may assist the process.

6   With the piston removed, the piston cup, spacer block and spring can be taken out; pressure from the spring may, however, cause the components to fly out; be ready for this occurence by covering the end with suitable lint-free cloth.

7   Wash all the dismantled parts in clean methylated spirits or brake fluid. **Never use petrol or paraffin**.

8   Examine the cylinder bore and the piston for signs of wear, roughness or scoring; if evident, it is best to purchase a replacement wheel cylinder complete with piston, piston cup, spring and spacer block.

9   Inspect the piston cup for wear, softening, swelling or damage of any kind. If any of these conditions exists, discard the cup and purchase a replacement item (usually supplied in a repair kit).

10   Ensure that all parts are clean, and commence reassembly by assembling the spring, spacer block and piston cup into the wheel cylinder. Coat each component with a little clean hydraulic fluid to aid assembly and lubricate the rubber piston cup,.

11   Fit the piston into the wheel cylinder, ensuring that it remains square to the wheel cylinder bore, thus preventing it from jamming in the bore.

12   Assemble the dust cover.

13   Refitting is a reversal of the removal sequence, but always ensure that the hydraulic connections have been secured correctly and are not cross-threaded in the wheel cylinders.

14   Refit the wheel drum, and set the wheel bearing preload as described in Chapter 11.

15   Adjust the brake linings and bleed the brakes; see Section 3 and 6 respectively for these operations.

16   Refit the roadwheel and lower the wheel to the ground.

17   Repeat the entire procedure for the other wheel if applicable. Road test the vehicle in a safe, convenient place to check for satisfactory brake operation.

14   Rear brake shoes – inspection, removal and refitting

1   Chock the front wheels and raise the rear end of the vehicle. Support the vehicle with axle stands or a secondary means of support. Release the handbrake and remove the roadwheel.

2   Loosen the adjustment anchor pin locknuts and turn each anchor pin anti-clockwise until the brake linings are completely clear of the drums.

3   Fig. 9.12 shows an exploded view of a rear brake assembly. Refer to this, and remove the two screws that retain the brake drum (1). Remove the brake drum (photo).

4   Brush all the dust from the backing plate and the interior of the brake drum. Check for any signs of brake fluid leakage at the wheel cylinder and signs of lubricant loss from the hub; refer to Section 15 of this Chapter, and Chapter 8 for corrective action for the wheel cylinder and hub respectively.

5   Inspect the linings for excessive wear or shoe damage, and contamination from brake fluid or hub grease. Linings that are worn to the specified limits must be renewed with exchange shoes; never attempt to fit new linings to the brake shoes yourself as this rarely proves successful.

6   Linings that have worn badly may have scored the brake drum. Light scoring can be removed professionally by machining, providing that the amount of material removed while machining does not increase the diameter of the brake drum beyond the specified figure. Heavily scored brake drums will have to be renewed.

14.3 The brake drum retaining screws

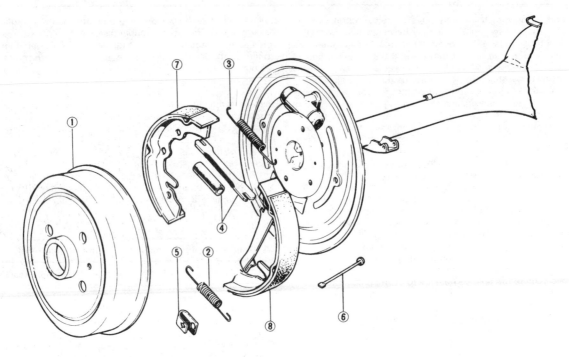

Fig. 9.12 Components of the rear brake shoe assembly (Sec 14)

1 Brake drum
2 Shoe return spring

3 Shoe return spring
4 Handbrake strut

5 Retaining spring
6 Guide pin

7 Front brake shoe
8 Rear brake shoe

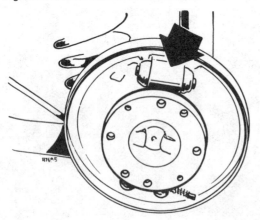

Fig. 9.13 Interior view of a rear brake drum (Sec 14)

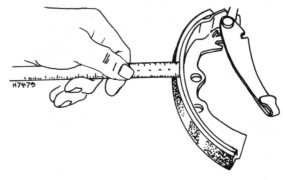

Fig. 9.14 Inspecting a brake lining for wear (Sec 14)

Fig. 9.15 A rear wheel brake cylinder (Sec 15)

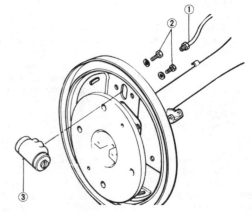

Fig. 9.16 Removing a rear wheel brake cylinder (Sec 15)

1 Hydraulic pipe
2 Retaining bolts and washers
3 Wheel cylinder

7 To remove the brake shoes, remove the shoe return spring (2) and (3) (photos).

8 Disconnect the handbrake cable from the operating arm on the rear brake shoe (8) and remove the handbrake strut (4) from between the two brake shoes.

9 Remove the retaining spring (5) and the guide pin (6) from each brake shoe, by holding the guide pin against the backing plate and pressing down the spring and turning it through 90° to release the guide pin (photo).

10 The brake shoes can now be removed from their locations in the wheel cylinder pistons and lifted out, but to prevent the wheel cylinder pistons from coming out, rubber bands can be used to hold them in.

11 With the brake shoes removed, it is wise to clean the backing plate area thoroughly with petrol or proprietary solvents.

12 Examine the adjustment anchor pins for signs of wear and/or damage to the eccentric cam and peg. To do this, remove the locknut and washer from the backing plate and lift out the anchor pin. Prise off the circlip and lift off the adjusting spacer. When reassembling the adjustment anchor pin, smear a little high melting point grease to the eccentric cam and peg and the adjusting spacer.

13 If the brake shoe return springs have been overheated - caused by incorrect adjustment, abuse or extremely heavy braking - it is a good policy to renew them.

14 Commence reassembly by refitting the adjustment anchor pins (if applicable).

15 Locate each brake shoe into the respective slots in the wheel cylinder pistons. If rubber bands were used to retain the wheel cylinder pistons, these should now be removed. Fit the retaining spring and guide pin to each brake shoe.

16 Assemble the handbrake strut between the two brake shoes and connect the handbrake cable to the operating arm on the rear brake shoe.

17 Hook the shoe return springs into their locating holes in the brake shoes and, using a pair of pliers, stretch them until they can be located into their respective holes in the opposite brake shoe. Ensure that the brake shoe tongues locate into the wheel cylinder slots at one end of the brake shoe, and at the other end they locate onto the adjusting spacer (photo).

18 Refit the brake drum and secure in position with the two retaining screws. Refit the roadwheel. Refer to Section 3, and adjust the brake shoes. Lower the vehicle to the ground, and repeat the entire procedure for the other rear brake assembly. After completion, road test the vehicle in a safe, convenient place to check for satisfactory brake operation.

15 Rear brake wheel cylinder – removal, servicing and refitting

1 Chock the front wheels and raise the rear end of the vehicle. Support the vehicle with axle stands or a secondary means of support.

14.7A The lower brake shoe return spring

14.7B The upper brake shoe return spring

14.9 A guide pin retaining spring

14.17 The brake shoe return springs correctly fitted

9

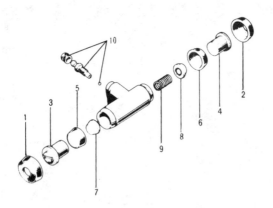

Fig. 9.17 Components of the rear brake wheel cylinder (Sec 15)

1	Dust boot	6	Piston cup
2	Dust boot	7	Spacer block
3	Piston	8	Spacer block
4	Piston	9	Spring
5	Piston cup	10	Bleed screw

16.1 A brake master cylinder with a separate hydraulic fluid reservoir

16.4 The flexible hose connection at the master cylinder

Release the handbrake and remove the roadwheel.
2   Remove the brake shoes as described in the previous Section.
3   Disconnect the hydraulic brake pipe at the rear of the wheel cylinder.
4   Undo and remove the two retaining bolts and washers and lift out the wheel cylinder.
5   Working on a clean surface, remove the dust boots from each end of the wheel cylinder.
6   To remove the piston, piston cup, spacer block and spring, press on either piston to force the components out of the wheel cylinder bore.
7   Wash all the parts in clean methylated spirits or hydraulic brake fluid. **Never use petrol or paraffin.**
8   Examine the cylinder bore and the pistons for signs of wear, roughness or scoring; if evident, it is best to purchase a replacement wheel cylinder assembly complete with all the component parts ready for fitting.
9   Inspect the piston cups for wear, softening, swelling or damage of any kind. If any of these conditions exists, discard the cups and purchase replacement items (usually supplied in a repair kit).
10   Ensure that all parts are clean, and commence reassembly by fitting the spring into the two spacer blocks.
11   Coat the wheel cylinder bore, pistons and piston cups with a little clean hydraulic fluid.
12   Assemble the piston cups over the pistons, ensuring that the small diameter of the taper is against the piston shoulder.
13   Press one piston and piston cup into the cylinder bore, and from the opposite end, fit the spacer blocks and spring, followed by the other piston and piston cup.
14   Assemble the dust boots to each end of the wheel cylinder.
15   Refitting of the wheel cylinder is a reversal of the removal sequence, but ensure that the hydraulic connection is secured correctly and is not cross-threaded in the wheel cylinders.
16   Refit the brake shoes as described in the previous Section.
17   Refit the roadwheel and adjust the brakes as described in Section 3; refer to Section 6 and bleed the hydraulic system. Lower the vehicle to the ground.
18   Repeat the entire procedure for the other wheel if applicable. Road test the vehicle in a safe, convenient place to check for satisfactory brake operation.

16   Master cylinder – removal and refitting

1   The master cylinder is of the tandem type, and in the case of vehicles fitted with vacuum servo-assistance, is mounted on the front face of the servo unit. On vehicles without vacuum servo assistance the cylinder is bolted to the engine bulkhead directly in line with, and located to, the brake pedal pushrod. The hydraulic fluid reservoir may be mounted directly above the master cylinder, or it may be mounted on a bracket behind the suspension strut tower, with flexible hoses to supply the fluid to the master cylinder (photo).
2   To remove the master cylinder with the fluid reservoir mounted separately, proceed as described in the following paragraphs.
3   First remove the hydraulic pipes from the master cylinder; disconnect one pipe at a time, catching the hydraulic fluid in a suitable drip tray. Plug the disconnected pipes to prevent dirt ingress. It is permissible to remove the two-way connection unions directly from the master cylinder, if desired.
4   It is now time to disconnect the flexible hoses supplying the master cylinder with hydraulic fluid. Peel back the rubber dust boot at each connection to the master cylinder (photo), and, firmly gripping each hose in turn, give them a sharp pull. Bearing in mind the pressures created in the hydraulic system by the master cylinder, the connections need to be very tight to effect a high pressure seal, so they will need considerable force to remove them.
5   Undo and remove the two nuts attaching the master cylinder to the servo unit (or the engine bulkhead, if applicable). Withdraw the cylinder from the two mounting studs on the servo unit (or the engine bulkhead).
6   To remove the master cylinder with the fluid reservoir mounted directly above the cylinder, proceed as described in the following paragraphs.
7   Disconnect the wire coupling to the brake fluid level sensor.
8   Disconnect the brake pipes from the master cylinder and allow the brake fluid to drain into a suitable drip tray.

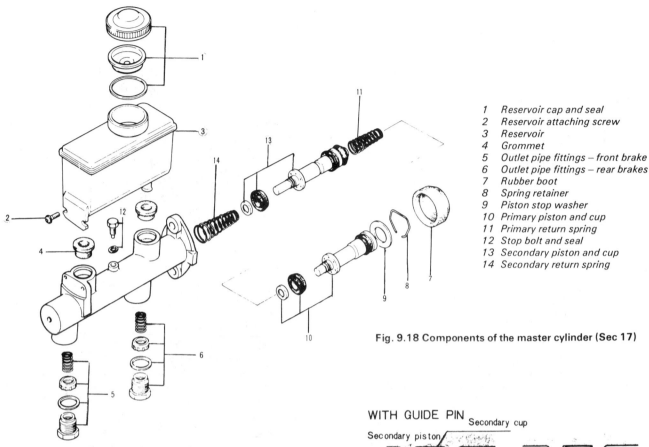

Fig. 9.18 Components of the master cylinder (Sec 17)

1   Reservoir cap and seal
2   Reservoir attaching screw
3   Reservoir
4   Grommet
5   Outlet pipe fittings – front brake
6   Outlet pipe fittings – rear brakes
7   Rubber boot
8   Spring retainer
9   Piston stop washer
10 Primary piston and cup
11 Primary return spring
12 Stop bolt and seal
13 Secondary piston and cup
14 Secondary return spring

9   Undo and remove the two nuts attaching the master cylinder to the servo unit (or the engine bulkhead, if applicable). Withdraw the cylinder from the two mounting studs on the servo unit (or the engine bulkhead).
10 Refitting is a reversal of the removal procedure in both cases. Check and, if necessary, adjust the brake pedal free travel, as described in Section 5, to ensure that the correct clearance exists between the operating rod and the piston in the servo unit (or the primary piston cup in the master cylinder).
11 Refer to Section 6, and bleed the hydraulic system.

17 Master cylinder – dismantling, servicing and reassembly

Note: *During any servicing operation on the brake master cylinder, great care must be taken to ensure that there is no contamination from dirt, oil or grease. As far as practically possible, hands should be clean. Do not use petrol or paraffin, or proprietary solvents to clean parts, but use only methylated spirits, alcohol or clean brake fluid.*
1   Separate the reservoir from the cylinder body by removing the two attaching screws. Lift the reservoir up, and out of the two grommets. Where the reservoir is mounted separately, this obviously won't be necessary.
2   Remove the two grommets (if applicable), and from the brake pushrod end, remove the rubber boot.
3   Using a small screwdriver, remove the spring retainer that holds the piston stop washer and primary piston and cup in position.
4   Remove the piston stop washer, primary piston and cup and the return spring.
5   Loosen, but do not remove, the secondary piston stop screw.
6   Make up a small guide pin, as shown in Fig. 9.19. Push in the secondary piston and cup with a screwdriver, remove the stop screw and seal, and insert the guide pin in its place. Remove the screwdriver and extract the secondary piston and cup, and the return spring.

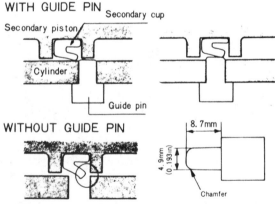

Fig. 9.19 The guide pin in use (Sec 17)

7   Remove the outlet port fittings, gaskets, check valves and check valve springs.
8   Remove the cups from the primary and secondary pistons. It is false economy to hope to use these again (unless they are known to have been in use for a very short period only) and they are best discarded.
9   Clean all the parts carefully (refer to the Note at the beginning of this Section) and dry them with compressed air, paper tissues or a clean lint-free cloth. Examine the cylinder bores and pistons for wear, scoring and signs of corrosion, renewing as necessary. Ensure that the compensating ports in the cylinder are not blocked.
10 When reassembling, dip all parts (except the cylinder body and reservoir) in clean brake fluid.
11 Insert the check valve springs and the valves into the cylinder ports and fit the gaskets and fittings.
12 Fit the cups to the secondary piston so that the flat sides of the cup are towards the piston (see Fig. 9.20).
13 Insert the secondary return spring, then fit the guide pin into the

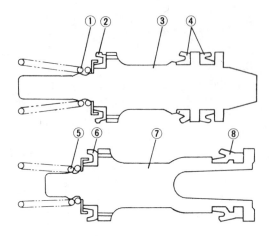

Fig. 9.20 Positioning the cups to the secondary and primary pistons

1 Secondary return spring
2 Primary cup
3 Secondary piston
4 Secondary cups

5 Primary return spring
6 Primary cup
7 Primary piston
8 Secondary cup

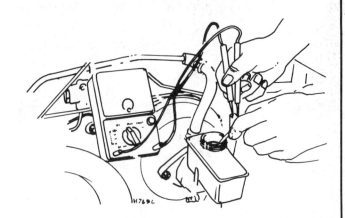

Fig. 9.21 Checking the brake fluid level sensor (Sec 18)

Fig. 9.22 The differential proportioning valve (Sec 19)

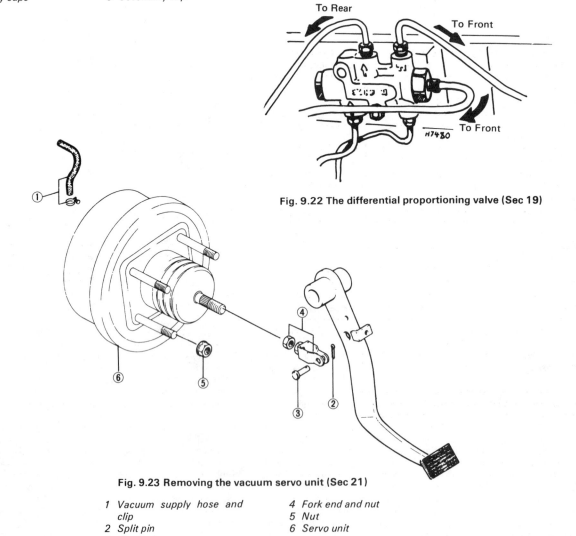

Fig. 9.23 Removing the vacuum servo unit (Sec 21)

1 Vacuum supply hose and clip
2 Split pin
3 Clevis pin

4 Fork end and nut
5 Nut
6 Servo unit

18.3 The master cylinder reservoir fluid level sensor float

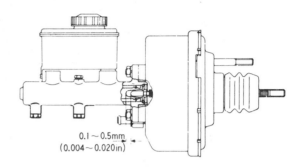

0.1 ~ 0.5mm
(0.004 ~ 0.020in)

Fig. 9.24 The clearance between the master cylinder and the servo unit (Sec 21)

stop screw hole and press in the secondary piston as far as it will go.
14 Remove the guide pin and fit the stop screw with a new seal.
15 Fit the primary cup to the primary piston, so that the flat side is towards the piston. Fit the secondary cup to the primary piston, with the open side facing the secondary piston.
16 Insert the return spring and the primary piston into the cylinder bore; fit the piston stop washer and the spring retainer to secure the stop washer in position.
17 If applicable, assemble the two grommets to the cylinder body and refit the reservoir. Fit the two reservoir attaching screws.
18 Fit the rubber boot to the brake pushrod end of the cylinder.

18 Brake fluid level sensor – checking

1 The fluid level sensor, which is fitted to the reservoir tank, is so designed to light the brake system warning lamp when the fluid in the reservoir falls below the minimum mark.
2 If the sensor is thought to be faulty, remove the reservoir cap and disconnect the coupler of the sensor.
3 Connect a circuit tester to the coupler and check the continuity by moving the float up and down (photo).
4 When the float is below the minimum mark, the tester should show continuity. When the float is above the minimum mark there should be no continuity shown on the circuit tester.
5 If the tests in paragraph 4 are not satisfactory, renew the fluid level sensor.

19 Differential proportioning valve – removal and refitting

1 The purpose of the differential proportioning valve is to prevent the rear brakes locking in advance of the front brakes under heavy braking.
2 To test the valve, drive the car in a straight line at 30 mph (50 kph) and apply the brakes hard to lock the wheels. The skid marks for the front wheels should be longer than those for the rear wheels, indicating that the front wheels locked first.
3 Where the test proves that the valve is faulty, renew it as an assembly.
4 Disconnect the hydraulic pipies from the valve body and unscrew and remove the valve mounting bolts.
5 When fitting the new valve, make sure that the fluid ports are correctly connected by reference to Fig. 9.22.

6 Refer to Section 6, and bleed the hydraulic system.

20 Vacuum servo unit – description

1 Most vehicles are fitted with a vacuum servo unit, which is fitted into the brake hydraulic circuit in series with the master cylinder to provide assistance to the driver when the brake pedal is depressed. This reduces the effort required by the driver to operate the brakes under all braking conditions.
2 The unit operates by vacuum obtained from the inlet manifold and comprises basically a booster diaphragm and a non-return valve. The servo unit and master cylinder are connected together so that the servo unit piston rod acts as the master cylinder pushrod.
3 Under normal operating conditions the vacuum servo unit is very reliable and requires no maintenance (except for periodic inspection of the vacuum hose and its connections). If a fault does occur, the unit should be renewed on an exchange basis and no attempt made to dismantle it.
4 It is emphasised that the servo unit assists in reducing the braking effort at the foot pedal and in the event of its failure, the hydraulic braking system is in no way affected except that the need for higher pedal pressures will be noticed.

21 Vacuum servo unit – removal and refitting

1 In order to remove the vacuum servo unit, refer to Section 16, and remove the brake master cylinder.
2 Loosen the hose clip, and disconnect the vacuum supply hose to the unit.
3 Working inside the vehicle, remove the split pin and push out the clevis pin to disconnect the brake pedal.
4 With the brake pedal disconnected, loosen the locknut and screw off the fork end servo to pedal connector, then remove the locknut.
5 Undo and remove the four nuts securing the unit to the engine bulkhead. Withdraw the vacuum servo unit from the engine bulkhead into the engine compartment.
6 Refitting is the reverse of the removal procedure but adjust the pedal free travel as described in Section 5 to ensure that the correct clearance exists between the primary piston in the master cylinder, and the pushrod in the vacuum servo unit.
7 Refer to Section 6, and bleed the hydraulic system.

9

22 Fault diagnosis – braking system

Symptom	Reason/s	Remedy
Pedal travels almost to floor before brakes operate	Brake fluid level too low	Top-up fluid reservoir
	Caliper leaking	Repair as necessary
	Master cylinder leaking	Remove and inspect piston cups etc
	Brake flexible hose leaking	Renew the hose
	Brake pipe fractured	Trace the fracture, renew as required
	Brake system unions loose	Tighten all unions
	Pad or shoe linings over 75% worn	Renew as necessary
Brake pedal feels springy	New linings not yet bedded-in	
	Brake disc or drums badly worn or cracked	Inspect and renew as necessary
	Master cylinder securing nuts loose	Examine and tighten
Brake pedal feels 'spongy' and 'soggy'	Caliper or wheel cylinder leaking	Locate and repair/renew as required
	Master cylinder leaking (bubbles in fluid)	Rectify leaks and bleed system
	Brake pipe or flexible hose leaking	Repair by renewal only
	Unions in brake system loose and leaking	Examine and tighten all unions
Excessive effort required to brake car	Faulty vacuum servo unit	Renew servo unit
	Pads or shoe linings badly worn	Renew as necessary
	Harder linings fitted than standard	Revert to original standard
	Linings and/or brake drums contaminated with oil, grease etc	Renew as required
Brakes uneven and pulling to one side	Linings or pads, discs or brake drums contaminated with oil, grease etc	Renew as required
	Tyre pressures unequal	Rectify tyre pressures
	Brake caliper loose	Tighten mounting bolts
	Brake pads or shoes incorrectly fitted	Assemble correctly
	Different type linings fitted to each wheel	Fit correct linings
Brakes tend to bind, drag or lock-on	Rear brakes overadjusted	Adjust correctly
	Air in hydraulic system	Bleed the hydraulic system
	Handbrake cables overtightened	Adjust handbrake cables

Chapter 10 Electrical system

See Chapter 13 for specifications and information applicable to 1979 thru 1983 North American models

Contents

Specifications

System type .... 12 volt, negative earth

Battery .... 35 or 45 amp hour, at 20 hour rate

Alternator

Earth polarity	Negative
Load test:	
Voltage	14 volt
Current	30 amp
Alternator revolutions	Not greater than 2500 rpm
Number of brushes	2
Brush length:	
New	0.71 in (18 mm)
Wear limit	0.31 in (8 mm)
Brush spring pressure	11 to 15 oz (310 to 430 gm)
Slip ring diameter:	
New	1.299 ± 0.008 in (33 ± 0.2 mm)
Wear limit	1.268 in (32.2 mm)
Pulley ratio	2.16 : 1

Regulator

Constant voltage relay:	
Air gap	0.028 to 0.051 in (0.7 to 1.3 mm)
Point gap	0.012 to 0.018 in (0.3 to 0.45 mm)
Back gap	0.028 to 0.059 in (0.7 to 1.5 mm)
Regulated voltage (no load) at 4000 rpm of alternator	14 to 15 volt
Ignition warning lamp relay:	
Air gap	0.039 to 0.059 in (1.0 to 1.5 mm)
Point gap	0.020 to 0.035 in (0.5 to 0.9 mm)
Back gap	0.028 to 0.059 in (0.7 to 1.5 mm)
Ignition warning light on	3.0 volts or less
Ignition warning light out	4.2 to 5.2 volts

Starter motor

Capacity	0.8 kw
Lock test:	
Voltage	5.0 volt
Current	310 amp or less
Torque	5.4 lbf ft (0.75 kgf m)

10

Free runing test:
Voltage	11.5 volt
Current	53 amp or less
Speed	6800 rpm or more
Number of brushes	3

Brush length:
New	0.67 in (17.0 mm)
Wear limit	0.45 in (11.5 mm)
Brush spring pressure	46 to 60 oz (1.3 to 1.7 gm)
Control switch	Solenoid
Voltage required to close solenoid contacts	8 volts or less

Bulbs

Headlights	45W/40W or 50W/40W *
Front turn signal light	21W, 23W or 27W *
Front parking light	5W or 8W *
Rear turn signal light	21W, 23W, or 27W *
Stop and tail light	21W/5W, 23W/5W or 27W/8W *
Number plate light	6W or 4W *
Reverse light	21W or 27W *
Interior light	5W

Indicator and warning lights (Instruments):
Turn signal light	3.4W
Hazard light	3.4W
Heat hazard light (USA)	1.4W
Rear window defogger light	1.5W
Brake system light	3.4W
Oil pressure light	1.4W
High beam light	3.4W
Generator light	1.4W
Instrument panel illumination light	3.4W

* The wattages given vary according to the vehicle type and market. In cases of difficulty, the advice of your vehicle main dealer should be sought.

1 General description

The electrical system is a 12 volt, negative earth type comprising an alternator and its associated regulator, the battery, starter motor, the ancillaries such as lighting equipment, windscreen wipers, etc, and the associated wiring and protective circuits.

When fitting electrical accessories to your vehicle it is important to ensure that they are suitable for negative earth vehicles (most vehicles today are negative earth, but accessories are still available for older positive earth systems). Equipment which incorporates semi-conductor devices may well be damaged if it is not suited to your particular system.

If the battery is to be charged from an external charging source, it is important to disconnect the battery positive cable in order to protect the semi-conductor devices in the alternator and any electrical accessories which may have been fitted. Similarly, when any electric (arc) welding or power tools are used the same precaution should be taken.

2 Battery – removal and refitting

1 The battery is in a special carrier fitted on the right-hand wing valance of the engine compartment. It should be removed once every three months for cleaning. Disconnect the leads from the battery terminals by slackening the clamp retaining nuts and bolts, or by un-screwing the retaining screws if terminal caps are fitted instead of clamps (photo).
2 Unscrew the clamp bar retaining nuts, then remove the clamp. Carefully lift the battery from its carrier. Hold the battery vertical to ensure that none of the electrolyte is spilled.
3 Refitting is a direct reversal of this procedure. **Note:** Fit the nega-tive lead after the positive lead and smear the terminals with petroleum jelly to prevent corrosion. NEVER use an ordinary grease as applied to other parts of the car.

3 Battery – maintenance

1 Normal weekly battery maintenance consists of checking the

2.1 The battery in position

electrolyte level of each cell to ensure that the separators are covered by $\frac{1}{4}$ inch of electrolyte. If the level has fallen, top up the battery using distilled or de-ionized water. Do not overfill. If the battery is overfilled or any electrolyte spilled, immediately wipe away the excess, as electrolyte attacks and corrodes any metal it comes into contact with very rapidly. In an emergency, where the electrolyte level is too low, it is permissible to use boiled drinking water which has been allowed to cool but this is not recommended as a regular practice.
2 If the battery terminals are showing signs of corrosion, brush or scrape off the worst taking care not to get the deposits on the vehicle paintwork or your hands. Prepare a solution of household ammonia, washing soda or bicarbonate of soda and water. Brush this onto all the corroded parts, taking care that none enters the battery. This will neutralize the corrosion and when all the fizzing and bubbling has stopped the parts can be wiped clean with dry, lint-free cloth. Don't

forget to smear the terminals and clamps with petroleum jelly afterwards to prevent further corrosion.

3   Inspect the battery clamp and mounting tray and treat these in the same way. Where the paintwork has been damaged, after neutralizing, the area can be painted with a zinc based primer and the appropriate finishing colour, or an underbody paint can be used.

4   At the same time inspect the battery case for cracks. If a crack is found, clean and plug it with one of the proprietary compounds marketed for this purpose. If leakage through the crack has been excessive then it will be necessary to refill the appropriate cell with fresh electrolyte as described later. Cracks are frequently caused at the top of the battery case by pouring in distilled water in the middle of winter *after* instead of *before* a run. This gives the water no chance to mix with the electrolyte and so the former freezes and splits the battery case.

5   If topping-up becomes excessive and the case has been inspected for cracks that could cause leakage, but none are found, the battery is being overcharged and the voltage regulator will have to be checked and reset.

6   With the battery on the bench, measure the specific gravity with a hydrometer to determine the state of charge and condition of the electrolyte. There should be very little variation between the different cells and, if a variation in excess of 0.025 is present, it will be due to either:

a)   *Loss of electrolyte from the battery at some time caused by spillage or a leak, resulting in a drop in the specific gravity of the electrolyte when the deficiency was replaced with distilled water instead of fresh electrolyte.*

b)   *An internal short circuit caused by buckling of the plates or similar malady pointing to the likelihood of total battery failure in the near future.*

7   The specific gravity of the electrolyte for fully charged conditions at the electrolyte temperature indicated, is listed in Table A. The specific gravity of a fully discharged battery at different temperatures of the electrolyte is given in Table B.

Table A
Specific gravity – battery fully charged
1.268 at 100°F or 38°C   electrolyte temperature
1.272 at 90°F or 32°C   electrolyte temperature
1.276 at 80°F or 27°C   electrolyte temperature
1.280 at 70°F or 21°C   electrolyte temperature
1.284 at 60°F or 16°C   electrolyte temperature
1.288 at 50°F or 10°C   electrolyte temperature
1.292 at 40°F or 4°C   electrolyte temperature
1.296 at 30°F or-1.5°C electrolyte temperature

Table B
Specific gravity – battery fully discharged
1.098 at 100°F or 38°C   electrolyte temperature
1.102 at 90°F or 32°C   electrolyte temperature
1.106 at 80°F or 27°C   electrolyte temperature
1.110 at 70°F or 21°C   electrolyte temperature
1.114 at 60°F or 16°C   electrolyte temperature
1.11,8 at 50°F or 10°C   electrolyte temperature
1.122 at 40°F or 4°C   electrolyte temperature
1.126 at 30°F or-1.5°C electrolyte temperature

4   Battery – electrolyte replenishment

1   If the battery is in a fully charged state and one of the cells maintains a specific gravity reading which is 0.025 or more lower than the others, and a check of each cell has been made with a voltage meter to check for short circuits (a four to seven second test should give a steady reading of between 1.2 to 1.8 volts), then it is likely that electrolyte has been lost from the cell with the low reading at some time.

2   Top-up the cell with a solution of 1 part sulphuric acid to 2.5 parts of water. If the cell is already fully topped-up draw some electrolyte out of it with a pipette.

3   When mixing the sulphuric acid and water **never add water to sulphuric acid** – always pour the acid slowly onto the water in a glass container. **If water is added to sulphuric acid it will explode.**

Continue to top-up the cell with the freshly made electrolyte and then recharge the battery and check the hydrometer readings.

5   Battery – charging

Note: If the battery is to remain in the vehicle when being charged, always disconnect the battery leads.

1   In the winter time when a heavy demand is placed on the battery, such as when starting from cold, and much electrical equipment is continually in use, it is a good idea to occasionally have the battery fully charged from an external source at a rate of approximately 4 amps.

2   Continue to charge the battery at this rate until no further rise in specific gravity is noted over a four hour period.

3   Alternatively, a trickle charger, charging at the rate of 1.5 amps can be safely used overnight.

4   Special rapid 'boost' charges which are claimed to restore the power of the battery in 1 to 2 hours are most dangerous unless they are thermostatically controlled as they can cause serious damage to the battery plates through overheating.

5   While charging the battery note that the temperature of the electrolyte should never exceed 100°F (37.8°C).

6   Alternator – general information and precautions

1   The use of alternators for generating the current required to operate the car electrical systems is now more commonplace. Their main advantage over the dynamo type of generator is that they provide a high output for lower revolutions and are light in weight/output ratio.

2   The alternator generates alternating current and this current is rectified by diodes into direct current which is the current required for battery storage.

3   The alternator output is controlled by a voltage regulator and a charge relay operates the charge warning lamp to indicate whether or not the alternator is charging the battery.

4   In service, a minimum amount of maintenance is required, the only items subject to wear being the brushes and bearings.

5   The brushes should be examined after 40 000 miles (65 000 km) and renewed if necessary.

6   The bearings are pre-packed with grease for life and should not require any further attention.

7   If there are indications that the charging system is malfunctioning in any way, care must be taken to diagnose faults properly, otherwise damage of a serious and expensive nature may occur to parts which in fact were quite serviceable.

8   The following basic precautions must be observed at all times:

a)   *All alternator systems use a negative earth. Even the simplest mistake of connecting a battery the wrong way round could burn out the alternator diodes in a few seconds.*

b)   *Before disconnecting any wiring in the system the engine ignition should be switched off. This will minimise accidental short circuits.*

c)   *The alternator must never be run with the output wire disconnected.*

d)   *Always disconnect the battery from the car's electrical system if an external charging source is being used.*

e)   *Do not use test wire connections that could move accidently and short circuit against nearby terminals. Short circuits will not blow fuses – they will blow diodes or transistors.*

f)   *Always disconnect the battery cables and alternator output lead before carrying out any electric (arc) welding work on the car.*

g)   *When checking the circuit between individual terminals, or when testing silicone diodes for continuity, never use a high voltage tester (such as a megger) as the diodes will be damaged. Use an ordinary tester.*

h)   *When using an extra battery as a starting aid, always connect it in parallel.*

7   Alternator – testing in the vehicle

1   In order to test the alternator properly, special test equipment will

10

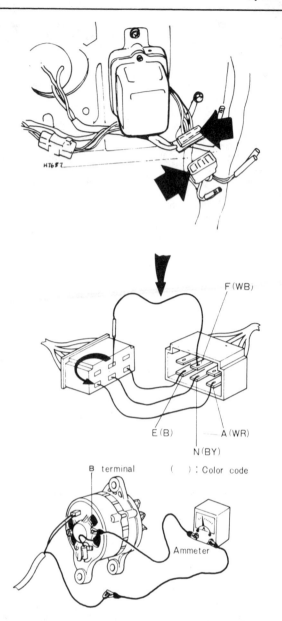

8.2 The alternator connections

Fig. 10.1 Alternator testing set-up (Sec 7)

8.3 The alternator adjustment slot bolt

8.5 Don't forget to recover any shims that may be fitted

be needed and it is considered outside the scope of a manual of this type. A simple check-out is given in the following paragraphs, but where this fails to isolate a fault it is essential that a vehicle main dealer is contacted or the job is entrusted to an automative electrical specialist.

2   Disconnect the wire from the alternator 'B' terminal and connect an ammeter into the lead. The ammeter positive lead connects to the 'B' terminal and the negative lead connects to the lead which was disconnected.

3   Remove the electrical connector from the regulator then remake each connection using suitable short flying leads.

4   Run the engine at approximately 2000 rpm and note the ammeter reading.

5   Disconnect the wire at the regulator 'F' terminal and short circuit this wire to the regulator 'A' terminal.

6   If there is a large increase in the charging current, the alternator is satisfactory and the regulator is at fault. If there is no change in current, the alternator is at fault.

8   Alternator – removal and refitting

1   Disconnect the battery terminals.

2   Disconnect the wires to the alternator at the multi-pin plug and

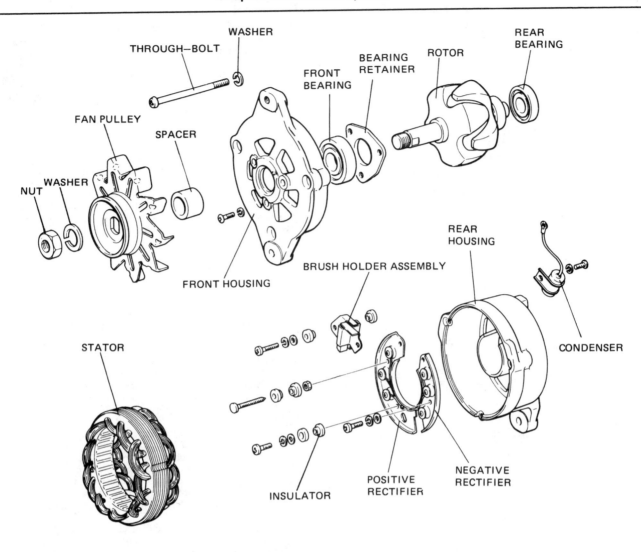

Fig. 10.2 Exploded view of the alternator (Sec 9)

the terminal post (photo).

3   Remove the bolt in the adjustment slot (photo).

4   Slacken the long pivot bolt and slide the alternator toward the engine. Remove the drivebelt.

5   Support the alternator with one hand, and with the other hand pull out the pivot bolt. Lift out the alternator taking care to recover any shim(s) that may be fitted to the mounting bracket (photo).

6   Refitting is a reversal of the removal procedure but ensure that the drivebelt tension is adjusted as described in Chapter 2.

9   Alternator – dismantling, testing, servicing and reassembly

Note: Before attempting to dismantle and repair the alternator, it must be appreciated that more harm than good can be done by anyone not familiar with repair techniques on semi-conductor devices. Also, before dismantling, ensure that there are spare parts available since it is unlikely that any individual components will be capable of being repaired. If parts are not available, or you feel that you do not have the knowledge and experience to do the job yourself, either purchase an exchange unit or entrust the job to a specialist in this type of work.

1   Remove the condenser (capacitor) from the rear of the alternator.

2   Insert a drift through the pulley fan fins then remove the nut, washer, pulley fan and spacer from the shaft.

3   Remove the three alternator through-bolts and remove the front housing and rotor.

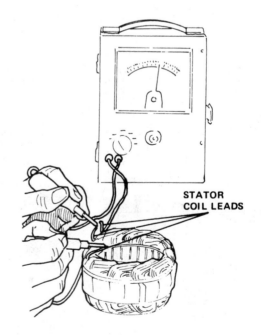

Fig. 10.3 Testing a stator for open circuit (Sec 9)

STATOR COIL LEADS

10

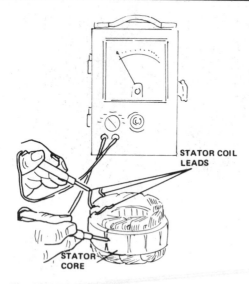

Fig. 10.4 Testing a stator for earthing (Sec 9)

STATOR COIL LEADS

STATOR CORE

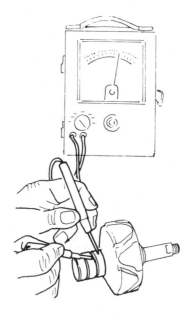

Fig. 10.5 Checking the resistance between the slip rings (Sec 9)

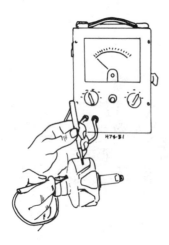

Fig. 10.6 Testing the rotor core for short circuits (Sec 9)

4   Remove the rotor from the front housing.

5   Remove the nut, washers and insulator from the 'B' terminal.

6   Remove the three screws attaching the rectifier plates to the housing.

7   Unscrew the brush holder attaching screws but do not pull them out.

8   Carefully remove the stator, rectifiers and brush holder from the rear housing.

9   Using a soldering iron of less than 100 watts, unsolder the stator leads from the rectifier and brush holder assembly. If necessary, unsolder the brush holder assembly wire from the positive rectifier heat sink.

10   Unsolder the rectifier assemblies using the minimum possible amount of heat.

11   If bearings are to be renewed, the rear one can be drawn off the rotor shaft. To remove the front one, remove the three attaching screws from the bearing retainer and press out the bearing.

12   Carefully clean the parts with a clean paintbrush, paper tissues or a lint-free cloth. Do not immerse them in proprietary solvents. Check the brushes and tension springs in accordance with the Specifications, renewing as necessary.

13   Using a suitable ohmmeter, check for continuity between the stator coil leads, or each pair of leads, as appropriate. If an open circuit exists, the stator must be renewed.

14   Using a suitable ohmmeter, connect between the stator core and each winding lead. If a short circuit is indicated, the stator must be renewed.

15   Using a suitable ohmmeter, check for a resistance of 5 to 6 ohms between the slip-rings. If this is not obtained, the rotor should be renewed.

16   Using a suitable ohmmeter, connect between the rotor core and one slip-ring. If a short circuit is indicated, the stator must be renewed.

17   Using a suitable ohmmeter, check between each diode terminal and the heat sink. An indication should be obtained when the leads are connected one way, but an open circuit should be indicated when the ohmmeter leads are reversed. If there is no indication, or a high indication, regardless of the polarity of the ohmmeter leads, the appropriate diodes and heat sink should be renewed as an assembly.

18   To commence assembly of the alternator, fit new bearings if they were removed. When fitting the rotor shaft bearing, press on the inner race only; when fitting the bearing to the front housing, press on the outer race only.

19   Assemble the stator, rectifiers and brush holder assembly, then using the minimum amount of heat possible, solder the wires.

20   Place the rear housing on the bench, open side up, then carefully

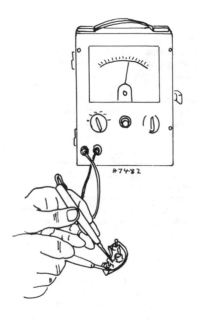

Fig. 10.7 Diode test (Sec 9)

lower the stator, rectifiers and brush holder assembly into the rear housing. Securely tighten the brush holder retaining screws.

21 Hold the rear housing and stator assembly with the rear of the housing upwards. Insert the insulator between the positive rectifier heat sink and the rear housing, allowing the shoulder of the insulator to drop into the large hole of the heat sink.

22 Fit the longest crosshead screw and washer through the flat insulator, heat sink and round insulator. Start the screw in the threaded hole in the rear housing but do not tighten it.

23 Insert another insulator between the positive rectifier heat sink and the rear housing so that the insulator shoulder fits into the 'B' terminal hole. Tighten the screw to hold the insulator in position against the rear housing.

24 Insert the square headed battery terminal screw through the heat sink, insulator and rear housing. Fit an insulator, shoulder towards the housing, washer and nut on the 'B' terminal screw. Ensure that the insulator shoulder recesses into the hole in the rear housing.

25 Fit the two screws that attach the negative rectifier heat sink to the rear housing.

26 Ensure that the insulator shoulders are recessed in the holes, then tighten the rectifier attaching screws and the 'B' terminal nut.

27 Push the brushes into the holder and fit a thin wire (eg a paper clip) into the small hole below the terminal connector at the rear of the alternator, and through the holes in the brushes to hold then in the retracted position.

28 Position the rotor to the front housing and fit the spacer, fan, pulley, washer and nut; tighten the nut securely.

29 Assemble the two parts ensuring that the notches are aligned, then fit and tighten the three through-bolts.

30 Fit the condenser with the lead connected to the 'B' terminal.

31 Remove the brush retracting wire then check that the pulley rotates freely.

10 Regulator – adjustment

1   A separate regulator is used to control alternator output, it being a single armature controlling current and voltage.

2   The regulator controls the current to the field coils, and responds to battery condition and the electrical load being drawn.

3   Whenever the battery is less than fully charged, the charging rate is relatively high, and extremely so immediately after starting up. As the battery becomes fully charged, the charge rate cuts back to a trickle. If an electric load is drawn, such as turning on the headlamps, the regulator responds and keeps the battery charge much as it was before.

4   If the regulator is set at too low a limiting voltage, the battery will be kept just off full charge, so will deteriorate. If set too high, on long journeys the battery will be overcharged. In the short term this calls for frequent topping up of the battery. In the long term it is not good for the battery.

5   As the alternator output rises the increased magnetic field in the regulator windings draws the armature down against its blade spring, so opening the contacts. This cuts the field current. The armature then vibrates as required to control the output. The setting depends on the mechanical adjustment of the points and the strength of the spring. Once the mechanical adjustments have been set, output is adjusted by tensioning the spring by bending its abutment stop. The adjustment of this stop is very delicate, a small bend giving a big effect.

6   After a long time the contacts will need cleaning. They should be polished with fine emery cloth. Then all dust must be cleaned away. Reset the mechanical clearances; point gap; air gap between core and armature; back gap between the frame and the contact arm.

7   For the electric setting the battery must be fully charged, so that little load is drawn. An old battery can never be fully recharged, so it may be necessary to buy or borrow a new one. Give it a final charge from a battery charger.

8   Connect a voltmeter between the 'A' and 'E' terminals of the regulator. Start the engine. Run it for about ten minutes at 2000 rpm to stabilise the temperature of the regulator, and to recharge the battery after starting up.

9   With the engine still at 2000 rpm, read the voltage. Allow the engine to slow down, and speed it up again, and watch for a kick on the voltmeter as regulation starts. The reading should be 14 to 15 volts.

10 If the car is used much for stop-start driving, the output wants to

Fig. 10.8 Fitting the alternator rotor (The arrow indicates the brush retracting wire) (Sec 9)

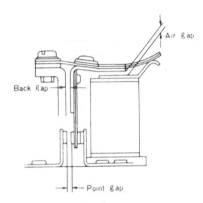

Fig. 10.9 The regulator gaps (Sec 10)

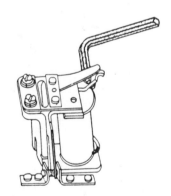

Fig. 10.10 Adjusting the regulator (Sec 10)

be up the top end of the tolerance. If the work is mainly long journeys, particularly in a hot climate, then it should be at the bottom end. To lower the voltage, the regulator spring must be weakened; the stop must be bent down, but only by a small amount. Bending it up raises the output.

11 Starter motor – general description

1   In the engine starting sequence the starter motor solenoid is energized from the ignition switch, which moves the starter drive gear into mesh with the flywheel ring gear. As this happens, the solenoid main contacts close and the motor is energized and starts to rotate.

2   As soon as the engine fires and the ignition key returns to the

10

13.5 Removing the starter motor

normal running position, the solenoid is de-energized. This cuts the supply to the motor and the solenoid return spring causes the shift fork to disengage the drive from the flywheel ring gear.

3   To prevent excessive motor speed as the engine fires, an over-running clutch is incorporated in the drive gear assembly. The drive gear can therefore run at a faster speed than the armature during this brief period of time and disengages itself from the flywheel ring gear.

12 Starter motor – maintenance

No routine maintenance is called for on the starter motor. It will normally work without attention for about the same mileage as the engine will go without overhaul. If should therefore be overhauled at the same time. All that should be needed is cleaning, renewal of the brushes, and lubrication.

13 Starter motor – removal and refitting

1   Disconnect the battery earth cable.
2   Remove the carburettor air cleaner for access; refer to Chapter 3 if necessary.
3   Disconnect the battery cable from the starter solenoid 'B' terminal

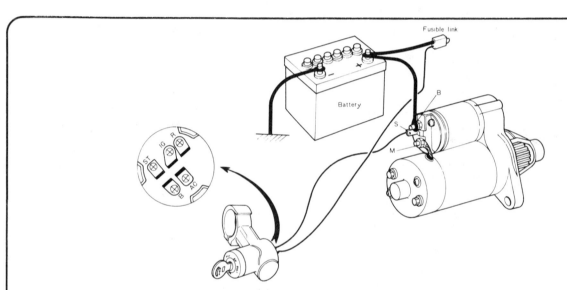

Fig. 10.11 The starter motor circuit (Sec 14)

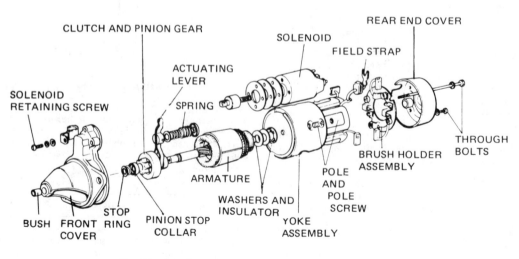

Fig. 10.12 An exploded view of the starter motor (Sec 14)

and the ignition switch wire from the 'S' terminal.
4   Raise the right-hand side of the vehicle if necessary, then working from below, remove the two starter attaching nuts, washers and bolts.
5   Tilt the drive end of the starter downwards then remove it from below (photo).
6   Refitting is a straightforward reversal of the removal procedure.

14 Starter motor – dismantling, testing, servicing and reassembly

1   Disconnect the solenoid field strap then remove the solenoid retaining screws.
2   Disengage the solenoid actuating lever from the solenoid plunger, then remove the solenoid.
3   Remove the two through-bolts and pull off the rear end cover.
4   Lift off the insulator and washers, then remove the brush and holder assembly.
5   Remove the yoke assembly from the front cover. Disengage the actuating lever and spring from the armature shaft.
6   To remove the pinion stop collar and the over-running clutch, hold the armature in protected vice jaws. Use a suitable size of tube to drive the stop collar towards the armature. The stop ring can now be removed along with the stop collar and the over-running clutch and pinion gear.
7   If it is necessary to remove the field coil, remove the pole shoe retaining screws and remove the pole shoes and field coil.
8   Check the clearance between the armature shaft and the bushes. If the side play exceeds 0.008 in (0.2 mm) press out the bushes and obtain replacement items. Before pressing in new bushes soak them overnight in clean engine oil, to allow the bush pores to absorb the oil.
9   Carefully clean the parts with a clean paintbrush, paper tissues or a lint-free cloth. Do not immerse them in proprietary solvents. Check the brush spring tension in accordance with the Specifications. Obtain new brushes, since it is false economy not to renew them if the starter is being stripped for major servicing.
10 If the commutator is dirty, discoloured or worn. it may be cleaned using a strip of fine emery paper. Where scoring or arcing has occurred it may be possible for it to be machined but in serious cases it will be necessary to obtain a replacement armature. After any repair operation to the commutator, undercut the mica segments as shown in Fig.10.14.
11 Using a suitable ohmmeter, check between the armature core and each segment of the commutator in turn. An infinite reading should be obtained in each case; if this is not obtained the armature must be renewed (see Fig. 10.15).
12 Using a suitable ohmmeter, test the commutator between all adjacent segments, checking for a similar resistance indication in each case. Provided that this test is satisfactory and there was no indication of arcing at the commutator, the armature can be assumed to be satisfactory. If the test proves satisfactory, and there was evidence of arcing at the commutator, it is advisable to have the armature checked by a specialist as there may be an intermittent fault.
13 Using a suitable ohmmeter test the field winding for a short circuit

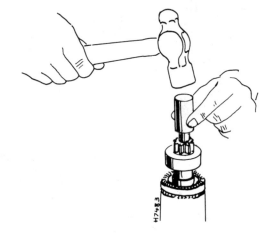

Fig. 10.13 Removing the stop ring (Sec 14)

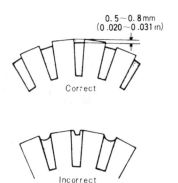

Fig. 10.14 Under-cutting dimensions for the commutator (Sec 14)

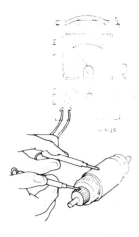

Fig. 10.15 Armature segment-to-core checks (Sec 14)

10

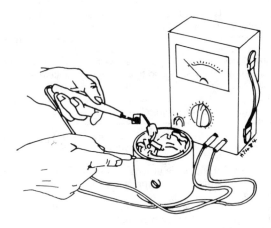

Fig. 10.16 Checking the field coils (Sec 14)

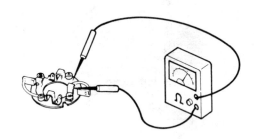

Fig. 10.17 Checking the brush holder assembly (Sec 14)

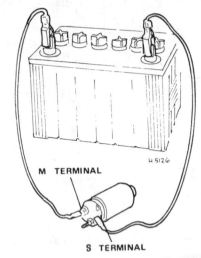

Fig. 10.18 Testing the solenoid pull-in (Sec 14)

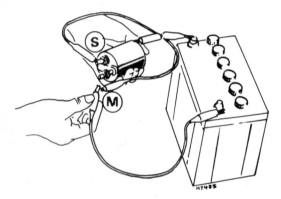

Fig. 10.19 Testing the holding coil (Sec 14)

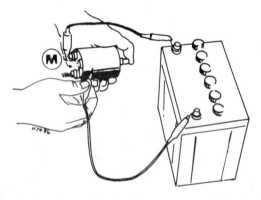

Fig. 10.20 Testing the solenoid for plunger return (Sec 14)

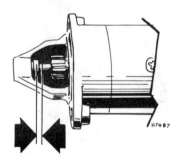

Fig. 10.21 The starter pinion-to-stop collar clearance (Sec 14)

to the frame. If a reading other than infinity is obtained, a new winding will be required.

14　Check the brush holder assembly for short circuits by touching one probe of the ohmmeter on the insulated brush holder, and the other probe to the brush holder frame. If the meter reading is other than infinite, the brush holder must be renewed.

15　To test the solenoid pull-in coil, apply 12 volts DC between the 'M' and 'S' terminals. The plunger should be drawn into the solenoid.

16　To test the solenoid holding coil, refer to Fig. 10.19 and connect a 12 volt positive lead to the 'M' terminal and solenoid body. Connect a 12 volt negative lead to the 'S' terminal then disconnect the lead from the 'M' terminal. Check that the plunger remains in the retracted position.

17　To check the plunger return, hold the solenoid plunger in and apply 12 volts DC between the 'M' terminal and the solenoid body. Release the plunger and check that it does **not** hold in.

18　To commence assembly, fit new bearings if applicable. If the field coils were removed, these too should now be fitted; ensure that the retaining screws are fully tightened.

19　Position the over-running clutch and pinion gear on the armature shaft followed by the stop ring collar and the stop ring. Slide the collar over the stop ring to secure it.

20　Engage the actuating lever to the armature shaft and fit the lever return spring.

21　Fit the armature shaft to the front cover and then assemble the yoke.

22　Fit the brush holder assembly followed by the washers and the insulator.

23　Assemble the rear end cover and fit the through-bolts.

24　Engage the solenoid plunger on the actuating lever. Secure the solenoid to the starter motor front cover and fit the field strap.

25　Apply 8 volts DC to the solenoid 'S' and 'M' terminals and check the clearance between the pinion and the stop collar. The clearance should be 0.02 to 0.08 in (0.5 to 2.0 mm). If the clearance is not as previously mentioned, shims are available to insert between the starter motor and the solenoid front face.

15　Headlights – removal and refitting

1　Remove the four cross-head screws that attach the turn signal and sidelight lens to the front of the vehicle (photo).

2　Carefully remove the lens to a safe place.

3　Remove the headlight surround retaining screw and lift away the surround (photo).

4　Loosen the headlight unit rim retaining screw, rotate the headlight unit anti-clockwise and remove it (photo).

5　Disconnect the wires from the rear of the headlight unit, and lift the unit from the vehicle (photo).

6　Refitting is a direct reversal of the removal procedure but it is advisable to have the headlight beam alignment checked at a garage.

16　Headlight aim – adjustment

1　Headlight aim is carried out by adjusting the three spring-mounted headlight retaining screws.

2　It is not advisable to adjust the aim yourself since not only may it be illegal, but it is not very easy to do it accurately.

3　For the small cost involved it is as well to take the vehicle to a garage which is specially equipped to carry out this adjustment.

17　Front parking and turn signal lamps

1　To renew either a flasher bulb or a parking lamp bulb, remove the four cross-head screws that secure the lens to the light unit. Remove the lens.

2　The bulbs are of the bayonet type, so removal is merely by pressing the bulbs inwards, and at the same time turning them anti-clockwise (photo).

3　If it is required to remove the reflector assembly after the lens has been removed, just pull it forward enough to be able to disconnect the wires at the rear of the reflector and then remove it from the vehicle.

4　Refitting of the reflector assembly and the lens is a direct reversal of the removal procedure.

15.1 Removing the sidelight/turn signal lens

15.3 The headlight surround retaining screw

15.4A Loosen the three rim retaining screws ...

15.4B ... and remove the headlight

15.5 Disconnect the wires from the rear of the headlight

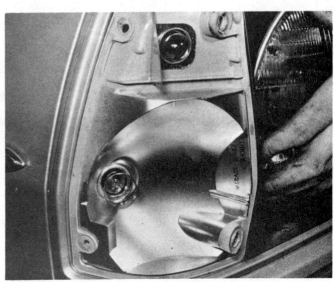
17.2 The front parking and turn signal bulbs

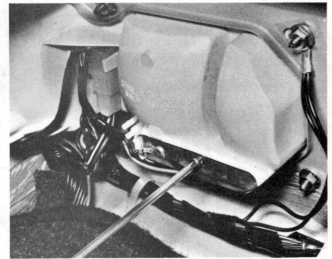

19.2 Remove the retaining screw ...

19.3 ... and pull out the bulb mounting plate, to gain access to the bulbs

20.1 The number plate lamp

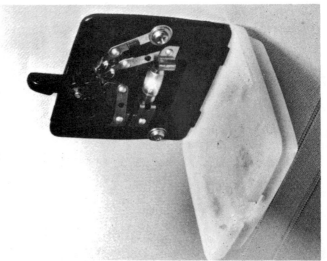

21.1 The interior lamp

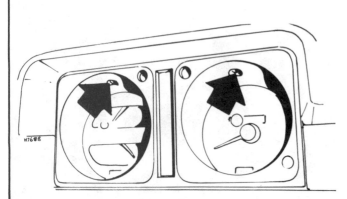

Fig. 10.22 The instrument hood retaining screws (Sec 23)

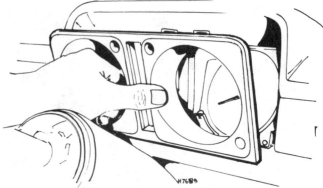

Fig. 10.23 Removing the instrument hood (Sec 23)

18 Rear turn signal lamp

1   Renewal of a rear flasher bulb is carried out from within the rear of the vehicle.
2   Remove the single screw retaining the bulb mounting plate to the light unit and pull out the bulb mounting plate.
3   Press the bulb inwards, and at the same time turn it anti-clockwise.
4   Refitting is a straightforward reversal of the removal procedure.
5   For removal of the lens, refer to the next Section.

19 Tail and reverse lamps

1   Renewal of the tail lamp bulb and the reverse lamp bulb is carried out from within the rear of the vehicle.
2   Remove the single screw retaining the bulb mounting plate to the light unit and pull out the bulb mounting plate (photo).
3   Press the bulb inwards, and at the same time turn it anti-clockwise (photo).
4   Refitting is the reverse of the removal procedure.
5   To remove the light lens, remove the reverse lamp bulb mounting plate as described in the previous Section.
6   Remove the tail and reverse lamp bulb mounting plate.
7   Undo and remove the six securing nuts and washers. The lens can now be removed from the vehicle.
8   When reassembling, ensure that the earth lead is fitted to its stud on the lens and securely retained with the nut and washer.

20 Number plate lamp

1   Removal of the number plate bulbs is carried out by removing the two screws that secure the lens and the lamp assembly to the vehicle. Remove the bulbs by pressing inwards, and at the same time turning them anti-clockwise (photo).
2   If it is required to remove the lamp assembly, pull it from the aperture, disconnect the wire and lift away the assembly.
3   Refitting of the lamp assembly and the bulb is a reversal of the removal procedure.

21 Interior light

1   To renew the festoon type bulb, press the lamp outer cover inwards, and at the same time apply a slight downward pressure. When the outer cover has been pressed in enough to clear its retaining recess, the cover can be hinged downward and unclipped from its rear mounting lugs (photo).
2   Remove the festoon bulb by pulling it from its two mountings.
3   Refitting is a straightforward reversal of the removal procedure.

22 Combination switch – removal

1   The removal of the combination switch is necessary when removing the steering wheel and is fully described in Chapter 11, Section 16.

23 Instrument panel – removal and refitting

1   Disconnect the battery.
2   Remove the two screws located in the top of each dial hood. Lift away the hood assembly.
3   Undo and remove the three screws that retain the unit to its mounting lugs.
4   Carefully pull the unit from the dash, then reach behind the dials and pull out the two multi-pin connectors.
5   Disconnect the speedometer drive cable from the speedometer by pressing the nylon lockplate downwards and pulling the cable from the speedometer.
6   Carefully lift the unit from the dashboard.
7   Refitting is a direct reversal of the removal procedure.

24 Instrument panel warning lights – renewal

1   Remove the instrument panel as described in the previous Section.
2   The instrument panel warning lights are located at the rear of the instrument panel.
3   Renewal of a warning light bulb is by way of pulling out the bulb holder.

25 Fuel gauge and water temperature gauge

1   Access for removal of these gauges is gained when the instrument panel has been removed (Section 23).

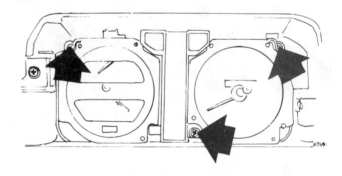

Fig. 10.24 The instrument panel retaining screws (Sec 23)

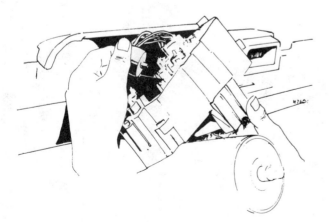

Fig. 10.25 Disconnecting a multi-pin plug (Sec 23)

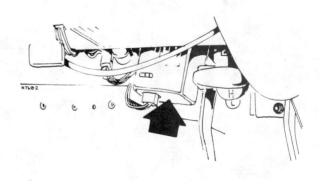

Fig. 10.26 The flasher and hazard unit (Sec 27)

10

28.2A Remove the retaining nut ...

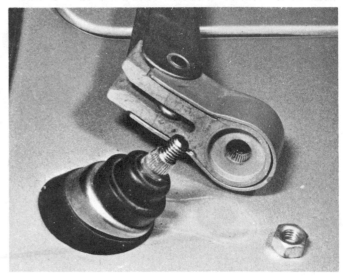

28.2B ... and pull off the wiper arm

28.5A A grille retaining screw

28.5B Remove the grille completely

29.1 The fuse panel is mounted under the dashboard

29.5 The fusible link

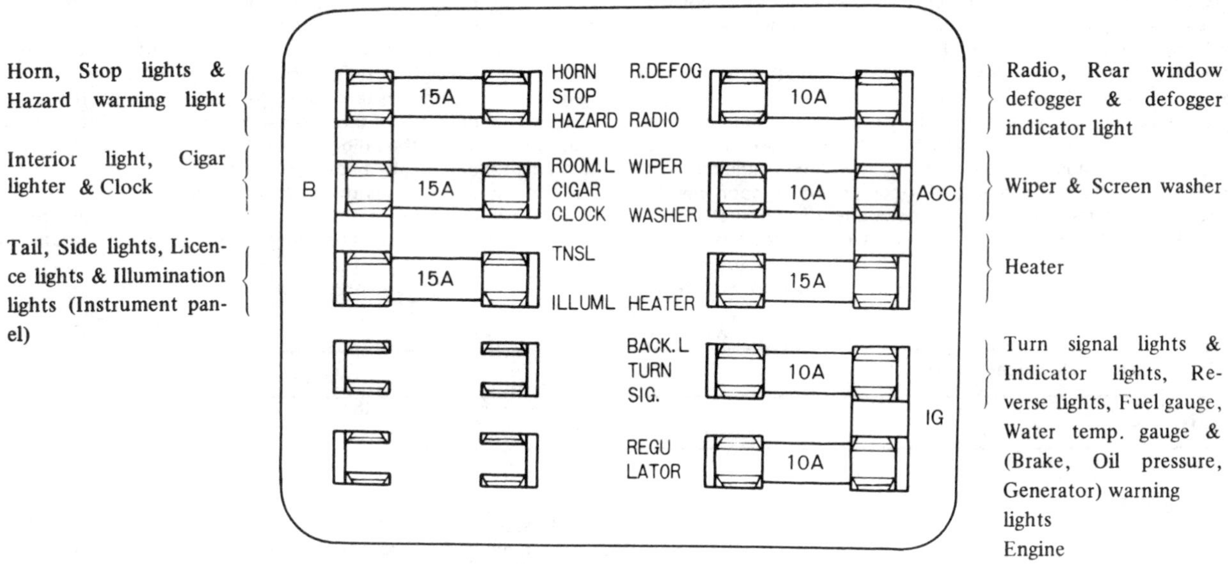

Horn, Stop lights & Hazard warning light

HORN
STOP
HAZARD   15A

R.DEFOG
RADIO   10A

Radio, Rear window defogger & defogger indicator light

Interior light, Cigar lighter & Clock

ROOM.L
CIGAR
CLOCK   15A

WIPER
WASHER   10A

Wiper & Screen washer

Tail, Side lights, Licence lights & Illumination lights (Instrument panel)

TNSL
ILLUML   15A

HEATER   15A

Heater

B

BACK.L
TURN
SIG.   10A

ACC

REGU
LATOR   10A

IG

Turn signal lights & Indicator lights, Reverse lights, Fuel gauge, Water temp. gauge & (Brake, Oil pressure, Generator) warning lights
Engine

Fig. 10.27 The fuse panel (Sec 29)

2 The gauges are connected to senders whose electrical resistance varies according to the circumstances they are recording.
3 The fuel gauge sender is located in the fuel tank, and the water temperature sender is screwed into the thermostat housing.

26 Speedometer drive cable – removal and refitting

1 Disconnect the battery.
2 Remove the instrument panel as described in Section 23.
3 Disconnect the speedometer cable from the speedometer by pressing down the nylon lockplate and pulling the inner drive cable from the speedometer.
4 Working beneath the vehicle, unscrew the cable retaining cup from the gearbox rear extension. Unhook the cable from its mounting clips along the route of the cable.
5 From under the bonnet, prise out the rubber grommet where the drive cable passes through the rear bulkhead.
6 Now tie a good length of string or flexible wire to the speedometer cable inside the vehicle and withdraw the cable into the engine compartment, drawing the string or wire along the exact route of the cable.
7 To refit the speedometer drive cable, slide on the rubber grommet from the speedometer end of the cable, then secure the cable to the previously fitted string or wire.
8 Pull the string inside the vehicle to feed the cable into its correct position. When the cable arrives at the dashboard, take off the string and connect it to the speedometer.
9 The rest of the refitting is the reverse of the removal procedure.

27 Turn indicator flasher and hazard unit – renewal

1 Failure of a bulb will be shown by a change in the speed of flashing, and the note of the audible clicking. Failure of the flasher itself is usually complete, sticking with the lamps unlit or lit.
2 The flasher unit is under the dash, and access for renewal is quite simple.
3 When fitting a replacement unit make sure to earth the flasher unit 'E'-terminal to the bulkhead.

28 Windscreen wiper motor – removal and refitting

1 Lift up the spring-loaded cover to gain access to the wiper arm retaining nut.
2 Undo and remove the retaining nut and pull off the wiper arm (photos).

3 Remove the other wiper arm as previously described.
4 Remove the rubber spacer, cup and flat rubber washer from each wiper spindle.
5 Pry back the rubber strip and remove the four cross-head screws and washers retaining the cover grille (photo). The grille can be removed completely by disconnecting the washer feed pipes (photos).
6 Disconnect the multi-pin plug and remove the rubber protection jacket from the motor.
7 Remove the single bolt and washer securing the wiper motor bracket to the bulkhead adjacent to the wiper spindle. Remove the two bolts and washers adjacent to the bonnet lock. The wiper motor can now be disengaged from the linkage and lifted out.
8 Refitting is a reversal of the removal procedure.

29 Fuses and fusible link

1 A fuse panel containing eight fuses is mounted under the dashboard. These fuses are rated as shown in the appropriate illustration (photo).
2 Fuses normally seldom blow. Sometimes when a lamp bulb fails it may blow the fuse as it goes, which can cause confusion in the fault finding process.
3 If a fuse consistently blows, it is indicating a short in that particular circuit, usually intermittent. Do not be tempted to renew the fuse with one of higher value than standard. The result will be that as the short gets worse, the wiring will overheat, causing widespread damage and creating the risk of burning the whole car.
4 Whilst searching for the cause of a blown fuse it is economical to use household fusewire laid in the clips and held in place by the burnt out fuse cartridge. The cause of the trouble is likely to be a frayed wire, chaffing where it passes through a hole in the sheet metal.
5 In addition to the fuse box for individual circuits there is a fusible link mounted on the front engine compartment next to the radiator (photo). This in effect is a master fuse protecting two basic circuits, namely, all circuits that pass through the ignition switch (with the exception of the starter motor), and the lighting circuits. The fusible link will burn out only in the event of a heavy overload caused probably by a direct short circuit. Always trace and rectify the fault before renewing the fusible link as an assembly.

30 Fault finding principles

1 Tracing an electrical fault follows the usual principle of a methodical check along the system.
2 First check for foolish errors, such as the wrong switch turned on,

10

or for such things as an over-riding control like the accessory position of the ignition/steering lock. Also, if other components have failed simultaneously, then a fuse appears to be the fault.

3 If the faulty component is a light, the next assumption to make is that a bulb has blown, so try changing it.

4 If there is still no cure have a brief glance for obvious faults such as a loose wire. On an old car used much in the wet, particularly with salty roads in winter, it is regretted that a good kick or a judicious blow is needed next! If the earth return of a component is poor this will strike a contact. If it does, then strip the component and derust it.

5 The proper systematic tracing of a more elusive fault requires a voltmeter or a test lamp. The latter is a small 12 volt bulb with two wires fitted to it, either using a bulb holder, or soldered direct. To test a circuit one wire is put to earth, and the other used to test the live side of a component. Work back from the component till the point is found where the bulb lights, indicating that the circuit is live until then To test the earth side, the test bulb is wired to a live source, such as the starter solenoid. Then if the other wire is put to a component properly earthed it will light the bulb.

6 Trace methodically back until a correct result is got. Inaccessible circuits can be bridged with a temporary wire to see if it effects a cure.

7 With the wiring in a loom, defective wiring must be renewed with a separate wire. This must be securely fixed so that it cannot chafe.

31 Fault diagnosis – electrical system

Symptom	Reason(s)
Starter motor fails to turn engine	Battery discharged
	Battery defective internally
	Battery terminal leads loose or earth lead not securely attached to body
	Loose or broken connections in starter motor circuit
	Starter motor solenoid switch faulty
	Starter motor pinion jammed in mesh with flywheel gear ring
	Starter brushes badly worn, sticking or brush wires loose
	Commutator dirty, worn or burnt
	Starter motor armature faulty
	Field coils earthed
Starter motor turns engine very slowly	Battery in discharged condition
	Starter brushes badly worn, sticking or brush wires loose
	Loose wires in starter motor circuit
Starter motor operates without turning engine	Pinion or flywheel gear teeth broken or worn
Starter motor noisy or engagement excessively rough	Pinion or flywheel teeth broken or worn
	Starter motor retaining bolts loose
Starter motor remains in operation after ignition key released	Faulty ignition switch
	Faulty solenoid
Charging system indicator on with ignition switch off	Faulty alternator diode
Charging system indicator light on – engine speed above idling	Loose or broken drivebelt
	Shorted negative diode
	No output from alternator
Charge indicator light not on when ignition switched on but engine not running	Burnt out bulb
	Field circuit open
	Lamp circuit open
Battery will not hold charge for more than a few days	Battery defective internally
	Electrolyte level too weak or too low
	Battery plates heavily sulphated
Horn will not operate or operates intermittently	Loose connections
	Defective switch
	Defective horn
Horns blows continually	Horn button stuck (earthed)
Lights do not come on	If engine not running, battery discharged
	Light bulb filaments burnt out or bulbs broken
	Wire connections loose, disconnected or broken
	Light switch shorting or otherwise faulty
Lights come on but fade out	If engine not running battery discharged
	Light bulb filament burnt out, or bulbs or sealed beam units broken
	Wire connections loose, disconnected or broken
	Light switch shorting or otherwise faulty
Lights give very poor illumination	Lamp glasses dirty
	Lamps badly out of adjustment

Lights work erratically – flashing on and off, especially over bumps	Battery terminals or earth connections loose Lights not earthing properly Contacts in light switch faulty
Wiper motor fails to work	Blown fuse Wire connections loose, disconnected, or broken Brushes badly worn Armature worn or faulty Field coils faulty
Wiper motor works slowly and takes excessive current	Commutator dirty, greasy or burnt Armature bearings dirty or unaligned Armature badly worn or faulty
Wiper motor works slowly and takes little current	Brushes badly worn Commutator dirty, greasy or burnt Armature badly worn or faulty
Wiper motor works but wiper blades remain static	Wiper motor gearbox parts badly worn or teeth stripped

Wiring diagrams on following pages

10

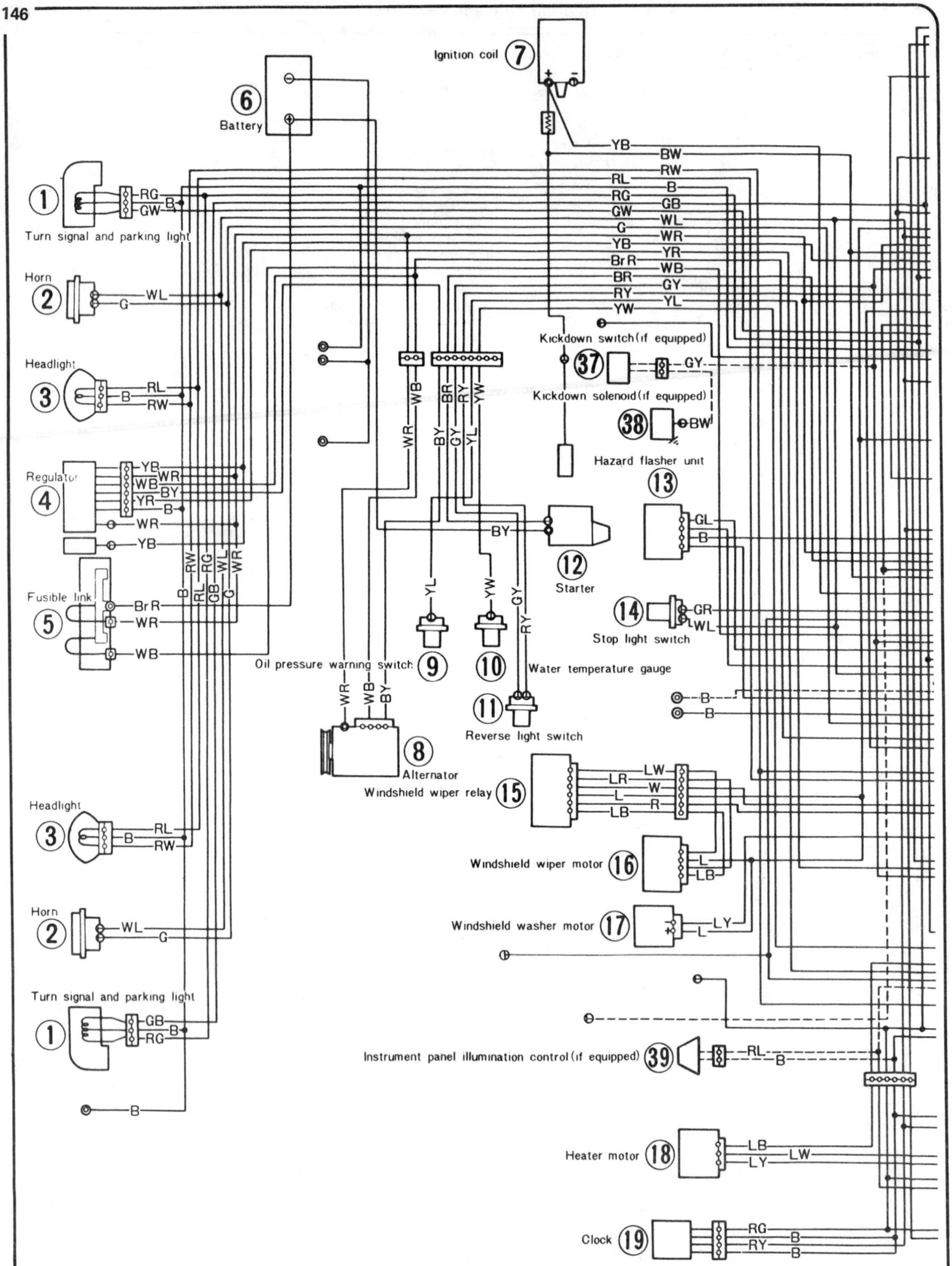

Fig. 10.28A Wiring diagram for all models

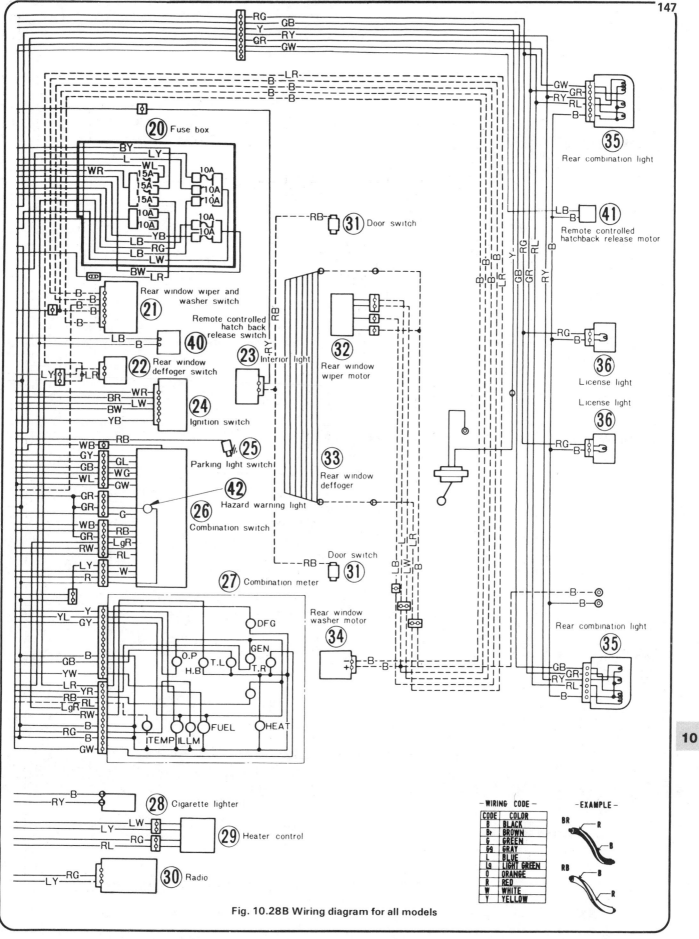

Fig. 10.28B Wiring diagram for all models

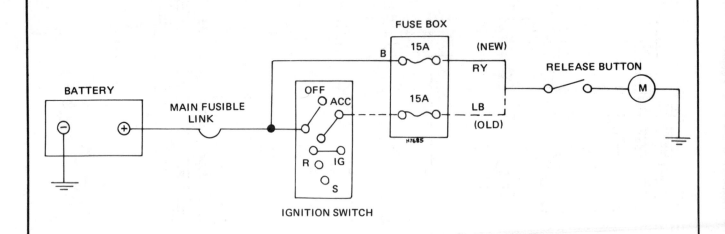

Fig. 10.29 Wiring diagram for rear door remote release button

Note: *At the time of compilation of this manual, no other information was available on the rear door remote release system*

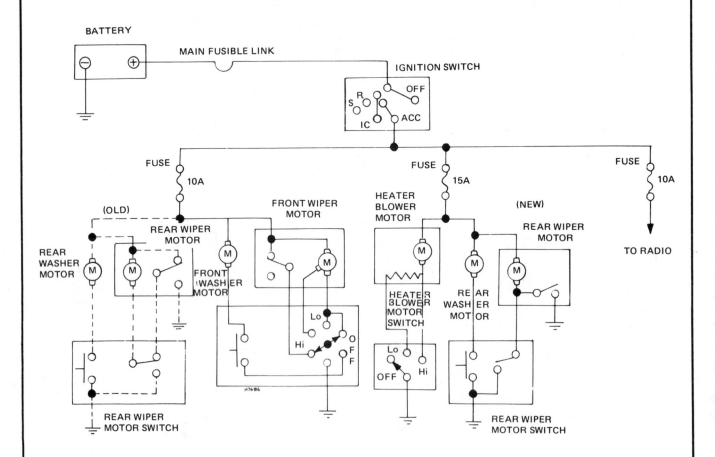

Fig. 10.30 Wiring diagram for rear wash/wipe unit

Note: *At the time of compilation of this manual, no other information was available on the rear wash/wipe system*

Chapter 11 Suspension and steering

See Chapter 13 for specifications and information applicable to 1979 thru 1983 North American models

Contents

Specifications

Front suspension

Type	Coil spring and strut
Spring free length:	
Hatchback/323	14.37 in (365 mm)
GLC	13.56 in (351.5 mm)

Rear suspension

Type	Four link with lateral rod and coil springs
Free length	9.90 in (251.5 mm)

Steering

Type	Recirculating ball and nut (varying ratio gear)
Reduction ratio	17 to 19 : 1
Free play of steering wheel	0.20 to 0.80 in (5 to 20 mm)
Wear limit	1.20 in (30 mm)
Clearance between rack and sector gear	0 to 0.004 in (0 to 0.1 mm)
Worm bearing preload:	
Without sector shaft and column bush	1.7 to 4.3 lbf in (2 to 5 kgf cm)
With sector shaft and column bush	5.2 to 10.4 lbf in (6 to 12 kgf cm)
Clearance between sector shaft and housing:	
New	0 to 0.0024 in (0 to 0.061 mm)
Wear limit	0.008 in (0.20 mm)
End clearance of adjusting screw and sector shaft	0 to 0.004 in (0 to 0.1 mm)
Lubricant type	SAE/90EP
Endplay of ball-studs of centre link and track-rods:	
New	0 to 0.010 in (0 to 0.25 mm)
Wear limit	0.039 in (1.0 mm)
Maximum wheel angle on full lock:	
Wheel on inside of turn	43° ± 2°
Wheel on outside of turn	31° ± 2°
Minimum turning radius	14 ft 6 in (4.4 mm)

Steering geometry

	Mazda Hatchback 323	Mazda GLC
King pin inclination *	8° 30'	8° 50'
Camber *	1° ± 1°	0° to 40' ± 1°
Caster *	1° 45' ± 45'	1° 35' ± 45'
Toe-in	0 to 0.24 in (0 to 6 mm)	0 to 24 in (0 to 6 mm)

*Non-adjustable

Wheels and tyres
Mazda Hatchback 323 (1000 cc)

Wheel size	4½J x 13, or 4J x 12C
Tyre size	6.15 - 13 - 4PR, 155SR13 or 6.00 - 12 - 4PR

11

Tyre pressure:

Front .... 24 lbf in^2 (1.7 kgf cm^2)

Rear .... 28 lbf in^2 (1.9 kgf cm^2), or 30 lbf in^2 (2.1 kgf cm^2) for 6.00 - 12 - 4PR

Mazda Hatchback 323 (1300 cc)

Wheel size .... $4\frac{1}{2}$J x 13 or $4\frac{1}{2}$J x 12

Tyre size .... 6.15 - 13 - 4PR, 155SR13 or 6.00 - 12 - 4PR

Tyre pressure:

Front .... 24 lbf in^2 (1.7 kgf cm^2)

Rear .... 28 lbf in^2 (1.9 kgf cm^2), or 30 lbf in^2 (2.1 kgf cm^2) for 6.00 - 12 - 4PR

Mazda GLC

Wheel size .... $4\frac{1}{2}$J x 13 or 5J x 13

Tyre size .... 155 - 13 - 4PR, 6.15 - 13 - 4PR, 155SR13 or 175/70 13

Tyre pressure:

Front and rear .... 26 lbf in^2 (1.8 kgf cm^2)

Note: For GLC Sport models, consult the driver's handbook or vehicle decal

Torque wrench settings

	lbf ft	kgf m
Front suspension		
Suspension arm-to-crossmember	29 to 40	4.0 to 5.5
Suspension arm- to-stabilizer bar	56 to 71	7.8 to 9.8
Knuckle arm-to-shock absorber	46 to 69	6.4 to 9.5
Suspension arm balljoint-to-knuckle arm	43 to 58	6.0 to 8.0
Front shock absorber:		
Piston rod-to-mounting block	47 to 59	6.5 to 8.2
Seal cap nut	36 to 43	5.0 to 6.0
Piston rod nut	10 to 12	1.35 to 1.65
Base valve nut	1	0.15
Rear suspension		
Lower arm-to-body	47 to 59	6.5 to 8.2
Lower arm-to-axle casing	47 to 59	6.5 to 8.2
Upper arm-to-body	47 to 59	6.5 to 8.2
Upper arm-to-axle casing	47 to 59	6.5 to 8.2
Lateral rod-to-body	47 to 59	6.5 to 8.2
Lateral rod axle casing	47 to 59	6.5 to 8.2
Shock absorber-to-body and axle casing	47 to 59	6.5 to 8.2
Steering		
Steering wheel nut	22 to 29	3.0 to 4.0
Steering gear housing-to-frame	32 to 40	4.4 to 5.5
Pitman arm-to-sector shaft	58 to 87	8.0 to 12.0
Idle arm bracket-to-frame	32 to 40	4.4 to 5.5
Idle arm-to-centre link	18 to 25	2.5 to 3.5
Pitman arm-to-centre link	22 to 33	3.0 to 4.5
Track-rod-to-centre link	22 to 33	3.0 to 4.5
Track-rod to knuckle arm	22 to 33	3.0 to 4.5
Track-rod locknut	51 to 58	7.0 to 8.0
Wheels		
Wheel bolts	65 to 79	9.0 to 11.0

1 General description

The front suspension is of independent MacPherson strut type. At the rear the axle is live, but the suspension is by coil spring and shock absorber with four links taking fore, aft and torque-reaction loads, whilst a lateral rod takes the side loads.

The steering uses a steering box, idler arm, and three piece track-rod.

The front hubs are conventional taper roller bearings. As the rear hubs are of the semi-floating type on the half-shaft these are dealt with in Chapter 8.

2 Suspension - general checks

1 The suspension and steering will normally last a long time and, unless damaged in an accident, wear and its effects in loss of precision or comfort will be so gradual as to be easy to ignore. Every 3000 miles (4800 km) a general suspension check should be carried out.

2 The car should be parked on level ground. Check that it sits level.

Compare it with the photographs at the front of this manual to see whether it has markedly sagged on the suspension. It will normally do so more on the driver's side.

3 Put the car in gear, and take off the handbrake. Grip the wheel at the top with both hands, and rock it to and fro vigorously. Listen for any squeaks, or metallic noises. Feel for any free play. If any is found, get an assistant to do the rocking whilst the source of the trouble is located. Repeat at the other wheels.

4 Check the shock absorbers. If they are damp on the outside from fluid leaks, they will definitely need renewal. Bounce on the car at the back. Work the car up and down in phase with its natural bouncing frequency to get it going to a large amplitude. It should feel stiff, damped by the shock absorbers. As soon as the bouncing is stopped, the car should come to rest, going only once past the static position. before stopping at the static height. Repeat at the front. The shock absorbers can be expected to last quite some considerable time. Do not renew single shock absorbers (unless there is an unusual failure at very low mileage) but do so in pairs, front and rear.

5 Check the appearance of the rubber bushes of the lateral rod and the links. The eye of the lateral rod should be central with its bolt. The rubber squeezed out at the back sides should be firm. though a certain

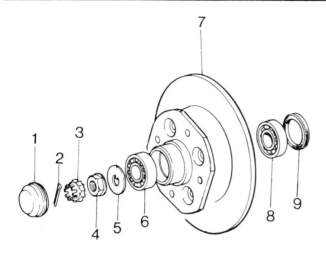

Fig. 11.1 The component parts of the front hub (disc) (Sec 3)

1 Grease cap
2 Split pin
3 Nut lock
4 Nut
5 Washer
6 Outer bearing (inner and

   outer race)
7 Brake disc
8 Inner bearing (inner and
   outer race)
9 Oil seal

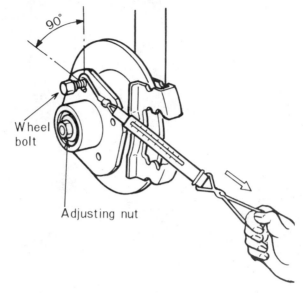

Fig. 11.2 Checking the wheel bearing preload (disc brake) (Sec 4)

amount of perishing must be accepted.

3 Front wheel hub (disc) - removal and inspection

1 Jack up the car and remove the wheels.
2 Take off the brake caliper by pulling out the spring clips and wedges, as described in Chapter 9. Lodge the caliper where the flexible brake hose cannot be strained. Remove the brake pads and shims. Remove the caliper carrier by undoing the two bolts holding it to the backplate.
3 Prise off the grease cap. Wipe off the excess grease from the stub axle.
4 Pull out the split pin from the end of the stub axle.
5 Take off the locking collar from the nut.
6 Undo the nut, and take it off.
7 Pull the hub towards you. It will come suddenly as the inner race of the outer bearing frees from the stub axle. Be careful that the oil seal at the far end of is not damaged. Be ready to catch the outer bearing and washer, which are now free in the hub.
8 Lift the hub off the stub axle, keeping it central so that the oil seal is not dragged along.
9 If further dismantling is necessary, a new oil seal will be needed. If the hubs have merely been dismantled to repack with grease, and the grease is quite clean, then this can be done by wiping away the old, and pushing in new past the oil seal. However, to check further, the inner bearing must come out, so the oil seal must be removed, and it will be damaged in the process.
10 Prise out the oil seal.
11 Take out the inner bearing.
12 For the present leave the bearing races in the hub.
13 Clean out the hub. Clean the stub axle. Wash both bearings very thoroughly in clean petrol. All traces of old grease must be removed, as they would lodge dirt.
14 Check the bearings run freely and smoothly, with an even rushing noise. Check the rollers are bright and unmarked. Check the bearing races are unmarked. Check the running surface on the stub axle for the oil seal is smooth and shiny.
15 If the bearings need renewal drive out of the hub the outer races, using a brass drift in the slots provided for this.

4 Front wheel hub (disc) - reassembly and adjustment

1 Fit the new outer races into the hub. Clean the hub. Put some

grease in the hub; it wants to be smeared liberally, but not full, so there will be still plenty of air space when assembled. Work grease thoroughly into the rollers of both bearings.
2 Put the inner bearing in the hub, then carefully and evenly tap in the new oil seal, lips towards the bearing.
3 Smear a little grease on the stub axle. Put the outer bearing in place in the hub. The lift the hub onto the stub axle, taking care the oil seal does not touch it. Slide the hub along the axle until the threads stick out of the outer bearing.
4 Fit the washer and nut. Tighten the nut, rotating the hub to and fro so that the rollers can climb into their proper position. If new bearings have been fitted this tightening will finally draw the races into their proper seated position. Tighten the nut fairly firmly to achieve this.
5 Now slacken the nut until the bearing goes free and easy to rotate, with endfloat just discernible; some play can be felt if the hub is pushed to and fro along the axis of the stub axle.
6 Retighten the nut to get the correct bearing preload. This will be just beyond the point where endfloat disappears, and some bearing drag can be felt. The correct preload is that the bearing should be 0.99 to 1.32 lbf (0.45 to 0.60 kgf) when pulling on the wheel bolts with a spring balance.
7 Fit the locking collar on the nut in such a position that one of the slots lines up with the split pin hole in the stub axle. Fit a new split pin, and bend the ends over.
8 Smear a little grease in the grease cap, and tap it squarely into place.
9 Reassemble the brakes (refer to Chapter 9 if necessary).

5 Front wheel hub (drum) - removal and inspection

1 Jack up the vehicle and remove the wheel.
2 At the rear of the brake drum, loosen the adjustment anchor pins, to ensure that the brake linings are clear of the brake drum (Refer to Chapter 9, if necessary).
3 Prise off the grease cap. Wipe off excess grease from the stub axle.
4 Pull out the split pin from the end of the stub axle.
5 Take off the locking collar from the nut.
6 Undo the nut, and take it off.
7 Using a suitable puller, draw the brake drum off the stub axle. When the brake drum starts to move, be ready to catch the outer bearing and washer, which are now free in the hub.
8 Lift the brake drum off the stub axle, keeping it central so that the oil seal is not dragged along the stub axle.

11

7.5 Remove the two bolts through the knuckle arm

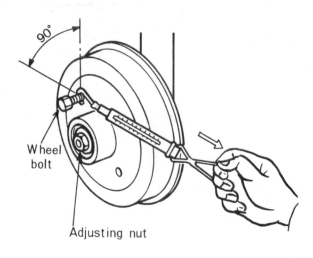

Fig. 11.3 Checking the wheel bearing preload (drum brake) (Sec 6)

90°

Wheel bolt

Adjusting nut

7.6 The strut-to-wheel arch mountings

7.7A Clear the knuckle arm from the bottom of the strut

7.7B Lift out the suspension strut

7.7C A view of the knuckle with the strut removed

9   If further dismantling is necessary, a new oil seal will be needed. If the brake drum hub has merely been removed to repack it with grease, and the existing grease is free from dirt, then this can be done by wiping away the old and pushing in new past the oil seal. However, to check further, the inner bearing must come out, so the oil seal must be removed, and it will be damaged in the process.

10   Prise out the oil seal.

11   Take out the inner bearing.

12   For the present, leave the bearing races in the hub.

13   Clean out the hub. Clean the stub axle. Wash both bearings very thoroughly in clean petrol. All traces of old grease must be removed. as they would lodge dirt.

14   Check that the bearings run freely and smoothly, with an even rushing noise. Check that the rollers are bright and unmarked. Check that the bearing races are unmarked. Check that the running surface on the stub axle for the oil seal is smooth and shiny.

15   If the bearings need renewal, drive their outer races out of the hub, using a brass drift in the slots provided for this.

---

6   Front wheel hub (drum) - reassembly and adjustment

1   Fit the new outer races into the hub. Clean the hub. Put some grease in the hub; it needs to be smeared liberally, but not full, so that there will still be plenty of air space when assembled. Work grease thoroughly into the rollers of both bearings.

2   Put the inner bearing into the hub, then carefully and evenly tap in the new oil seal, lips towards the bearing.

3   Smear a little grease on the stub axle. Put the outer bearing in place in the hub. Then lift the hub onto the stub axle, taking care that the oil seal does not touch it. Slide the hub along the axle until the threads protrude from the inner bearing.

4   Fit the washer and nut. Tighten the nut, rotating the hub to and fro so that the rollers can climb into their proper positions. If new bearings have been fitted this tightening will finally draw the races into their proper seated positions. Tighten the nut fairly firmly to achieve this.

5   Now slacken the nut until the bearing goes free and easy to rotate, with the endfloat just discernible; some play can be felt if the hub is pushed to and fro along the axis of the stub axle.

6   Retighten the nut to get the correct bearing preload. Bearing preload has been fully described in Section 4, paragraph 6.

7   Fit the locking collar on the nut in such a position that one of the slots lines up with the split pin hole in the stub axle. Fit a new split pin, and bend the ends over.

8   Smear a little grease into the grease cap, and tap it squarely into place.

9   Adjust the brakes, as described in Chapter 9. Refit the roadwheel, and lower the vehicle to the ground.

---

7   Front suspension strut - removal and refitting

1   Raise the front of the vehicle, and remove the appropriate roadwheel.

2   Remove the brake and hub as described in Section 3, or Section 5, if applicable.

3   On vehicles fitted with drum brakes, the brake linings will have to be removed. Refer to Chapter 9 for details of this. It is not necessary to disconnect the hydraulic pipes at the rear of the backplate. Remove the bolts securing the backplate to the stub axle. It is advisable to retain the pistons in the wheel cylinders with rubber bands to prevent them accidentally falling out of their cylinder bores and necessitating hydraulic bleeding. Prise off the spring retaining clips securing the flexible brake hose to the suspension strut. Lift the backplate from the stub axle and tie it to a convenient anchorage point, making sure that the brake flexible hose is not strained in any way.

4   On vehicles fitted with disc brakes, remove the bolts securing the backplate to the stub axle. Lift off the backplate.

5   All vehicles: Remove the two bolts in the bottom of the suspension strut through the knuckle arm from below. (photo).

6   In the engine compartment, undo the three nuts holding the top of the strut to the wheel arch. (photo).

7   Push down on the suspension arm and knuckle assembly to clear the knuckle arm outwards away from the bottom of the strut. When it is clear, lower it away from the top mounting. (photos).

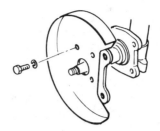

Fig. 11.4 Removing a backplate (disc brake) (Sec 7)

Fig. 11.5 Removing a backplate (drum brake) (Sec 7)

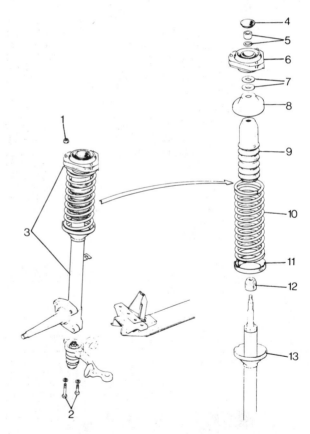

Fig. 11.6 The front suspension (Sec 7)

1 Upper mounting nut	assembly
2 Knuckle arm bolts and	7 Washers
washers	8 Upper spring seat
3 Shock absorber and	9 Rubber boot
spring assembly	10 Spring
4 Cap	11 Spring seat
5 Nut and washer (piston	12 Rubber stopper
rod)	13 Shock absorber and lower
6 Upper mounting block	spring seat

11

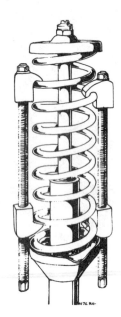

Fig. 11.7 Typical spring compressors in use (Sec 8)

Fig. 11.8 Fitting the piston rod assembly into the reservoir tube (Sec 9)

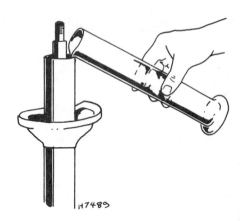

Fig. 11.9 Filling the reservoir tube with shock absorber fluid (Sec 9)

8   Keep the strut upright, so that the air in the shock absorber will stay at the top, clear of the valves.
9   When reassembling, grease the bolts, nuts and studs, and make sure no dirt gets between the mating surfaces at the top and bottom of the strut.
10 Reassemble the backplate to the stub axle. On vehicles fitted with drum brakes, reassemble the brake linings and the drum. Adjust the brake linings as described in Chapter 9.
11 Refer to Section 4, or Chapter 6, and set the wheel bearing preload.

8   Front suspension strut - removal of the spring

1   With the suspension strut off the car, the spring load is held by the piston rod at full extension. In order to remove it, special spring compressors must be purchased or borrowed. In desperation, these could be made up using suitable steel bars and bolts, but due to the danger of serious injury occurring should they fail, this is not recommended. Also the cost of the materials and the problem of bending, drilling and shaping would not be worthwhile. Mazda spring compressors can be purchased under the part numbers 49 0223 640A and 49 0370 641, but universal types are also available as an off-the-shelf item at large motor factors and spring specialists at very moderate prices.
2   Refer to Fig 11.7 which shows typical spring compressors in use. Mount the assembly vertically in a vice then compress the spring to release the load from the upper seat.
3   Hold the upper end of the piston rod with a small spanner. Remove the locknut and washer. Take off the upper mounting block assembly, the washers and the upper seat. Pull off the rubber boot.
Note: *When removing the upper mounting block assembly, place the components in their correct disassembled order in a safe place.*
4   Lift off the spring, still compressed. Do not jar it, or leave it where it could fly apart and hit anybody. Only ease the compressor if it is necessary to check the spring, or fit a new one. This will be necessary if the car has sagged, or there is visible damage to the spring.
5   When reassembling use new rubbers. Lubricate them with a vegetable lubricant, ie rubber grease, or if nothing else, a little brake fluid. Check the condition of the swivel bearing and pack it with grease.

9   Front suspension - overhaul of the shock absorber

1   It is not possible to fit new parts of the front shock absorber; it is only available as a complete assembly, so the information in this Section has been limited to just changing the shock absorber.
2   Having removed the spring, the front strut contains only the shock absorber. Compress the shock absorber. Remove the cap nut at the top of the tube. A special spanner will probably need to be made from some steel water pipe. Remove the O-ring fitted on the piston guide. Pull the piston rod assembly from the reservoir tube.
3   To reassemble, insert the new piston rod assembly into the reservoir tank.
4   Fill the reservoir tube with 270 cc (16.5 cu in) of shock absorber fluid. Too much will burst the shock absorber. Too little will give aeration.
5   Apply grease to the lip of the oil seal. Fit the O-ring and push the cap nut slowly down the piston rod. Tighten the nut temporarily with the piston rod extended. Get it as tight as possible, to make an air seal at the top. Then compress the shock absorber, so that the proper tightening can be done with the rod out of the way. Tighten the nut finally to the specified torque.
6   From now on, keep the shock absorber upright, to keep the air at the top.

10 Suspension arm and balljoint assembly – removal, inspection and refitting

1   It is not possible to lubricate the balljoint, as it is filled with grease and sealed during production.
2   To check the joint, jack the car up and make it secure with auxiliary supports. Then push and pull the bottom of the suspension strut to see if there is any free movement. Inspect the condition of the rubber boot,

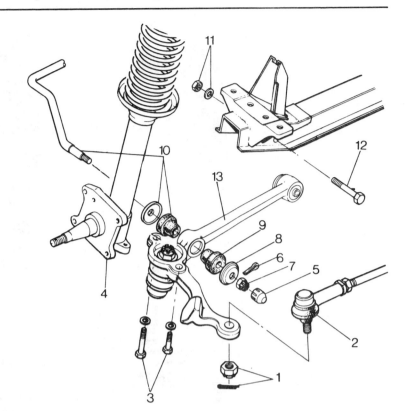

1 Nut and split pin (track-rod)
2 Track-rod end balljoint
3 Knuckle arm to lower strut securing bolts and washers
4 Suspension strut
5 Rubber plug
6 Split pin
7 Nut (stabilizer bar)
8 Washer
9 Rubber bush
10 Stabilizer bar, washer and rubber bush
11 Nut and washer (crossmember)
12 Bolt (crossmember)
13 Suspension arm and knuckle arm assembly

Fig. 11.10 The suspension arm assembly and its associated components (Sec 10)

which must not be perished or damaged in any way. Check the condition of the rubber mounting bushes at the inner and outer end of the suspension arm. Failure of any of these parts can create excessive movement at the balljoint. If the free movement in the balljoint is nearing 0.040 in (1.0 mm) the suspension arm and balljoint will have to be renewed as an assembly.

3   Note that, depending on market destination, the suspension arm may either be manufactured of pressed steel, or be of the cast type. In either case the procedures for removal and refitting are the same.

4   To remove the suspension arm and ball-joint assembly, remove the split pin from the outer track-rod end balljoint and remove the nut.

5   Disconnect the outer track-rod end balljoint from the knuckle arm. This can be done by using two hammers of equal weight to strike the outer diameter of the knuckle arm. Aim the hammer blows to arrive at the knuckle arm together; this will shock the balljoint pin from its taper. Alternatively, purchase or borrow a balljoint separator.

6   Remove the two bolts and washers securing the knuckle arm to the lower suspension strut.

7   Now pull off the rubber plug and remove the split pin from the end of the stabilizer bar.

8   Undo and remove the nut and washer securing the stabilizer bar to the suspension arm.

9   Moving to the inner end of the suspension arm, undo and remove the nut and washer that secures the suspension arm to the crossmember.

10 Push out the bolt from the crossmember, releasing the suspension arm.

11 Now push down on the outer end of the stabilizer bar and, at the same time, pull the suspension arm from the stabilizer bar. Retrieve the rubber bush that will have come away from the stabilizer bar whilst pulling off the suspension arm. It is advisable to remove the inner rubber bush and washer from the stabilizer bar for safe keeping.

12 Now that the suspension arm and knuckle have been removed from the vehicle, it is time to separate them.

13 Remove the split pin from the end of the balljoint pin. Undo and remove the nut. With the assembly now free from the vehicle, the ball-

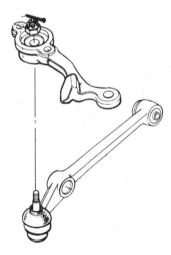

Fig. 11.11 The suspension arm and knuckle arm separated (Sec 10)

joint can be separated from the knuckle by resting the knuckle across the jaws of a vice and, using a soft faced hammer, striking the end of the balljoint pin.

14 Commence reassembly by fitting the knuckle arm to the balljoint. Screw on the nut and tighten to the specified torque. Fit a new split pin and bend the ends over.

15 Assemble the washer and the rubber bush to the inner end of the stabilizer bar. Smear a little grease to the inside and outside diameters of the rubber bush to aid the assembly.

16 Now locate the suspension arm onto the stabilizer bar. Push the stabilizer bar down and, at the same time, push the suspension arm onto the stabilizer bar and under the lower strut.

17 Now locate the inner end of the suspension arm into the

11

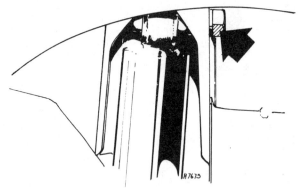

Fig. 11.12 The upper mounting point of the rear shock absorber (Sec 11)

Fig. 11.13 The lower mounting point of the rear shock absorber (Sec 11)

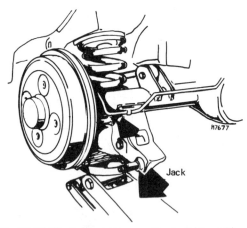

Fig. 11.14 The lower arm mounting bolt (Sec 12)

Fig. 11.15 Removing the rear spring (Sec 12)

11.4 A rear shock absorber

crossmember and, aligning the through-holes, fit the bolt. Assemble the washer and nut, tightening the nut to the specified torque.

18 Grease the other rubber bush and assemble it to the stabilizer bar, followed by the washer. Screw on the nut and tighten it to the torque specified. Fit a new split pin and bend over the ends. Fit the rubber plug.

19 Fit the two bolts and washers that secure the knuckle arm to the lower strut. Tighten these bolts to the specified torque.

20 Assemble the outer track-rod end balljoint to the knuckle arm. Screw on the nut, and tighten to the specified torque. Fit a new split pin, and bend over its ends.

21 Provided that the track rod end has not been revolved on its track-rod, the vehicle's tracking should not need checking.

22 Refit the roadwheel and lower the vehicle to the ground.

11 Rear shock absorber - removal, inspection and refitting

1 Raise the rear end of the vehicle and support it under the bodyframe side rails with suitable stands.

2 Remove the appropriate roadwheel.

3 Undo and remove the bolt that secures the shock absorber to its upper mounting point.

4 Remove the nut and bolt that secures the lower end of the shock absorber to the lower arm. The shock absorber can then be removed from the vehicle (photo).

5 To test the shock absorber, hold it in an upright position, and work the piston rod up and down its full length of travel four or five times. If a strong resistance is felt due to hydraulic pressure, the shock absorber is functioning properly. If no resistance is felt, or there is a sudden free movement in travel, the shock absorber is obviously suspect, and should be renewed as a complete assembly.

6 Fitting the rear shock absorber is a reversal of the removal procedure. Tighten the securing bolts to the specified torque.

7 Refit the roadwheel and lower the vehicle to the ground.

8 It is unwise to renew a single shock absorber. Unless failure occurs at a very low mileage due to an unusual fault, always renew them in pairs.

12 Rear spring - removal and refitting

1 Raise the rear end of the vehicle and support it under the bodyframe side rails with suitable stands.

2 Remove the appropriate roadwheel.

3 Refer to the previous Section and remove the shock absorber.

4 Place a jack under the lower arm. The best jack for this task is a trolley type jack; but a bottle jack will do the job, provided that a piece of heavy duty wood is placed on the jack head. This will, in effect, spread the load of the spring tension on the lower arm and prevent the

jack from kicking away.

5   Remove the nut and bolt that secure the lower arm to the axle housing.

6   Carefully and slowly lower the jack until the spring pressure on the lower arm is relieved. The spring can now be removed.

7   Refitting is a reversal of the removal procedure, but before tightening the nut and bolt to the specified torque the vehicle should be lowered to the ground.

13 Rear suspension upper arm - removal and refitting

1   Raise the rear end of the vehicle and support it under the bodyframe side rails with suitable stands.

2   Remove the appropriate roadwheel.

3   Remove the nut and bolt securing the upper arm to the axle housing.

4   At the other end of the upper arm, remove the nut and bolt that secure the arm to the bodyframe. The upper arm can be removed from the vehicle.

5   Reassembly is the reverse of the removal sequence, but before finally tightening the nuts and bolts to the specified torque, lower the vehicle to the ground. This will ensure that the arm is in its normal working position.

14 Rear suspension lower arm - removal and refitting

1   Raise the rear end of the vehicle and support it under the bodyframe side rails with suitable stands.

2   Remove the appropriate roadwheel.

3   Refer to Section 11, and remove the rear shock absorber.

4   Refer to Section 12, and remove the rear spring.

5   Undo and remove the nut and bolt securing the lower arm to the vehicle's bodyframe. Lift away the lower arm.

6   To refit the lower arm, reverse the removal procedure, but before finally tightening the nuts and bolts to the specified torque, the vehicle should be lowered to the ground.

15 Lateral rod - removal and refitting

1   Raise the rear end of the vehicle and support it under the bodyframe side rails with suitable stands.

2   Remove both rear roadwheels.

3   Remove the nut and bolt at either end of the lateral rod. One nut and bolt secures the lateral rod to the axle housing, and the other is attached to vehicle's bodyframe. The lateral rod can now be removed from the vehicle.

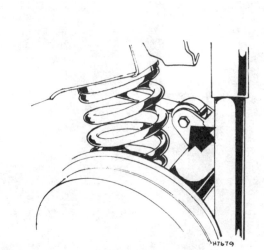

Fig. 11.16 The axle housing mounting point for the upper suspension arm (Sec 13)

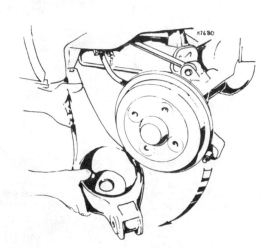

Fig. 11.17 Removing the lower suspension arm (Sec 14)

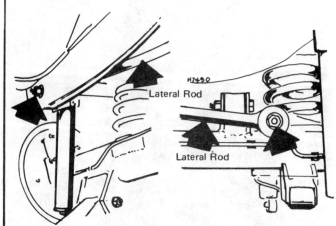

Fig. 11.18 The lateral rod mounting points (Sec 15)

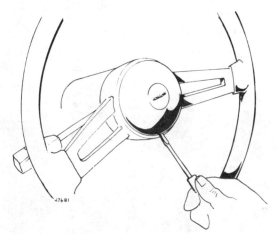

Fig. 11.19 Removing the horn cap (Sec 16)

11

16.3A Prise out the ignition switch surround

4   Refitting of the lateral rod is a straightforward reversal of the removal procedure. Tighten the retaining nuts and bolts to the specified torque.
5   Refit the roadwheels and lower the vehicle to the ground.

16 Steering gear - removal and refitting

1   Disconnect the battery leads then prise off the horn cap.
2   Mark the relative positions of the steering wheel and column shaft then remove the wheel attaching nut and washer. Pull off the steering wheel.
3   Remove the screws that secure the two halves of the steering column shroud together. Prise out the ring that surrounds the ignition switch. Remove the single nut at the lower end of the shroud assembly. Pull the shroud assembly from the lower mounting stud, then carefully remove the two halves of the shroud (photos).
4   Identify the wire couplers and disconnect them. From the end of the steering column shaft, remove the stop ring. Remove the screw(s) attaching the combination switch to the column shaft. Lift the combination switch away from the column shaft. Lift out the bush that will be revealed after the combination switch has been removed (photo).
5   Remove the column jacket support bracket bolts and remove the bracket from the vehicle.

16.3B Remove the two halves of the shroud

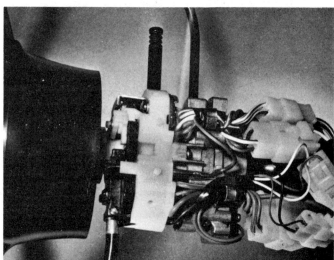

16.4 The wire couplers inside the shroud assembly

16.8 The split pin and nut that secure the centre link to the pitman arm

16.10 The steering box securing bolts

6   Turn the ignition switch to the 'on' position, then pull the column jacket and steering lock assembly from the steering shaft.
7   Raise the front end of the vehicle and support it with stands. The vehicle needs to be high enough to allow clearance for the steering gear and shaft, which will be removed from under the vehicle.
8   Remove the split pin and nut securing the centre link to the Pitman arm (photo). In view of the limited accessibility, it is advisable to obtain a balljoint separator for this task. Proceed to remove the Pitman arm from the centre link.
9   Now remove the three nuts and bolts that secure the steering box to the vehicle's chassis (photo). When the steering box bolts are removed, recover any shims that may be between the box and the chassis. Where applicable, unclip the speedometer cable from the steering box.

10   With the steering gear assembly now loose, revolve and incline the steering box as necessary, and feed it over the crossmember and under the radiator until it is clear of the vehicle (photo).
11   Refitting is essentially the reverse of the removal procedure, but remember to refit any shims that were between the steering box and the chassis.

17 Steering gear - dismantling and reassembly

1   Having removed the steering gear, as described in the previous Section, remove the filler plug and drain out the lubricant.
2   Support the gear in a vice. Remove the nut and washer and draw off the Pitman arm using a suitable extractor.

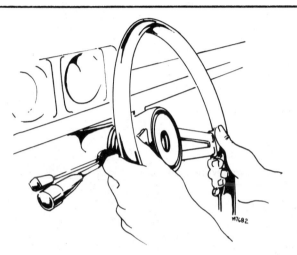

Fig. 11.20 Removing the steering wheel (Sec 16)

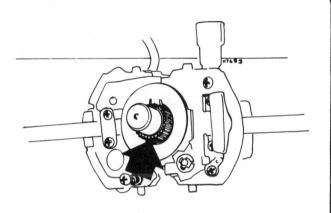

Fig. 11.21 The stop ring (Sec 16)

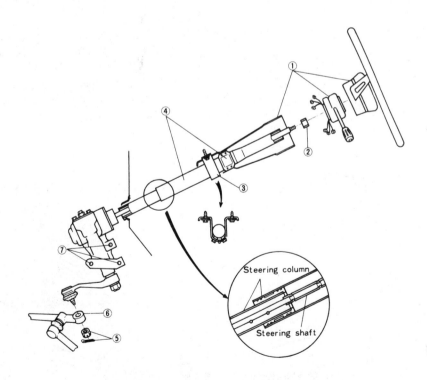

Fig. 11.22 A sectional view of the steering gear (Sec 16)

1 Steering wheel, column shrouds and combination switch
2 Steering shaft bush
3 Support bracket
4 Column jacket and steering lock assembly
5 Split pin and nut (Pitman arm)
6 Centre link
7 Steering box mounting holes

11

160

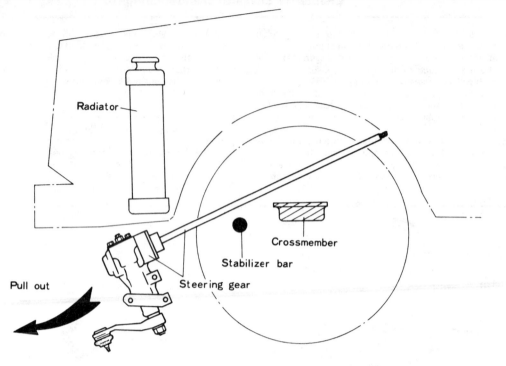

Fig. 11.23 Removing the steering gear assembly (Sec 16)

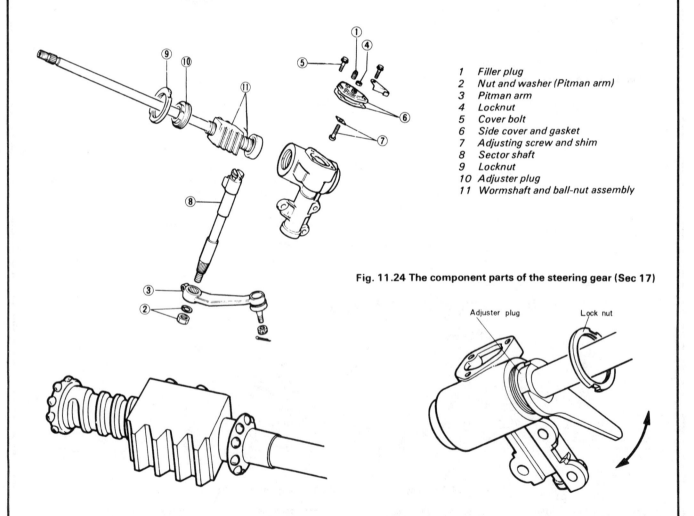

1   Filler plug
2   Nut and washer (Pitman arm)
3   Pitman arm
4   Locknut
5   Cover bolt
6   Side cover and gasket
7   Adjusting screw and shim
8   Sector shaft
9   Locknut
10 Adjuster plug
11 Wormshaft and ball-nut assembly

Fig. 11.24 The component parts of the steering gear (Sec 17)

Fig. 11.25 The wormshaft and ball-nut assembly (Sec 17)

Fig. 11.26 Adjusting the worm bearing preload (Sec 17)

3 Remove the locknut from the sector shaft adjusting screw.

4 Remove the bolts from the side cover, then turn the adjusting screw clockwise through the cover so that the cover and gasket can be removed.

5 Remove the adjusting screw and shim from the end of the sector shaft.

6 Remove the sector shaft from the gear housing, taking care not to damage the bushings and oil seal.

7 Using a suitable spanner, remove the locknut from the adjuster plug. Lift the locknut from the steering shaft. Then undo and remove the adjuster plug from the gear housing. Remove the wormshaft and ball-nut assembly from the gear housing.

8 Clean all the parts carefully using petrol and inspect for wear and damage. Check the operation of the ball-nut assembly on the wormshaft; if it does not travel freely and smoothly, the ball-nut and wormshaft must be renewed as an assembly. Check for wear and damage in the bushings and bearings. Renew parts as necessary.

9 When reassembling, first fit the housing oil seal if it was removed.

10 Insert the wormshaft and ball-nut into the housing, and adjust the worm bearing preload as follows: Fit the adjuster plug and tighten it until it takes 1.7 to 4.3 lbf in (2 to 5 kgf cm) to turn the steering shaft or, if the sector shaft is fitted, 5.2 to 10.4 lbf in (6 to 12 kgf cm). This is best measured by clamping a bar to the end of the steering shaft and at 90° to it. With 4 in (10 cm) between the centre of the steering shaft and the end of the bar, a spring balance attached to the end of the bar should, when pulled, turn the shaft within the figures previously given.

11 When a satisfactory worm bearing preload has been achieved, assemble the locknut and firmly tighten it. Before moving on to the next stage of assembly, it is advisable to check the worm bearing preload with the locknut firmly tightened.

12 Fit the adjusting screw into the slot at the end of the sector shaft and check the end clearance with a feeler gauge. By the use of shims, adjust this clearance to 0 to 0.004 in (0 to 0.1 mm), replacement shims are available in the following sizes: 0.077 in (1.95 mm), 0.079 in (2.00 mm), 0.081 in (2.05 mm) and 0.083 in (2.10 mm).

13 Turn the wormshaft to place the rack in the centre of the worm thread.

14 Insert the sector shaft and adjusting screw into the housing, taking care not to damage the oil seal. Ensure that the centre of the sector gear aligns with the centre of the rack.

15 Place the side cover and gasket on the adjusting screw, and turn the screw anti-clockwise until the side cover seats. Tighten the cover bolts.

16 Now it is time to adjust the clearance between the sector gear and the rack. To do this proceed as follows: with the gear housing mounted in a vice, turn the shaft through its full movement a couple of times; finally, stop it in the half-way position. Assemble the Pitman arm temporarily and use a dial test indicator to check the movement of the Pitman arm before the wormshaft begins to move. Screw the adjusting screw clockwise until clearance of 0 to 0.004 in (0 to 0.1 mm) is obtained; then tighten the locknut. Finally, check that the sector shaft turns smoothly to the right and left without any tendency to bind.

17 Remove the Pitman arm (that was temporarily fitted to check the clearance) and align the identification marks on the sector shaft and the Pitman arm. Fit the Pitman arm, assemble the washer and tighten the nut to the specified torque.

18 Fill the gear housing with the correct grade of oil to just below the filler neck. Fit and tighten the filler plug. The unit is now ready for refitting.

18 Steering linkage

1 If excessive play is suspected in the steering linkages, set the front roadwheels in the 'straight ahead' position and check that there is no play in the balljoints, idler arm bushes and wheel bearings. This is best done by grasping the centre link, track rods, idler arm and Pitman arm, and feeling for any movement. A large tyre lever may help in detecting free play, but care should be taken not to lever against parts which might suffer damage. Any excessive movement in the part(s) previously mentioned will necessitate renewal of the offending part(s).

2 Due to the varied accessibility of the steering joints, it is advisable to obtain a balljoint separator. However, if one is not available, suitable split wedges or even a cold chisel will suffice. If these are not available, it should be possible to free the joint by striking the centre link eye

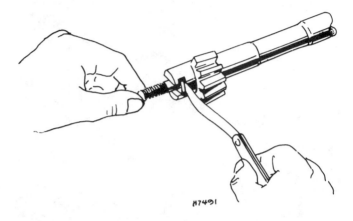

Fig. 11.27 Measuring the clearance between the adjuster screw and the sector shaft (Sec 17)

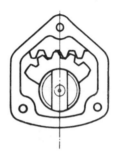

Fig. 11.28 Correct alignment of the rack and the sector shaft gear (Sec 17)

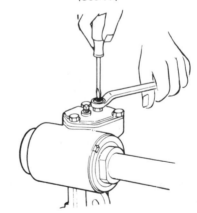

Fig. 11.29 Adjusting the clearance (Sec 17)

11

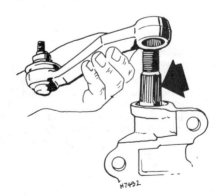

Fig. 11.30 The Pitman arm alignment mark (Sec 17)

18.3 The idler arm viewed from beneath the vehicle

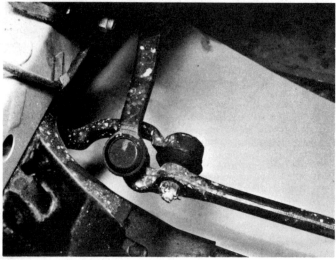

18.15 An inner track-rod end

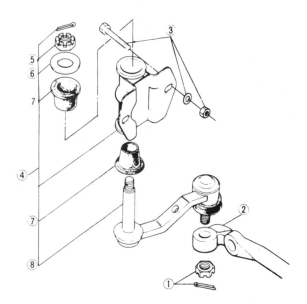

Fig. 11.31 The component parts of the idler arm (Sec 18)

1 Nut and split pin	5 Nut and split pin
2 The centre link	6 Washer
3 Nut, washer and bolt	7 Rubber bushes
4 Idler arm assembly	8 Idler arm shaft

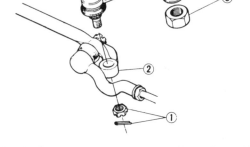

Fig. 11.32 The Pitman arm location points (Sec 18)

1 Split pin and nut
2 The centre link
3 Nut and washer
4 Pitman arm

sharply whilst it is being supported by a heavy weight.

Renewing the idler arm

3   Pull out the split pin and remove the nut. Separate the idler arm from the centre link using one of the methods suggested in paragraph 2. Remove the idler arm by unscrewing the attaching nuts and bolts, and lifting it from its mounting position on the side of the engine compartment (photo). Hold the assembly in a vice and pull out the split pin. Undo and remove the nut, then lift off the washer. The idler arm can then be removed from its mounting bracket. Pull out the rubber bushes and inspect them for perishing or undue wear.

4   When reassembling, liberally grease the internal bore of each rubber bush and fit them to their mounting bracket.

5   Push the idler arm shaft through the rubber bushes until its thread protrudes through the top rubber bush.

6   Assemble the washer and nut. Firmly tighten the nut, and fit a new split pin.

7   Bolt the assembly to its mounting position in the engine compartment. Tighten the bolts to the specified torque.

8   Locate the idler arm balljoint through the eye in the centre link. Screw on the retaining nut and torque tighten the nut to the specified torque. Fit a new split pin and bend over its ends.

Renewing the Pitman arm

9   Pull out the split pin and remove the nut retaining the Pitman arm balljoint in the centre link.

10 Separate the joint, using one of the suggested methods in paragraph 2.

11 Undo and remove the nut, lift off the washer, and using a suitable extractor, draw the Pitman arm from the sector arm.

12 Refitting is essentially the reverse of the removal procedure, but

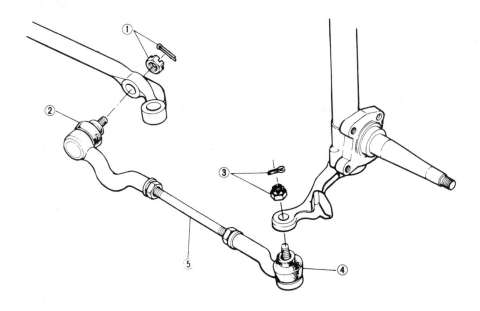

Fig. 11.33 The track-rod ends (Sec 18)

1 Split pin and nut	4 Track-rod end
2 Track-rod end	5 Track-rod
3 Split pin and nut	

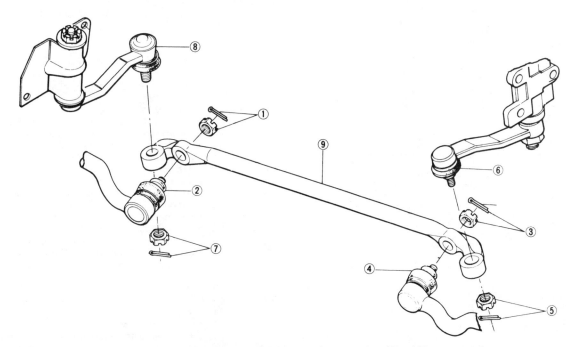

Fig. 11.34 The centre link attachment points (Sec 18)

1 Nut and split pin (track-rod end)	3 Nut and split pin (track-rod end)	5 Nut and split pin (Pitman arm)	7 Nut and split pin (idler arm)
2 Track-rod end	4 Track-rod end	6 Pitman arm	8 Idler arm
			9 Centre link

11

ensure that the master splines align when fitting the Pitman arm to the sector shaft. The Pitman arm retaining nut and the balljoint retaining nut should be tightened to the torques specified.

Renewing a track-rod and track-rod ends

13 Before removing the track-rod, carefully measure the distance between the inner and outer track-rod end locknuts; the track-rod can then be refitted without fear of the toe-in settings being upset.

14 Pull out the split pin and remove the nut retaining the outer track-rod and balljoint to the knuckle arm. Separate this joint using one of the methods suggested in paragraph 2.

15 If it is only required to renew the outer track-rod end, this can now be removed by loosening the locknut and unscrewing the track-rod end from the track-rod. **Note**: *The outer track-rod end has a left-hand thread.*

16 Alternatively, if the outer track-rod end will not unscrew due to

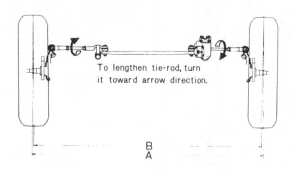

To lengthen tie-rod, turn it toward arrow direction.

B
A

Fig. 11.35 Checking the front wheel alignment (toe-in). Dimension B should be less than A by the specified toe-in (Sec 19)

Fig. 11.36 Adjusting a track-rod (Sec 19)

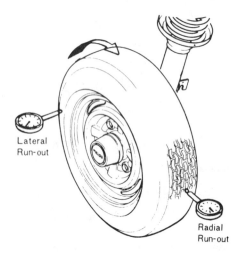

Lateral Run-out

Radial Run-out

Fig. 11.37 Checking a tyre for run-out (Sec 20)

rusting, loosen the locknut at the inner track-rod end and remove the outer track-rod end, together with the track-rod. The track-rod can then be held in a vice and, with the aid of some penetrating oil, easily be separated. (photo).

17 In the unlikely event of both track-rods ends seizing, the only alternative is to break the joint with the centre link at the inner end of the track-rod, and remove the entire track-rod assembly.

18 When reassembling, ensure that the measurement between the inner and outer track-rod end locknuts is obtained, before connecting the last track-rod end to the centre link or the knuckle arm. Tighten the nuts retaining the balljoints to the specified torque.

Renewing the centre link

19 The centre link is held by four balljoints, namely the idler arm, the Pitman arm, and the inner track-rod ends, at each side of the vehicle. After disconnecting the two inner track-rod ends from the centre link, described in paragraph 16, disconnect the centre link from the idler arm and then the Pitman arm. The centre link can then be removed from the vehicle. To refit, simply reverse the removal procedure.

19 Steering geometry checks

1   The camber, caster, and king pin inclination are not adjustable. These are set during the vehicle's production, and will not be altered during normal driving, unless the vehicle is involved in a serious crash. Whenever camber, caster or king pin inclination is suspect, check all parts of the front suspension and body alignment. This will obviously involve the use of specialised equipment, and can only be entrusted to your local Mazda dealer.

2   In order to obtain satisfactory tyre wear and good steering stability, the vehicle's tracking should be checked from time-to-time. This is not really a do-it-yourself job and it is always preferable to have the job done by your vehicle main dealer, who will have the correct alignment gauges. However, if you feel competent to do the job yourself, the basic procedure is given below.

3   Before commencing the check, the vehicle must be standing on a level floor, the tyres must be inflated to the correct pressure (cold), excessive mud must be removed from the vehicle underframe, and the fuel tank level, engine oil and coolant level must be correct.

4   Raise the front end of the vehicle so that the wheels are clear of the ground.

5   Turn each wheel by hand and scribe a chalk line in the centre of the tyre tread around its circumference. Lower the car to the ground and set the front wheels in the 'straight ahead' position.

6   Measure the distance between the chalk marks at the front and rear of the tyre, at equal heights from the ground. If the toe-in is correct, the distance between the chalk marks at the rear should be greater than the distance between them at the front, by the amount given in the Specifications.

7   If adjustment is required, loosen the track-rod locknuts and rotate the track-rod as necessary. Each track-rod is threaded, right-hand at one end and left-hand at the other end, in order to retain the correct balance of the linkage; the two track-rods must be the same length after any adjustment. On completion, ensure that the locknuts are tightened.

20 Wheels and tyres

1   A wheel and/or a tyre can get damaged if a kerb is struck. Therefore during maintenance tasks, spin them round and check that they are within 0.080 in (2.0 mm) on radial run-out, and within 0.10 in (2.5 mm) on lateral run-out

2   Include in your car cleaning programme washing the backs of the wheels , as these get caked with mud. Always do this before having them balanced.

3   The wheels require repainting occasionally, either after fitting new tyres, or after several years use. Rusting is liable to start where the wheel rim is joined to the centre. If this becomes severe the wheel may fail.

4   The tyres are very important for safety. The tread should not be allowed to wear near the legal limit. Not only is there fear of police action, but it is dangerous, particularly in wet weather, or at speed. The recommended pressures must be used to ensure that the tyres last, and to preserve the handling of the car.

5 Radial ply tyres should be used on all four wheels.
6 Fitting tubeless tyres is best left to a tyre factor, because they have the equipment to force the tyre into position on the rim easily and quickly.
7 Changing tyres round to even out wear is not recommended. Apart from the expenditure when all five tyres have to be renewed in one batch, it masks any aberrations of wear which might be happening on one wheel. This can give advanced warning of suspension defects.
8 Excessive wear of the front tyres indicates incorrect toe-in alignment.

9 If the thread wear in the centre is worse than that at the edges it indicates over-inflation, vice versa indicating under-inflation.
10 It will pay to have the wheels balanced. Wheel out-of-balance gives shaking of the steering, which is unpleasant, and causes much wear to the suspension and tyres. In bad cases it will have adverse effects on the roadholding.
11 Examine the tyres for cracks. Any cracks or bulges should be dealt with by a tyre repairer; though seldom can they be cured, as the carcase will have been damaged. Such tyres are dangerous and illegal, and must be renewed.

21 Fault diagnosis – suspension and steering

Symptom	Reason(s)	Remedy
Steering feels vague, car wanders, and floats at speed		
General wear or damage	Tyre pressures uneven	Check pressure and adjust as necessary
	Shock absorbers worn	Test, and renew if worn
	Steering gear balljoint badly worn	Fit new balljoints
	Suspension geometry incorrect	Check and rectify
	Steering mechanism free play excessive	Adjust or overhaul steering mechanism
	Front suspension and rear suspension pickup points out of alignment	Normally caused by poor repair work after a serious accident. Extensive rebuilding necessary
	Mixed radial and crossply tyres	Fit all radial
Stiff and heavy steering		
Lack of maintenance or accident damage	Tyre pressure too low	Check pressures and inflate tyres
	No oil in steering gear	Top up steering gear
	Front wheel toe-in incorrect	Check and reset toe-in
	Suspension geometry incorrect after accident	Realign body
	Steering gear incorrectly adjusted: too tight	Check and readjust steering gear
	Steering column badly misaligned	Determine cause and rectify (usually due to bad repair after severe accident damage and difficult to correct)
Wheel wobble and vibration		
General wear or damage	Wheel nuts loose	Check and tighten as necessary
	Wheels and tyres out of balance	Balance wheels and tyres and add weights as necessary
	Steering balljoints badly worn	Renew steering balljoints
	Hub bearings badly worn	Remove and fit new hub bearings
	Steering gear free play excessive	Adjust and overhaul steering gear
Rattles		
General wear	Suspension rubber bushes worn	Renew
	Steering balljoints worn	Renew
	Excessive play in steering box	Adjust the sector shaft endfloat

11

Chapter 12 Bodywork and fittings

See Chapter 13 for specifications and information applicable to 1979 thru 1983 North American models

Contents

1 General description

1   The body, which is of the welded pressed steel type, has been extensively rust-protected during manufacture. There are no metal to metal contacts between the bodyshell and the various body panels. All screws and clips to secure trim and other equipment to the body are in plastic sleeves. The engine compartment bonnet, the doors and the rear tailgate are fitted with adjustment points to ensure correct fitting and to eliminate rattles. The engine compartment bonnet release catch is operated from inside by a release lever beneath the dash panel. Both front doors can be locked from the outside by a key, or from the inside by pressing the door lock knob down. Interior trim for the deluxe versions includes full floor carpeting and cloth covered seats. The 1000 cc basic version has a vinyl floor covering and black mesh vinyl seats. All models are equipped with laminated windscreens. In the event of a breakdown the front and rear of all models are equipped with towing hooks.

2 Maintenance – bodywork and underframe

1   The general condition of a vehicle's bodywork is the one thing that significantly affects its value. Maintenance is easy but needs to be regular and particular. Neglect, particularly after minor damage, can lead quickly to further deterioration and costly repair bills. It is important also to keep watch on those parts of the vehicle not immediately visible, for instance the underside, inside all the wheel arches and the lower part of the engine compartment.
2   The basic maintenance routine for the bodywork is washing - preferably with a lot of water from a hose. This will remove all the loose solids which may have stuck to the vehicle. It is important to flush these off in such a way as to prevent grit from scratching the finish.
3   The wheel arches and underbody need washing in the same way to remove any accumulated mud which will retain moisture and tend to encourage rust. Paradoxically enough, the best time to clean the underbody and wheel arches is in wet weather when the mud is thoroughly wet and soft. In very wet weather the underbody is usually cleaned of large accumulations automatically and this is a good time for inspection.
4   Periodically it is a good idea to have the whole of the underside of the vehicle steam cleaned, engine compartment included, so that a thorough inspection can be carried out to see what minor repairs and renovations are necessary. Steam cleaning is available at many garages and is necessary for removal of accumulation of oily grime which sometimes is allowed to cake thick in certain areas near the engine, gearbox and back axle. If steam facilities are not available, there are one or two excellent grease solvents available which can be brush applied. The dirt can then be simply hosed off.

5   After washing paintwork, wipe off with a chamois leather to give an unspotted clear finish. A coat of clear protective wax polish will give added protection against chemical pollutants in the air. If the paintwork sheen has dulled or oxidised, use a cleaner/polisher combination to restore the brilliance of the shine. This requires a little effort, but is usually caused because regular washing has been neglected. Always check that the door and ventilator opening drain holes and pipes are completely clear so that the water can drain out. Bright work should be treated the same way as paintwork. Windscreen and windows can be kept clear of smeary film which often appears, if a little ammonia is added to the water. If they are scratched, a good rub with a proprietary metal polish will often clear them, Never use any form of wax or other body or chromium polish on glass.

3 Maintenance – upholstery and carpets

1   Mats and carpets should be brushed or vacuum cleaned regularly to keep them free of grit. If they are badly stained remove them from the vehicle for scrubbing or sponging and make quite sure they are dry before refitting. Seats and interior trim panels can be kept clean by a wipe over with a damp cloth. If they do become stained (which can be more apparent on light coloured upholstery) use a little liquid detergent and a soft nail brush to scour the grime out of the grain of the material. Do not forget to keep the head lining clean in the same way as the upholstery. When using liquid cleaners inside the car do not over-wet the surfaces being cleaned. Excessive damp could get into the seams and padded interior causing stains, offensive odours or even rot. If the inside of the vehicle gets wet accidentally it is worthwhile taking some trouble to dry it out properly particularly where carpets are involved. **Do not** leave oil or electric heaters inside the car for this purpose.

4 Minor body damage – repair

See photo sequences on pages 174 and 175

Repair of minor scratches in the vehicle's bodywork

If the scratch is very superficial, and does not penetrate to the metal of the bodywork - repair is very simple. Lightly rub the area of the scratch with a paintwork renovator or a very fine cutting paste, to remove the loose paint from the scratch and to clear the surrounding bodywork of wax polish. Rinse the area with clean water.

Apply touch-up paint to the scratch using a thin paint brush; continue to apply thin layers of paint until the surface of the paint in the scratch is level with the surrounding paintwork. Allow the new paint at least two weeks to harden, then, blend it into the surrounding paintwork by rubbing the paintwork in the scratch area with a paintwork renovator, or a very fine cutting paste. Finally apply wax polish.

An alternative to painting over the scratch is to use a paint transfer. Use the same preparation for the affected area, then simply, pick a patch of a suitable size to cover the scratch completely. Hold the patch against the scratch and burnish its backing paper; the paper will adhere to the paintwork, freeing itself from the backing paper at the same time. Polish the affected area to blend the patch into the surrounding paintwork.

Where a scratch has penetrated right through to the metal of the bodywork, causing the metal to rust, a different repair technique is required. Remove any loose rust from the bottom of the scratch with a penknife; then apply rust inhibiting paint to prevent the formation of rust in the future. Using a rubber or nylon applicator, fill the scratch with bodystopper paste. If required, this paste can be mixed with cellulose thinners to provide a very thin paste which is ideal for filling narrow scratches. Before the stopper paste on the scratch hardens, wrap a piece of smooth cotton rag around the top of a finger. Dip the finger in cellulose thinners and then quickly sweep it across the surface of the stopper-paste in the scratch; this will ensure that the surface of the stopper-paste is slightly hollowed. The scratch can now be painted over as described earlier in this Section.

Repair of dents in the vehicle's bodywork

When deep denting of the vehicle's bodywork has taken place the first task is to pull the dent out, until the affected bodywork almost attains its original shape. There is little point in trying to restore the original shape completely, as the metal in the damaged area will have stretched on impact and cannot be reshaped fully to its original contour. It is better to bring the level of the dent up to a point which is about 1/8 inch (3 mm) below the level of the surrounding bodywork. In cases where the dent is very shallow anyway, it is not worth trying to pull it all out.

If the underside of the dent is accessible, it can be hammered out gently from behind, using a mallet with a wooden or plastic head. Whilst doing this, hold a suitable block of wood firmly against the outside of the dent. This block will absorb the impact from the hammer blows and thus prevent a large area of bodywork from being 'belled-out'.

Should the dent be in a section of the bodywork which has a double skin or some other factor making it inaccessible from behind, a different technique is called for. Drill several small holes through the metal inside the dent area - particularly in the deeper sections. Then screw long self-tapping screws into the holes just sufficiently for them to gain a good purchase in the metal. Now the dent can be pulled out by pulling on the protruding heads of the screws with a pair of pliers.

The next stage of the repair is the removal of the paint from the damaged area, and from an inch or so of the surrounding 'sound' bodywork. This is accomplished more easily by using a wire brush or abrasive pad on a power drill, although it can be done just as effectively by hand using sheets of abrasive paper. To complete the preparations for filling, score the surface of the bare metal with a screwdriver or the tang of a file, or alternatively, drill small holes in the affected area. This will provide a really good 'key' for the filler paste. To complete the repair see the Section on filling and re-spraying.

Repair of rust holes or gashes in the vehicle's bodywork

Remove all paint from the affected area and from an inch or so of the surrounding 'sound' bodywork, using an abrasive pad or a wire brush on a power drill. If these are not available a few sheets of abrasive paper will do the job just as effectively. With the paint removed you will be able to gauge the severity of the corrosion and therefore decide whether to renew the whole panel (if this is possible) or to repair the affected area. New body panels are not as expensive as most people think and it is often quicker and more satisfactory to fit a new panel than to attempt to repair large areas of corrosion.

Remove all fittings from the affected area except those which will act as a guide to the original shape of the damaged bodywork (eg. headlamp shells etc). Then, using tin snips or a hacksaw blade, remove all loose metal and any other metal badly affected by corrosion. Hammer the edges of the hole inwards in order to create a slight depression for the filler paste.

Wire brush the affected area to remove the powdery rust from the surface of the remaining metal. Paint the affected area with rust inhibiting paint; if the back of the rusted area is accessible treat this also.

Before filling can take place it will be necessary to block the hole in some way. This can be achieved by the use of one of the following materials: Zinc gauze, Aluminium tape or Polyurethane foam.

Zinc gauze is probably the best material to use for a large hole. Cut a piece to the approximate size and shape of the hole to be filled, then position it in the hole so that its edges are below the level of the surrounding bodywork. It can be retained in position by several blobs of filler paste around its periphery.

Aluminium tape should be used for small or very narrow holes. Pull a piece off the roll and trim it to the approximate size and shape required, then pull off the backing paper (if used) and stick the tape over the hole; it can be overlapped if the thickness of one piece is insufficient. Burnish down the edges of the tape with the handle of a screwdriver, or similar, to ensure that the tape is securely attached to the metal underneath.

Polyurethane foam is best used where the hole is situated in a section of bodywork of complex shape, backed by a small box section (eg. where the sill panel meets the rear wheel arch – most vehicles). The usual mixing procedure for this foam is as follows: put equal amounts of fluid from each of the two cans provided in the kit, into one container. Stir until the mixture begins to thicken, then quickly pour this mixture into the hole, and hold a piece of cardboard over the larger apertures. Almost immediately the polyurethane will begin to expand, gushing frantically out of any small holes left unblocked. When the foam hardens it can be cut back to just below the level of the surrounding bodywork with a hacksaw blade.

Having blocked off the hole the affected area must now be filled and sprayed - see Section on bodywork filling and respraying.

Bodywork repairs - filling and re-spraying

Before using this Section, see Sections on dent, deep scratch, rust hole and gash repairs.

Many types of bodyfiller are available, but generally speaking those proprietary kits which contain a tin of filler paste and a tube of resin hardener are best for this type of repair. A wide, flexible plastic or nylon applicator will be found invaluable for imparting a smooth and well contoured finish to the surface of the filler.

Mix up a little filler on a clean piece of card or board - use the hardener sparingly (follow the maker's instructions on the packet) otherwise the filler will set very rapidly.

Using the applicator, apply the filler paste to the prepared area; draw the applicator across the surface of the filler to achieve the correct contour and to level the filler surface. As soon as a contour that approximates the correct one is achieved, stop working the paste - if you carry on too long the paste will become sticky and begin to 'pick-up' on the applicator.

Continue to add thin layers of filler paste at twenty-minute intervals until the level of the filler is just 'proud' of the surrounding bodywork.

Once the filler has hardened, excess can be removed using a plane or file. From then on, progressively finer grades of abrasive paper should be used, starting with a 40 grade 'wet or dry' paper. Always wrap the abrasive paper around a flat rubber cork or wooden block - otherwise the surface of the filler will not be completely flat. During the smoothing of the filler surface the 'wet or dry' paper should be periodically rinsed in water - this will ensure that a very smooth finish is imparted to the filler at the final stage.

At this stage the 'dent' should be surrounded by a ring of bare metal, which in turn should be encircled by the finely 'feathered' edge of the good paintwork. Rinse the repair area with clean water, until all of the dust produced by the rubbing-down operation is gone.

Spray the whole repair area with a light coat of grey primer; this will show up any imperfections in the surface of the filler. Repair these imperfections with fresh filler paste or bodystopper, and once more smooth the surface with abrasive paper. If bodystopper is used, it can be mixed with cellulose thinners to form a really thin paste which is ideal for filling small holes. Repeat this spray and repair procedure until you are satisfied that the surface of the filler, and the feathered edge of the paintwork are perfect. Clean the repair area with clean water and allow to dry fully.

The repair area is now ready for spraying. Paint spraying must be carried out in a warm, dry, windless and dust free atmosphere. This condition can be created artificially if you have access to a large indoor working area, but if you are forced to work in the open, you will have to pick your day very carefully. If you are working indoors, dousing the floor in the work area with water will 'lay' the dust which would otherwise be in the atmosphere. If the repair area is confined to one body panel, mask off the surrounding panels; this will help to minimise

12

Fig. 12.1 The door hinge adjustment points (Sec 7)

Fig. 12.2 The striker plate adjusting screws (Sec 7)

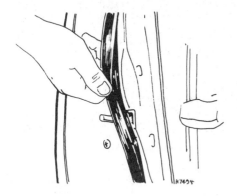

Fig. 12.3 Removing the weatherstrip (Sec 7)

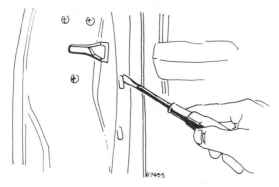

Fig. 12.4 Removing the weatherstrip retaining clips (Sec 7)

the effects of a slight mis-matching in paint colours. Bodywork fittings (eg chrome strips, door handles etc) will also need to be masked off. Use genuine masking tape and several thicknesses of newspapers for the masking operation.

Before commencing to spray, agitate the aerosol can thoroughly, then spray a test area (an old tin, or similar) until the technique is mastered. Cover the repair area with a thick coat of primer; the thickness should be built up using several thin layers of paint rather than one thick one. Using 400 grade 'wet or dry' paper, rub down the surface of the primer until it is really smooth. While doing this, the work area should be thoroughly doused with water. Allow to dry before spraying on more paint.

Spray on the top coat, again building up the thickness by using several thin layers of paint. Start spraying in the centre of the repair area and then using a circular motion, work outwards until the whole repair area and about 2 inches of the surrounding original paintwork is covered. Remove all masking material 10 to 15 minutes after spraying on the final coat of paint. Allow the new paint at least two weeks to harden fully; then, using a paintwork renovator or a very fine cutting paste, blend the edges of the new paint into the existing paintwork. Finally, apply wax polish.

5   Major body damage – repair

Where serious damage has occurred or large areas need renewal due to neglect, it means certainly that completely new sections or panels will need welding in and this is best left to professionals. If the damage is due to impact it will also be necessary to completely check the alignment of the bodyshell structure. Due to the principle of construction the strength and shape of the whole can be affected by damage to a part. In such instances the services of the official agent with specialist checking jigs are essential. If a body is left misaligned it is first of all dangerous as the car will not handle properly and secondly, uneven stresses will be imposed on the steering, engine and transmission, causing abnormal wear or complete failure.

6   Maintenance – hinges and locks

1   Periodically lubricate the hinges of the doors, bonnet and tailgate with a few drops of light oil.
2   Similarly lubricate the door catches, bonnet release mechanism and tailgate, and the release mechanism for the spare wheel.
3   Apply a smear of general purpose grease to lock strikers, striker plates and door pillar blocks.

7   Doors – rattles and their rectification

1   Check first that the door is not loose at the hinges and that the latch is holding the door firmly in position. Check also that the door lines up with the aperture in the body.
2   If the hinges are loose, or the door is out of alignment, it will be necessary to reset the hinge positions. This is a straightforward matter after slackening the hinge retaining screws slightly, following which the door can be repositioned in/out or up/down.
3   If the latch is holding the door properly it should hold the door tightly when fully latched and the door should line up with the body. If adjustment is required, slacken the striker plate screws slightly and reposition the plate in/out or up/down as necessary.
4   Other rattles from the door could be caused by wear or looseness in the window winder, or the glass channels, seal strips and interior lock mechanism.
5   To remove the door weatherstrip, pull it from the retaining clips. Using a suitable screwdriver, prise the retaining clips from the door.
6   To refit the weatherstrip, first fit the retaining clips, then fit the weatherstrip to them.

8   Door window regulator and glass – removal and refitting

Front door
1   Lower the glass to its full extent and remove the regulator handle.
2   Pull off the inset trim and remove the two screws that attach the arm rest to the door. Remove the arm rest.

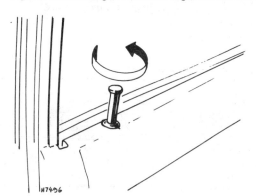

Fig. 12.5 Removing the window regulator (Sec 8)

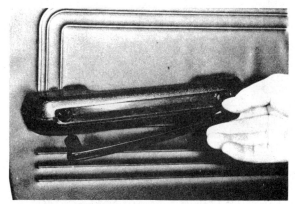

Fig. 12.6 Removing the arm rest (Sec 8)

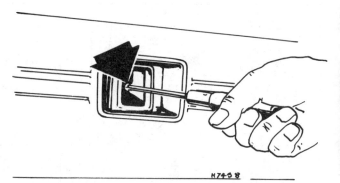

Fig. 12.7 Unscrewing the door lock knob (Sec 8)

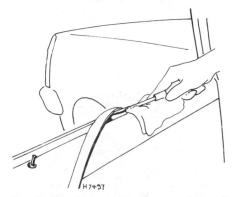

Fig. 12.8 Removing the weatherstrip (Sec 8)

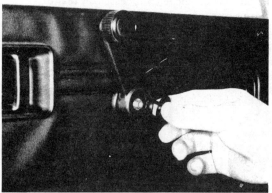

Fig. 12.9 Removing the inner handle cover (Sec 8)

Fig. 12.10 Removing the door trim (Sec 8)

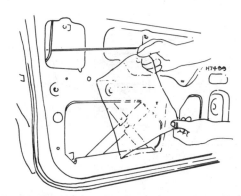

Fig. 12.11 Rolling away the polythene covers (Sec 8)

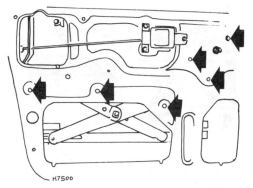

Fig. 12.12 The regulator mechanism securing bolts (Sec 8)

12

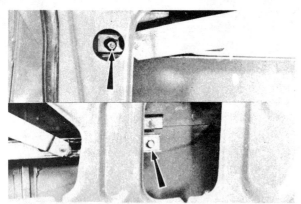

Fig. 12.13 The glass to regulator attaching bolts at each end of the regulator (Sec 8)

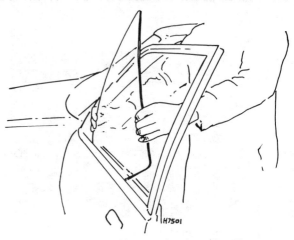

Fig. 12.14 Lifting out the glass (Sec 8)

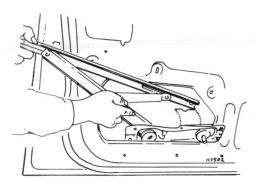

Fig. 12.15 Lifting out the regulator assembly (Sec 8)

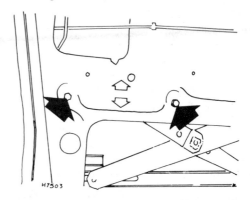

Fig. 12.16 Adjusting the regulator position (Sec 8)

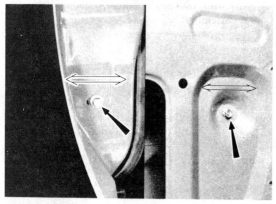

Fig. 12.17 The front and rear glass guide securing bolts (Sec 8)

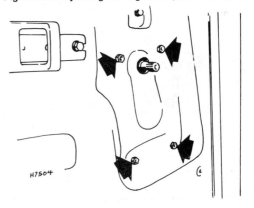

Fig. 12.18 The regulator mechanism attaching bolts (Sec 8)

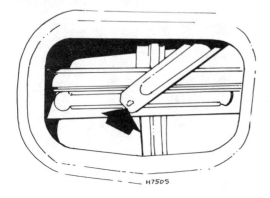

Fig. 12.19 The regulator arm roller (Sec 8)

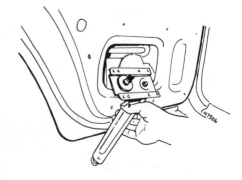

Fig. 12.20 Lifting out the regulator assembly (Sec 8)

3 Unscrew and remove the door lock knob.

4 Using a suitable screwdriver, carefully prise the weatherstrip from its retaining clips. To avoid damage to the door's paintwork, use a piece of rag or soft wood between the screwdriver and the point of leverage.

5 Unscrew and remove the door inner handle cover, and carefully, using the flat blade of a knife, lever the door trim panel away from the door.

6 Carefully remove the polythene covers over the apertures in the door.

7 Undo and remove the six bolts securing the window regulator to the door.

8 At either end of the glass, remove the bolt attaching the glass to the window regulator. The glass can now be removed from the door. Exercise great care when removing the glass.

9 The regulator can be removed now by manipulating it through the large aperture.

10 Refitting is a reversal of the removal procedure, but before finally fitting the door trim the following adjustments should be made: Operate the regulator to ensure that the glass is sealing at the top of the door frame. Adjustment is carried out, by loosening the six regulator attaching bolts, and moving the window regulator up or down as required. Adjust the front and rear glass guides by loosening their retaining bolts, and moving them in the required direction to eliminate any side play.

Rear door

11 Lower the glass to its full extent and remove the regulator handle.

12 Using a suitable flat-bladed knife, carefully prise the door trim from the door. Remove the polythene cover from the access hole.

13 Undo and remove the four bolts attaching the regulator mechanism to the door.

14 Disconnect the roller on the regulator arm from the glass channel. This can be carried out by placing the regulator arm at one end of the channel, which will allow the regulator arm roller to pass through the large hole at the end of the channel.

15 The regulator assembly can now be manipulated through the access hole.

16 Using a suitable screwdriver, lever the weatherstrip from the door. Exercise great care in this operation, and place a piece of rag or wood between the screwdriver and the leverage point to avoid damaging the paintwork.

17 At the top edge of the door, prise back the rubber sealing strip to reveal the two screws retaining the upper end of the division sash. Remove the two screws.

18 Remove the two bolts attaching the division sash to the door panel. Remove the division sash together with the quarter light glass.

19 The rear door window can now be removed from the door.

20 To refit the rear door glass and regulator is a reversal of the removal procedure.

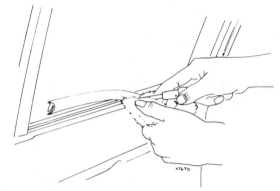

Fig. 12.21 Removing the weatherstrip (Sec 8)

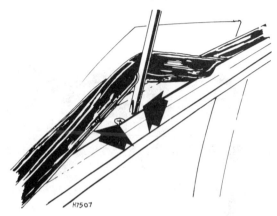

Fig. 12.22 The division sash retaining screws (Sec 8)

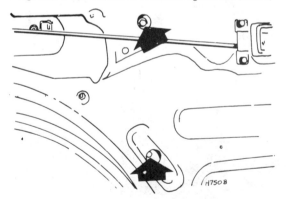

Fig. 12.23 The division sash securing bolts (Sec 8)

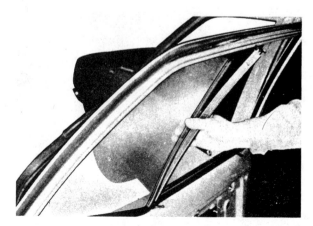

Fig. 12.24 Lifting out the quarter light with the division sash (Sec 8)

Fig. 12.25 Lifting out the glass (Sec 8)

12

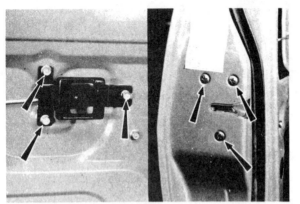

Fig. 12.26 The inner handle retaining bolts and the door lock screws (Sec 9)

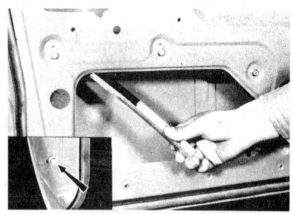

Fig. 12.27 Removing the rear glass guide (Sec 9)

Fig. 12.28 The outer handle attaching nuts and the remote control rods (Sec 9)

Fig. 12.29 Removing the door lock assembly (Sec 9)

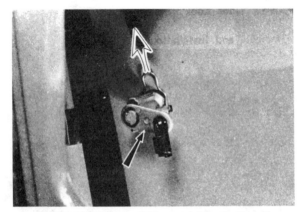

Fig. 12.30 The key cylinder retaining clip (Sec 9)

Fig. 12.31 Adjusting the inner handle (Sec 9)

9 Door lock – removal and refitting

1   Carry out the operations described in paragraphs 1, 2, 5 and 6, Section 8, to remove the trim panel.

2   Remove the three bolts attaching the inner handle to the door, and allow the handle to hang on the operating rod.

3   At the door edge, remove the three screws attaching the door lock to the door.

4   Remove the rear glass guide attaching screw and lift out the guide.

5   Remove the two nuts attaching the outer handle. Disconnect the remote control rod from the key cylinder and the outer handle.

6   The door lock and inner handle can now be removed as an assembly.

7   The key cylinder can now be removed, if so desired, by prising out its retaining clip.

8   Refitting of the door lock is a reversal of the removal procedure,

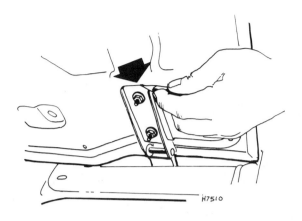

Fig. 12.32 Marking the hinge positions (Sec 10)

Fig. 12.33 The hinge attaching nuts (Sec 10)

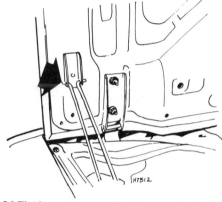

Fig. 12.34 The bonnet support location on the bonnet (Sec 10)

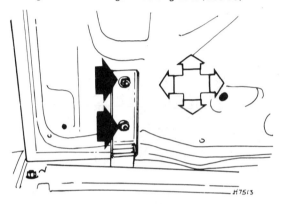

Fig. 12.35 Adjusting the position of the bonnet (Sec 10)

Fig. 12.36 The bonnet lock mechanism attaching bolts (Sec 10)

Fig. 12.37 The hatchback securing bolts (Sec 11)

but if necessary, adjust the inner handle free play.

10 Bonnet – removal and refitting

11   Open the bonnet, and using a fibre tip pen, mark the position of the hinges in relation to the bonnet.
2   The bonnet support can be removed by either removing the two retaining clips at the bonnet location bracket and pressing the two rods inwards until they clear the bracket, or by removing the single retaining screw and washer on the top of the inner ring.
3   At this stage it is wise to seek the assistance of a second person to help support the bonnet. Remove the two nuts securing each hinge to the bonnet. Lift the bonnet away from the vehicle.
4   Refitting is a reversal of the removal procedure, but ensure that

the hinge positions, marked previously, are aligned. Fore and aft movement and side to side movement is possible, as the hinge location holes are slotted to allow this. The bonnet lock mechanism is also adjustable when the attaching bolts are loose.

11 Hatchback door – removal and refitting

1   Before any work is commenced, place rags along the roof of the vehicle to prevent the door marking the paintwork when the hinge bolts are removed.
2   Remove the two bolts that attach each hinge to the hatchback door. It is advisable to have a second person to help support the door when removing these bolts, and also when removing the hatchback

12

These photos illustrate a method of repairing simple dents. They are intended to supplement *Body repair - minor damage* in this Chapter and should not be used as the sole instructions for body repair on these vehicles.

1   If you can't access the backside of the body panel to hammer out the dent, pull it out with a slide-hammer-type dent puller. In the deepest portion of the dent or along the crease line, drill or punch hole(s) at least one inch apart . . .

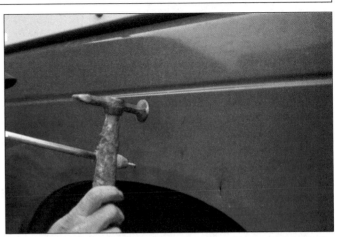

2   . . . then screw the slide-hammer into the hole and operate it. Tap with a hammer near the edge of the dent to help 'pop' the metal back to its original shape. When you're finished, the dent area should be close to its original contour and about 1/8-inch below the surface of the surrounding metal

3   Using coarse-grit sandpaper, remove the paint down to the bare metal. Hand sanding works fine, but the disc sander shown here makes the job faster. Use finer (about 320-grit) sandpaper to feather-edge the paint at least one inch around the dent area

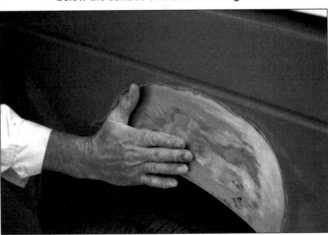

4   When the paint is removed, touch will probably be more helpful than sight for telling if the metal is straight. Hammer down the high spots or raise the low spots as necessary. Clean the repair area with wax/silicone remover

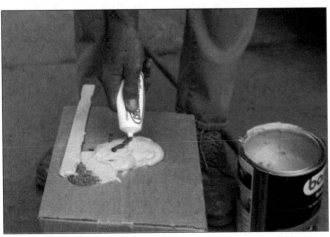

5   Following label instructions, mix up a batch of plastic filler and hardener. The ratio of filler to hardener is critical, and, if you mix it incorrectly, it will either not cure properly or cure too quickly (you won't have time to file and sand it into shape)

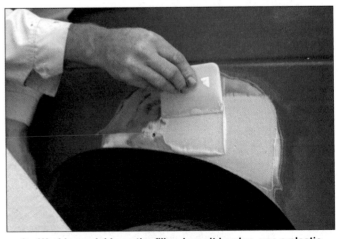

6   Working quickly so the filler doesn't harden, use a plastic applicator to press the body filler firmly into the metal, assuring it bonds completely. Work the filler until it matches the original contour and is slightly above the surrounding metal

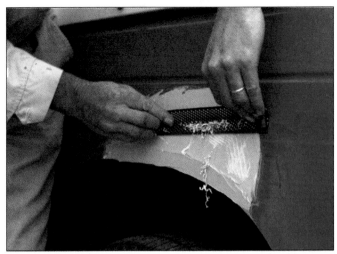

7   Let the filler harden until you can just dent it with your fingernail. Use a body file or Surform tool (shown here) to rough-shape the filler

8   Use coarse-grit sandpaper and a sanding board or block to work the filler down until it's smooth and even. Work down to finer grits of sandpaper - always using a board or block - ending up with 360 or 400 grit

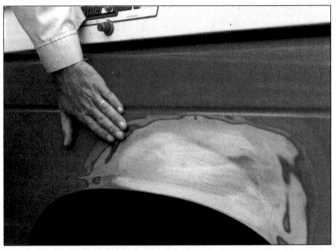

9   You shouldn't be able to feel any ridge at the transition from the filler to the bare metal or from the bare metal to the old paint. As soon as the repair is flat and uniform, remove the dust and mask off the adjacent panels or trim pieces

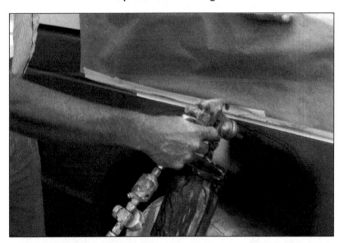

10   Apply several layers of primer to the area. Don't spray the primer on too heavy, so it sags or runs, and make sure each coat is dry before you spray on the next one. A professional-type spray gun is being used here, but aerosol spray primer is available inexpensively from auto parts stores

11   The primer will help reveal imperfections or scratches. Fill these with glazing compound. Follow the label instructions and sand it with 360 or 400-grit sandpaper until it's smooth. Repeat the glazing, sanding and respraying until the primer reveals a perfectly smooth surface

12   Finish sand the primer with very fine sandpaper (400 or 600-grit) to remove the primer overspray. Clean the area with water and allow it to dry. Use a tack rag to remove any dust, then apply the finish coat. Don't attempt to rub out or wax the repair area until the paint has dried completely (at least two weeks)

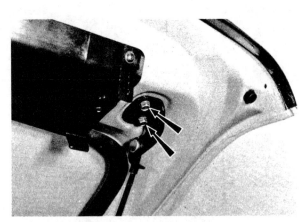

Fig. 12.38 The stay attaching bolts (Sec 11)

Fig. 12.39 Adjusting the striker plate (Sec 11)

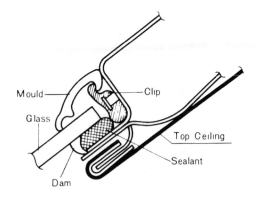

Fig. 12.40 A sectional view of the windscreen location components (Sec 12)

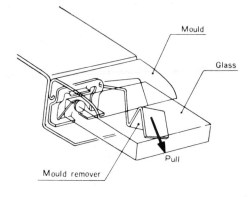

Fig. 12.41 Removing the windscreen mould (Sec 12)

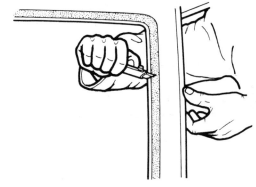

Fig. 12.42 Cutting the sealant with piano wire (Sec 12)

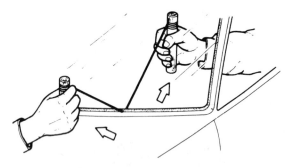

Fig. 12.43 Trimming the sealant (Sec 12)

Fig. 12.44 Repairing the sealant (Sec 12)

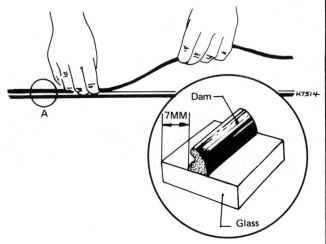

Fig. 12.45 Bonding the dam into position (Sec 12)

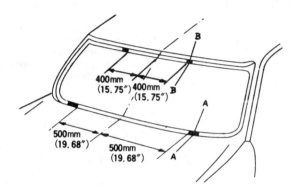

Fig. 12.46 The spacer setting dimensions (Sec 12)

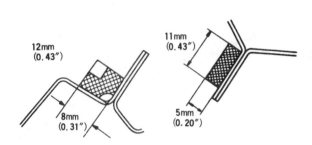

Fig. 12.47 A sectional view of the two different types of spacers (Sec 12)

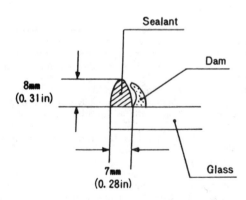

Fig. 12.48 A mould retaining clip and a spacer in position (Sec 12)

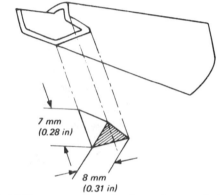

Fig. 12.49 The correct shape for the sealant nozzle (Sec 12)

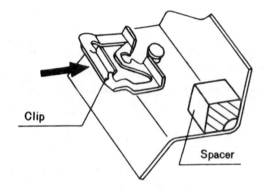

Fig. 12.50 The glass ready for fitting (Sec 12)

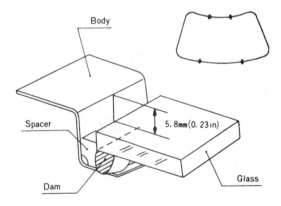

Fig. 12.51 The step height of the glass in relation to the body (Sec 12)

stay.

3   Remove the two bolts attaching the stay to the door and, very carefully, lift away the door.

4   Refitting is a straightforward reversal of the removal procedure but, if necessary, loosen the door lock striker plate attaching bolts to adjust the locking position.

12 Windscreen glass – removal and refitting

1   Windscreen renewal is a job that requires the use of special sealant, and if it is at all possible, it should be entrusted to a specialist. However, if the owner has the necessary competence, and time to spare so that the job is not rushed, proceed as described in the following paragraphs.

2   Remove the interior mirror, and the two side pillar trims. Take off the wiper arms. Spread padding material over the front of the car, and tape newspaper over the top of the dash.

3   Insert a bent strip of metal under the trim mould round the outer edge of the glass, and prise it out of the clips. Remove the clips from their pins.

4   Now remove the remains of the old glass. Drill a small hole through the sealant between the glass and the flange. A needle pushed hard should do this. Pass a length of piano wire through the hole. Wind each end of the piano wire around a stick to form a suitable grip. Using a sawing motion, go all around the windscreen separating the sealant from the windscreen. Take care to ensure that the wire does not rub on the paintwork. Lift out the old glass.

5   With a sharp knife cut away the old sealant, so that about 0.04 to 0.08 in(1 to 2 mm) remains around the circumference of the flange. If

12

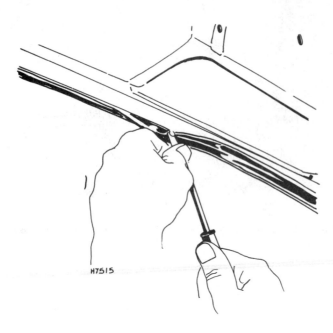

Fig. 12.52 Removing the service hole clip (Sec 13)

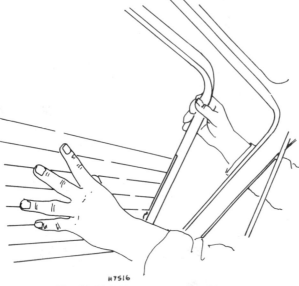

Fig. 12.53 Lifting out the glass (Sec 13)

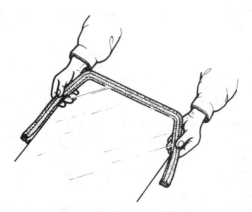

Fig. 12.54 Assembling the weatherstrip to the glass (Sec 13)

all the sealant has come off in places, apply some primer, and allow it 30 minutes to dry. Then paint on more sealant to build up to the required thickness.

6   Lay the new glass flat, inside uppermost, on some soft padding such as a blanket. Degrease with solvent the perimeter of the glass for about 2 inches (50.8 mm) in from the edge, the whole way around.

7   Bond the new dam to the glass with bonding agent. It should be positioned with its outer edge 0.028 in (7 mm) from the glass edge, and the lip outermost.

8   Apply a thin coat of primer to the bonding area of the body flange and the glass, and allow 30 minutes for it to dry. Keep it clear of dust. Do not touch the surface. If any gets on the hands, wipe it off quickly.

9   Bond the spacers to the body 15.75 in (400 mm) either side of the centre line of the top of the windscreen flange, and 19.68 in(500 mm) either side of the centre line of the bottom edge of the windscreen flange (Fig. 12.46).

10 Insert the moulding clips on their pins. If any are loose, renew them.

11 Prepare the nozzle of a sealant gun so that it has a flange which can run along the edge of the glass, and a 'V' out of which the sealant can flow.

12 Open the car windows so that the glass will not be pushed out by air pressure when a door is closed.

13 Apply the sealant around the whole circumference to fill the gap between the dam and the edge of the glass with a ridge of sealant 0.31 in (8 mm) high. Keep the run of the sealant smooth and even, reshaping it where necessary with a suitable spatula.

14 Now lift the glass into position. A person is needed at each end of the glass, and a third inside the car to check that all is in order.

15 Push the glass lightly in towards the car to squeeze the sealant. Measure the amount the glass is in from the body on the top and bottom edges at the positions shown in Fig. 12.51. The step height from the glass to the body should be 0.23 in (5.8 mm). Smooth away any sealant that oozes out with a spatula. If there are any points of poor contact, add more sealant.

16 Leave trhe sealant time to harden without disturbing the glass. This will take 5 hours at 20°C(68°F) and 24 hours at 5°C(41°F).

17 Recheck for any points of poor sealing that might allow a leak, and rectify.

18 Clip in the mould. Refit the interior mirror, and the pillar trims.

13 Hatchback door glass – removal and refitting

1   If the vehicle is equipped with a heated rear window, disconnect the terminals to it.

2   Prise off the service hole clip.

3   Using a suitable wooden spatula, carefully lever the weatherstrip from the body; work the spatula all the way around the weatherstrip to break the cemented surfaces.

4   Using a putty knife or similar flat blade tool, prise the lips of the weatherstrip away from the body flange.

5   Working from inside the car, with an assistant outside, start pushing the glass out.

6   With the glass removed, pry the moulding out of the weatherstrip.

7   To commence refitting, fit the weatherstrip to the glass, making sure it is properly seated and positioned.

8   Press the moulding into the recess in the weatherstrip.

9   Now insert a piece of good quality string into the channel of the weatherstrip, and feel it all the way around the weatherstrip.

10 To aid refitting, apply soapy water to the weatherstrip and the body flange.

11 Now press the glass assembly against the body, aligning the weatherstrip lip with the flange.

12 From inside the vehicle, start pulling the string out of the weatherstrip channel, locating the weatherstrip lip over the flange. Whilst the string is being pulled out, have someone to tap the glass with the flat of the hand near the weatherstrip. This will aid the lip to enter the flange and locate over the lip. Continue pulling the string all around the glass, until the entire length of the lip has located over the flange, and the string has been removed from the weatherstrip channel.

13 Using a suitable adhesive gun, apply adhesive between the weatherstrip and the body, both inside and outside the vehicle. Apply adhesive between the weatherstrip and the glass inside and outside

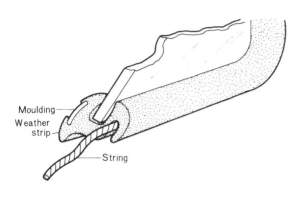

Fig. 12.55 A sectional view of the weatherstrip showing the string (Sec 13)

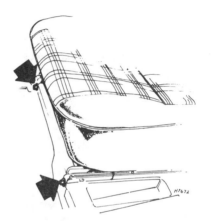

Fig. 12.56 Locating the weatherstrip lip by removing the string – part A (Sec 13)

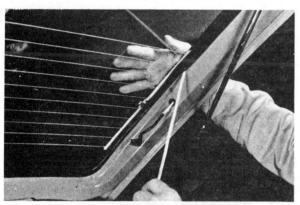

Fig. 12.57 Locating the weatherstrip lip by removing the string – part B (Sec 13)

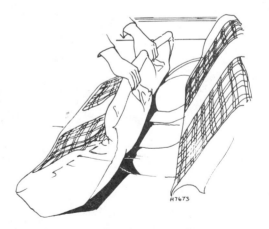

Fig. 12.58 The front seat attaching bolts (Sec 14)

Fig. 12.59 The front seat attaching nuts (Sec 14)

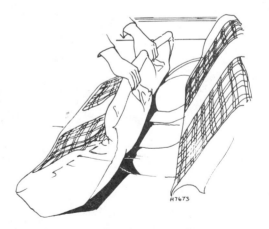

Fig. 12.60 Lift up the back seat (Sec 14)

the vehicle. Use a spatula to remove any excess adhesive that may ooze out.

14 Fit the service hole clip and reconnect the heated rear window terminals (if applicable).

14 Seats – removal and refitting

Front seat

1 At the front end of the seat, remove the bolt that attaches each seat runner to the mounting bracket.

2 At the rear end of the seat, remove the nut that attaches each seat runner to the mounting brackets.

3 The seat can now be lifted out of the vehicle, together with the seat runners.

4 To refit a front seat, reverse the removal procedure.

Rear seat

5 Lift up the back of the rear seat.

6 Now unsnap the clips of the seat and remove the seat.

12

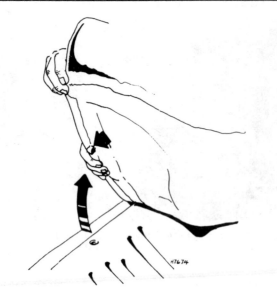

Fig. 12.61 Unsnap the clips to remove the rear seat (Sec 14)

7   Refitting is the reverse of the removal procedure.

15 Front bumper and shock absorber – removal, checking and reassembly

1   Undo and remove the three bolts that attach the shock absorber bracket to the bodyframe. Repeat this at the other shock absorber.
2   Lift the bumper, together with the two shock absorbers, away from the vehicle.
3   Remove the two bolts and washers that secure each shock absorber to the bumper.
4   To check the shock absorber, measure the distance between the inside of the bumper mounting flange and the stop bracket. The distance should be 1.18 ± 0.08 in (30 ± 2 mm).
5   A replacement shock absorber can be fitted to the original bracket, provided that the bracket is not damaged, by removing the nut securing it to the bracket, and fitting the replacement item into place.
6   To reassemble the front bumper, fit both shock absorbers to the bumper, then fit the bumper to the vehicle in the reverse of the removal procedure.

16 Rear bumper and shock absorber – removal, checking and reassembly

1   Remove the nut and washer securing the inner end of the shock

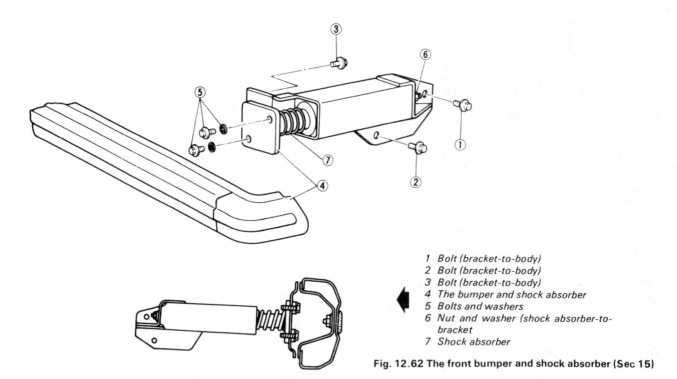

1 Bolt (bracket-to-body)
2 Bolt (bracket-to-body)
3 Bolt (bracket-to-body)
4 The bumper and shock absorber
5 Bolts and washers
6 Nut and washer (shock absorber-to-bracket
7 Shock absorber

Fig. 12.62 The front bumper and shock absorber (Sec 15)

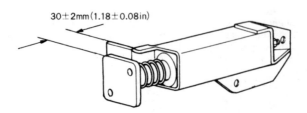

30±2mm(1.18±0.08in)

Fig. 12.63 The front shock absorber checking dimension (Sec 15)

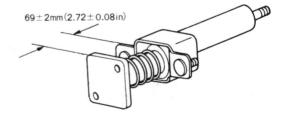

69±2mm(2.72±0.08in)

Fig. 12.64 The rear shock absorber checking dimension (Sec 16)

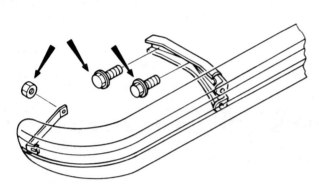

Fig. 12.65 The front bumper mounting points (Sec 17)

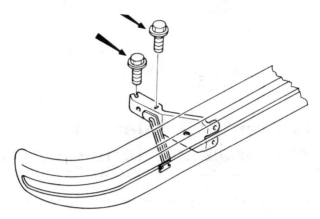

Fig. 12.66 The rear bumper mounting points (Sec 18)

absorber to the inside of the box frame. Repeat this at the other shock absorber.

2   Remove the two nuts and washers securing the shock absorber to the outside of the box section. Repeat this at the other shock absorber.

3   Lift the bumper, together with the shock absorbers, away from the vehicle.

4   Remove the two bolts anbd washers securing each shock absorber to the bumper.

5   To check the shock absorber, measure the distance between the inside of the bumper mounting flange and the shock absorber body. The distance should be 2.72 $\pm$ 0.08 in (69 $\pm$ 2 mm).

6   Commence reassembly by fitting both shock absorbers on the bumper, then fit the bumper assembly to the vehicle in the reverse order of removal.

17 Front bumper (non shock absorber type) – removal and refitting

1   At each end of the bumper, remove the nut attaching the stay to

the vehicle body.

2   Remove the two bolts securing the bumper bracket to the vehicle frame. Repeat this at the other bracket.

3   The bumper together with the brackets, can now be removed from the vehicle.

4   To refit the bumper, reverse the removal procedure.

18 Rear bumper (non shock absorber type) – removal and refitting

1   Remove the two bolts attaching the bumper bracket to the vehicle frame. Repeat this at the other bumper bracket.

2   Lift the bumper, together with the brackets, away from the vehicle.

3   Refitting is the reverse of the removal procedure.

12

Chapter 13 Supplement:
Revisions and information on later North American models

Contents

1 Introduction

This supplementary Chapter covers changes made to the Mazda GLC hatchback during the 1979 and 1980 model years and the Mazda GLC wagon during the 1979 through 1983 model years, and the procedures affected by those changes.

Operations that are not included in this Chapter are the same or similar to those described for the 1978 1300cc US model found in the first 12 Chapters of this manual.

The recommended way of using this supplement is, prior to any operation, check here first for any relevant information pertaining to your model. After noting any model differences, particularly in the Specifications Section, you can then follow the appropriate procedure, either in this Chapter or one of the preceding 12.

2 Specifications

Dimensions

Mazda GLC hatchback

Overall length (1979)	154 in (3920 mm)
Overall length (1980)	156 in (3975 mm)
Overall width	63 in (1605 mm)
Overall height	54 in (1370 mm)
Wheelbase	91 in (2315 mm)
Front track	51 in (1295 mm)
Rear track	52 in (1310 mm)

Mazda GLC Station wagon

Overall length (1979)	163 in (4145 mm)
Overall length (1980 thru 1983)	166 in (4205 mm)
Overall width	63 in (1605 mm)
Overall height	56 in (1425 mm)
Wheelbase	91 in (2315 mm)
Front track	51 in (1295 mm)
Rear track	52 in (1310 mm)

Engine — 1415 cc

Bore	3.03 in (77 mm)
Stroke	2.99 in (76 mm)
Displacement	1415 cc (86.4 cu in)
Compression ratio	9.0 : 1
Cylinder block bore	3.0315 to 3.0322 in (77.0 to 77.019 mm)
Piston diameter (measured at 90° to pin bore axis and 0.71 in	
[18.0 mm] below oil ring groove)	3.031 ± 0.0004 in (76.964 ± 0.010 mm)
Piston-to-cylinder clearance (standard)	0.0021 to 0.0026 in (0.054 to 0.067 mm)
Piston ring groove width	
Top	0.0599 to 0.0604 in (1.520 to 1.534 mm)
Second	0.0599 to 0.0604 in (1.520 to 1.534 mm)
Oil	0.1583 to 0.1588 in (4.020 to 4.034 mm)
Piston ring groove depth	0.1461 to 0.1516 in (3.71 to 3.85 mm)
Piston ring width	
Top	0.0579 to 0.0587 in (1.47 to 1.49 mm)
Second	0.0579 to 0.0587 in (1.47 to 1.49 mm)
Oil	0.1583 to 0.1588 in (4.020 to 4.032 mm)
Piston ring thickness	
Top	0.1260 ± 0.0039 in (3.2 ± 0.1 mm)
Second	0.1260 ± 0.0039 in (3.2 ± 0.1 mm)
Oil	0.1260 in (3.2 mm)
Piston ring side clearance	
Top	0.0012 to 0.0025 in (0.030 to 0.064 mm)
Second	0.0012 to 0.0025 in (0.030 to 0.064 mm)
Oil ring end gap	0.012 to 0.035 in (0.3 to 0.9 mm)
Piston pin diameter	0.7865 to 0.7869 in (19.076 to 19.088 mm)
Piston pin length	2.5591 in (65 mm)
Piston-to-pin clearance	0 to 0.0009 in (0 to 0.024 mm)
Main journal diameter	1.9661 to 1.9667 in (49.040 to 49.055 mm)
Main bearing oil clearance (standard)	0.009 to 0.0017 in (0.023 to 0.042 mm)
Crankpin diameter	1.5724 to 1.5730 in (39.040 to 39.055 mm)
Connecting rod bearing oil clearance (standard)	0.0009 to 0.0019 in (0.024 to 0.048 mm)
Connecting rod length (center-to-center)	5.3544 ± 0.0020 in (136.0 ± 0.05 mm)
Crankshaft endplay	0.004 to 0.006 in (0.10 to 0.15 mm)

Engine — 1490 cc

Bore	3.03 in (77 mm)
Stroke	3.15 in (80 mm)
Displacement	1490 cc (90.9 cu in)
Compression ratio	9.0 : 1
Valve seat angle (intake and exhaust)	45°
Valve face angle (intake and exhaust)	45°
Valve spring free length	
New spring	1.705 in (43.3 mm)
Service limit	1.654 in (42.0 mm)
Valve spring installed length	1.319 in (33.5 mm)
Cylinder block bore	3.0315 to 3.0322 in (77.0 to 77.019 mm)
Piston diameter (measured at 90° to pin bore axis and 0.71 in	
[18.0 mm] below oil ring groove)	3.031 ± 0.0004 in (76.964 ± 0.010 mm)
Piston-to-cylinder clearance	0.0015 to 0.0020 in (0.039 to 0.052 mm)
Piston ring groove width	
Top	0.0480 to 0.0488 in (1.22 to 1.24 mm)
Second	0.0599 to 0.0607 in (1.52 to 1.54 mm)
Oil	0.1583 to 0.1591 in (4.02 to 4.04 mm)
Piston ring groove depth	0.1461 to 0.1516 in (3.71 to 3.85 mm)
Piston ring width	
Top	0.0460 to 0.0468 in (1.197 to 1.199 mm)
Second	0.0579 to 0.0587 in (1.497 to 1.499 mm)
Piston ring thickness	
Top	0.1299 ± 0.0039 in (3.3 ± 0.1 mm)
Second	0.1299 ± 0.0039 in (3.3 ± 0.1 mm)
Oil	0.1220 ± 0.0079 in (3.1 ± 0.2 mm)
Piston ring side clearance	
Top	0.0012 to 0.0028 in (0.030 to 0.070 mm)
Second	0.0012 to 0.0028 in (0.030 to 0.070 mm)
Piston ring end gap	
Top	0.006 to 0.012 in (0.15 to 0.30 mm)
Second	0.006 to 0.012 in (0.15 to 0.30 mm)
Oil	0.012 to 0.035 in (0.30 to 0.90 mm)
Piston pin diameter	0.7864 to 0.7866 in (19.974 to 19.980 mm)
Piston pin length	2.5591 in (65 mm)
Piston-to-pin clearance	0.0003 to 0.0010 in (0.008 to 0.026 mm)
Main journal diameter	1.5724 to 1.5731 in (39.940 to 39.956 mm)

13

Main bearing oil clearance	0.0009 to 0.0017 in (0.024 to 0.042 mm)
Crankshaft endplay	0.004 to 0.006 in (0.10 to 0.15 mm)
Connecting rod length (center-to-center)	5.3544 ± 0.0020 in (136 ± 0.05 mm)
Connecting rod side clearance	0.004 to 0.010 in (0.11 to 0.26 mm)
Connecting rod bearing oil clearance	0.0009 to 0.0026 in (0.024 to 0.066 mm)
Engine oil capacity	3.9 US qt (3.7 liters)

Torque wrench settings

1415 cc engine	lbf ft	kgf m
Connecting rod cap bolts	22 to 25	3.0 to 3.5
Exhaust manifold bolts	14 to 19	1.9 to 2.6
1490 cc engine	**lbf ft**	**N-m**
Main bearing cap bolts	47.7 to 51.4	66 to 71
Connecting rod cap bolts	22 to 25	30 to 35
Oil pump sprocket bolts	22 to 25	30 to 35
Cylinder head bolts	56 to 59	78 to 82
Exhaust manifold bolts	14 to 19	19 to 26

Fuel, exhaust and emission control systems

Carburetor — 1979

Main jet	No.
Primary (manual transmission — Canada)	110
Primary (automatic transmission — Canada)	108
Primary (US except California)	110
Primary (manual transmission — California)	112
Primary (automatic transmission — California)	114
Secondary (all)	150
Main air bleed	
Primary	80
Secondary	140
Slow jet	
Primary	
US	48
Canada	46
Secondary	130
Slow air bleed	
Primary	190
Secondary	180
Power jet	40
Fast idle cam clearance	0.054 in (1.37 mm)
Choke valve opening	0.043 in (1.10 mm)
Choke diaphragm	0.050 in (1.27 mm)
Unloader system	0.09 in (2.28 mm)
Secondary throttle valve	0.236 in (6.0 mm)
Idle speed	
Manual transmission	700 to 750 rpm
Automatic transmission (in Drive)	600 to 650 rpm
CO concentration at idle	2.0 ± 0.5% (without air injection)

Carburetor — 1980

Main jet	No.
Primary (California with auto. trans.)	116
Primary (except California with auto. trans.)	114
Secondary (California with auto. trans.)	145
Secondary (except California with auto. trans.)	150
Main air bleed	
Primary	80
Secondary	140
Slow jet	
Primary	50
Secondary	150
Slow air bleed	
Primary	190
Secondary	180
Power jet	40
Fast idle cam clearance	0.041 in (1.03 mm)
Choke valve opening	0.043 in (1.10 mm)
Choke diaphragm	0.050 in (1.27 mm)
Unloader	0.09 in (2.28 mm)
Secondary throttle valve	1.236 in (6.0 mm)
Idle speed	
Automatic transmission (in Drive)	700 ± 50 rpm
Manual transmission	700 ± 50 rpm
CO concentration at idle	2.5 ± 1.5% (without air injection)

Carburetor — 1981 thru 1983

	No.
Main jet	
Primary	110
Secondary	155
Main air bleed	
Primary	70
Secondary	140
Slow jet	
Primary	48
Secondary	160
Slow air bleed	
Primary	140
Secondary	180
Power jet	40
Fast idle cam clearance	
1981	0.033 in (0.85 mm)
1982 and 1983	0.031 to 0.038 in (0.78 to 0.96 mm)
Choke valve opening (1981 and 1982)	0.042 in (1.08 mm)
Choke diaphragm	
1981	0.050 in (1.27 mm)
1982 and 1983	0.100 ± 0.010 in (2.54 ± 0.25 mm)
Unloader (1981 and 1982)	0.088 in (2.23 mm)
Secondary throttle valve (1981 and 1982)	0.287 to 0.335 in (7.3 to 8.5 mm)
Secondary throttle valve (1983)	0.277 to 0.325 in (7.05 to 8.25 mm)
Idle speed	
Automatic transmission (in Drive)	750 rpm
Manual transmission	800 rpm

General

Fuel pump feed capacity	0.7 US qt/min. (0.6 Imperial qt/min.) (700 cc/min.) at idle
Fuel tank capacity — station wagons	11.9 US gallons (45 liters) (9.9 Imperial gallons)

Ignition system

Spark plug type	
NGK	BPR-6ES or BPR-5ES
Nippon Denso	W20-EXR-U or W16-EXR-U
Distributor	
Type	
US	Transistorized
Canada (1979)	Single points on vehicles with auto. trans.
	Dual points on vehicles with manual trans.
Canada (1980 thru 1983)	Transistorized
Air gap	
1979 and 1980	0.010 to 0.014 in (0.25 to 0.35 mm)
1981 thru 1983	0.012 to 0.018 in (0.30 to 0.45 mm)
Centrifugal advance (transistorized ignition)	
Starts	0° at 700 rpm
Maximum	10° at 2350 rpm
Vacuum advance (transistorized ignition)	
Starts	0° at 2.95 in-Hg (75 mm Hg)
Maximum	7° at 13.8 in-Hg (350 mm Hg)
Centrifugal advance (contact breaker ignition)	
Starts	0° at 600 rpm
Maximum	8° at 2000 rpm
Vacuum advance (contact breaker ignition)	
Starts	0° at 11.81 in-Hg (300 mm Hg)
Maximum	4° at 16.14 in-Hg (410 mm Hg)
Ignition timing	
1979 conventional ignition with manual trans.	8 ± 1° ATDC
1979 conventional ignition with auto. trans.	8 ± 1° BTDC
1979 transistorized ignition - except California	7 ± 1° BTDC
1979 transistorized ignition - California	5 ± 1° BTDC
1980 (all)	5 ± 1° BTDC
1981 thru 1983 (all)	8° BTDC

Automatic transmission

Stall speed	
During break-in period	1850 to 2100 rpm
After break-in period	1900 to 2150 rpm

Rear Axle

Ratio	
Automatic transmission	3.909 : 1
Manual transmission	3.727 : 1

13

Rear axle (continued)

Oil capacity

1979	1.1 US qt (1.0 liter)
1980	1.1 US qt (1.0 liter)
1980 hatchback, 1981 thru 1983	0.8 US qt (0.8 liter)

Braking system

Power brake pushrod-to-master cylinder clearance

1979 thru 1982	0.004 to 0.020 in
1983	0.004 to 0.012 in

Electrical system

Battery (1982)	60 amp/hour at 20 hour rate
Alternator pulley ratio (except 1979)	2.4 : 1

Suspension and steering

Front wheel bearing preload

1979 and 1980	0.33 to 1.32 lb (0.15 to 0.60 kg)
1981 thru 1983	0.99 to 1.43 lb (0.45 to 0.65 kg)

Kingpin inclination

1979 and 1980 hatchback	8° 45'
1979 and 1980 wagon	8° 30'
1981	8° 17'
1982 and 1983	8° 43'

Camber

1979 and 1980 hatchback	0° 45' ± 30'
1979 and 1980 wagon	1° 00' ± 30'
1981	1° 13' ± 30'
1982 and 1983	0° 47' ± 30'

Caster

1979 and 1980 hatchback	1° 40' ± 45'
1979 and 1980 wagon	1° 45' ± 45'
1981	1° 32' ± 45'
1982 and 1983	1° 36' ± 45'
Toe-in	0 to 0.24 in (0 to 6 mm)

Torque wrench settings

Rear suspension — station wagons	lbf ft	Kgf m
U-bolt	23 to 34	3.2 to 4.7
Spring pin nut	23 to 34	3.2 to 4.7
Spring pin-to-frame bracket	12 to 17	1.6 to 2.3
Shackle pin nuts	23 to 34	3.2 to 4.7

Wheels		
Wheel lug bolts (1983 models)	65 to 87	9.0 to 12.0

---

3 Engine

General information

The engines used on vehicles produced between 1979 and 1983 are of the same basic design as those described in Chapter 1. The 1979 and 1980 models used a 1415 cc engine, while the 1981 thru 1983 models used a 1490 cc engine. The major differences between these engines are the bore and stroke (see Specifications). There may be minor design changes between components of the different sized engines but repair procedures are essentially the same as described in Chapter 1.

---

4 Fuel, exhaust and emission control systems

Carburetor — general information

The carburetors used on vehicles produced between 1979 and 1983 are basically the same as the one described in Chapter 3. The major difference being the introduction of an electrically operated automatic choke on 1979 vehicles sold in the US. From 1980 through 1983, all vehicles used this type of choke. To comply with emission control standards, 1981 thru 1983 vehicles also have an accelerator switch, a slow fuel cut solenoid and a tamper-proof mixture adjustment screw. See the accompanying illustrations for minor component differences.

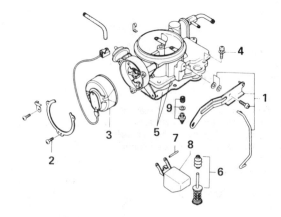

Fig. 13.1 Typical carburetor air horn with an automatic choke — exploded view (Sec 4)

1 Accelerator pump lever/rod/screw
2 Screw/bi-metal cover plate
3 Bi-metal cover
4 Screw
5 Air horn/gasket
6 Accelerator pump piston/boot
7 Float retaining pin
8 Float
9 Needle valve

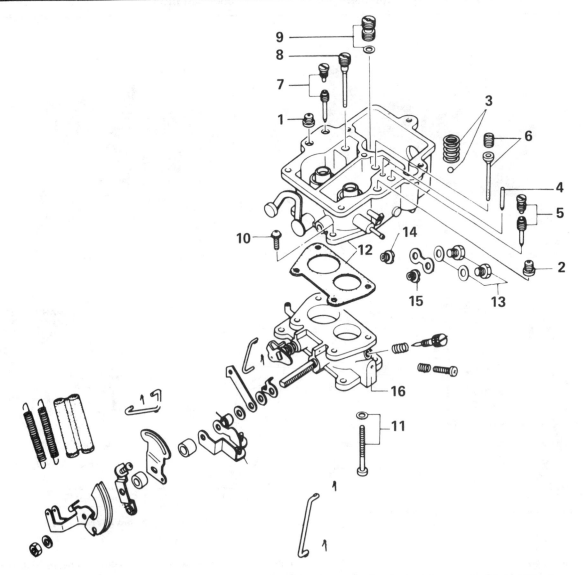

Fig. 13.2 Typical main body and throttle body of carburetor with automatic choke — exploded view (Sec 4)

1 Secondary slow air bleed	5 Primary slow jet	9 Power valve	13 Plug
2 Primary slow air bleed	6 Primary main air bleed	10 Screw	14 Secondary main jet
3 Accelerator pump spring/inlet check ball	7 Secondary slow jet	11 Screw	15 Primary main jet
4 Injector weight	8 Secondary main air bleed	12 Main body/gasket	16 Throttle body

Fig. 13.3 Slow fuel cut solenoid, spring and plunger (Sec 4)

Fig. 13.4 Location of the accelerator switch on the carburetor (Sec 4)

13

Fig. 13.5 Location of the diaphragm mounting screws (1)
and linkage pin (2) (Sec 4)

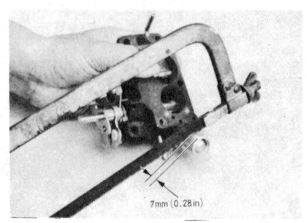

Fig. 13.6 Cutting the mixture adjusting screw shell off (Sec 4)

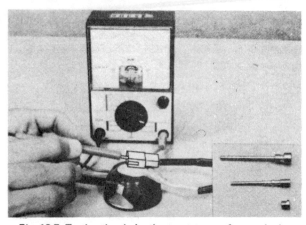

Fig. 13.7 Testing the choke thermostat cover for continuity
with an ohmmeter (Sec 4)

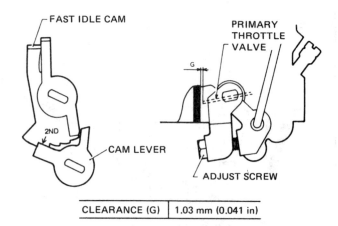

CLEARANCE (G)	1.03 mm (0.041 in)

Fig. 13.8 Adjusting the fast idle cam (Sec 4)

Carburetor — overhaul

1 On carburetors with an automatic choke, disconnect the electrical
lead to the choke. On 1979 and 1980 models, unscrew the bi-metal
cover plate and lift off the bi-metal cover. 1981 thru 1983 models have
the cover plate riveted on. Therefore, to remove the choke cover and
the choke thermostat cover, the rivets must be drilled out using a 1/8 in
diameter drill bit.

2 For 1981 thru 1983 models, continue disassembling the carburetor
as described in Chapter 3. After removing the accelerator pump spring
discharge weight (outlet valve), remove the slow fuel cut solenoid and
gasket.

3 Disconnect the spring and remove the screws attaching the accelerator
switch to the carburetor. Pry open the bracket holding the electrical
lead and remove the switch assembly.

4 Remove the screws attaching the diaphragm to the carburetor,
disconnect the linkage pin and remove the diaphragm.

5 To remove the mixture adjustment screw, either cut the shell open
with a hacksaw 0.28 in (7 mm) from the outside edge or pull out on the
whole shell while turning it. Count the number of turns required to
seat the mixture adjustment screw so it can be reinstalled to approximately the same setting.

6 Be careful not to submerge any plastic or electrical parts in carburetor
cleaner.

7 Inspect the thermostat cover for continuity by connecting an ohmmeter to the choke heater body and the coupler. If there is no
continuity, replace the choke heater with a new one.

8 Install a new shell, spring and mixture adjustment screw in the
throttle body. Turn the screw in the same number of turns it took to
remove it. Do not install the blind cap until the final adjustments have
been made as described in Chapter 3.

9 When installing the main body to the throttle body, position the

mixture adjustment screw shell so it will not interfere with the two
mating surfaces.

Carburetor — adjustment (except 1979 Canada vehicles)

Fast idle cam

10 Remove the bi-metal cover. Close the choke valve with your finger
and check that the fast idle cam is on the second step. Turn the adjusting screw until the primary throttle valve clearance is as specified.

Choke valve opening angle

11 With the fast idle cam still on the second step, check the clearance
between the choke valve and the wall of the choke bore. Bend the
fast idle cam for minor adjustments and the choke rod for major
adjustments.

Choke diaphragm

12 Place the fast idle cam on the first step. Apply vacuum to the choke
diaphragm (about 15.7 in Hg [400 mm Hg]). Measure the choke valve
clearance. If adjustment is necessary, bend the choke lever.

Unloader

13 Completely close the choke valve. Next, open the primary throttle
valve as far as possible.

14 Measure the choke valve clearance in its bore. Bend the tab in the
housing as required to get the specified clearance.

Secondary throttle valve

15 Install the bi-metal cover on the choke housing. For 1979 and 1980
vehicles, connect the bi-metal hook to the choke shaft lever and turn
the thermostatic cover to see if the choke valve operates correctly.

16 Align the center index mark of the bi-metal cover with the center
mark on the choke housing and tighten the attaching screws.

17 On 1981 thru 1983 vehicles, install the thermostatic cover on the

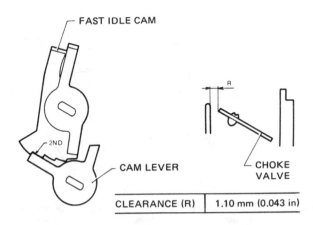

CLEARANCE (R)	1.10 mm (0.043 in)

Fig. 13.9 Adjusting the choke valve opening angle (Sec 4)

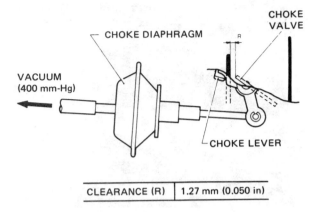

CLEARANCE (R)	1.27 mm (0.050 in)

Fig. 13.10 Apply vacuum to the choke diaphragm and measure the choke valve clearance (Sec 4)

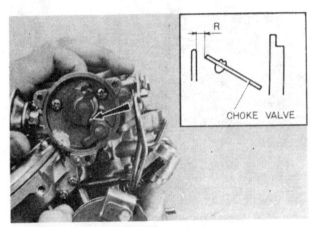

Fig. 13.11 The unloader is adjusted by bending the tab (arrowed) in the choke housing (Sec 4)

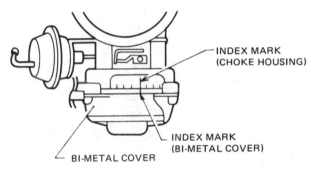

Fig. 13.12 Align the index marks before securing the bi-metal cover (Sec 4)

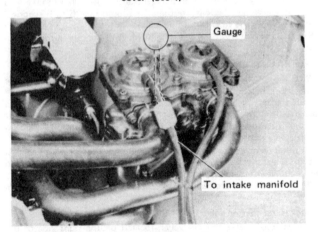

Fig. 13.13 Connect a vacuum gauge as shown to test the vacuum delay valves (Sec 4)

choke housing so the set pin is aligned with the alignment hole. Rivet the cover on with a pop rivet gun.

18 Open the primary throttle valve until the secondary throttle valve begins to open. Measure the clearance between the wall of the throttle bore and the primary throttle valve. If necessary, bend the connecting rod until the correct clearance is obtained.

Emission control system — general information

Most of the emission control devices described in Chapter 3 are used on vehicles produced between 1979 and 1983. Some devices that were only used in certain geographic areas are now used on all cars sold in North America. Also, many new devices are incorporated in the emission control systems for vehicles made between 1979 and 1983. To be sure what emission control devices your car should have, check the emission control service label located in the engine compartment.

Emission control system — testing and adjustment

Air switching valve (1981 thru 1983)

19 The air switching valve is connected to the air control valve, mounted on the left shock tower.

20 Disconnect the air hoses from the outlet on the front of the air control valve (A) and the outlet on the bottom of the air switching valve (B).

21 Disconnect the vacuum hose(s) from the relief valve(s) (1981 and 1983 vehicles have two relief valves). Disconnect the vacuum line from the air switching valve and plug it.

22 With the engine idling, attach a hose, with manifold vacuum, to the number one relief valve. Feel the two air control valve outlets to be sure no air is flowing.

23 Attach another hose, with intake manifold vacuum, to the air

switching valve diaphragm. Place your finger over the air control valve outlet (B) to be sure air is flowing. **Note:** *On 1983 models, apply vacuum to both relief valves and make sure air flows out of air control valve outlet A.*

Vacuum delay valves (1980 thru 1983)

24 Connect a vacuum gauge to the side of the vacuum delay valve that the arrow points away from.

25 Apply intake manifold vacuum to the other end of the vacuum delay valve with the engine idling. Note the reading on the vacuum gauge.

26 Disconnect the hose used to apply the intake manifold vacuum and see how many seconds it takes before the vacuum gauge drops 11.81 in

13

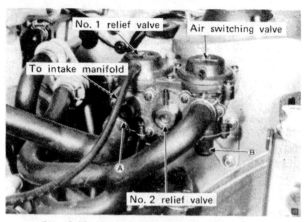

Fig. 13.14 Air control valve components (Sec 4)

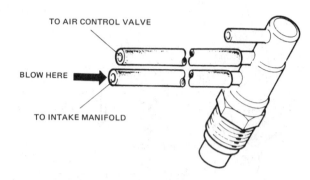

Fig. 13.15 Test the 1981 thru 1983 radiator water thermo valves by blowing into the bottom port and checking for air flow at various temperatures (Sec 4)

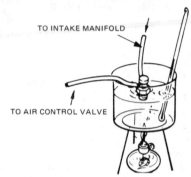

Fig. 13.16 Testing the upper intake manifold water thermo valve (Sec 4)

Fig. 13.17 Location of the No. 2 three-way solenoid valve on the 1980 models (Sec 4)

Hg (300 mm Hg). Vehicles with manual transmissions should take 4 to 6 seconds while vehicles with automatics should take 6 to 9 seconds. On 1980 vehicles the time should be 7.5 seconds.

27 Replace the valve with a new one if it doesn't operate in the specified time.

Radiator water thermo valve (1980 only)

28 With the engine below 64°F (18°C), disconnect the vacuum hose (white) from the air control valve.

29 With the engine idling, check for vacuum in the hose. California vehicles with automatic transmissions should not have vacuum until the coolant temperature reaches 64°F (18°C). There should be vacuum present above that temperature.

30 All other vehicles produced in 1980 should have the opposite results. That is, they should have vacuum below 64°F (18°C) and have no vacuum above that temperature.

31 If the valve needs replacing, follow the instructions in the next subsection.

Radiator water thermo valve (1981 thru 1983)

32 The radiator water thermo valve, located on the right side of the radiator, must be removed to test it.

33 With the engine cool, drain the coolant into a clean container so it can be reused. Disconnect and label the vacuum hoses to ease reassembly.

34 Remove the valve and place a vacuum hose on the bottom (closest to threads) port.

35 Immerse the threaded part of the valve in a pot of water and gradually heat the water. Place a thermometer in the water. When the temperature reaches approximately 84°F (21°C), blow into the hose connected to the bottom port. If air flows out of the valve at 84°F or above, the valve is working correctly.

36 Apply thread sealing tape to the threads of the valve and reinstall it. Reconnect the hoses and fill the radiator with coolant.

Upper intake manifold water thermo valve (1981 thru 1983)

37 The upper intake manifold water thermo valve is connected to the front of the intake manifold.

38 Remove the valve and blow air through the hose that goes to the intake manifold to make sure the air passes through the valve.

39 Place the threaded part of the valve in a pot of water along with a thermometer. Gradually heat the water and watch the temperature.

40 Blow into the hose that leads to the intake manifold. Air should not flow after the temperature gets above 138°F (59°C). If air flows, the valve should be replaced with a new one.

No. 2 three-way solenoid valve (1980 only)

41 This valve is located on the right shock tower on the same bracket as the No.3 three-way solenoid valve on California cars with automatic transmissions.

42 Disconnect the brown vacuum hose from the No.2 three-way solenoid valve and the white vacuum hose from the No. 2 EGR valve.

43 Unplug the yellow/black wire from the No.2 three-way solenoid valve. Use a jumper wire and ground the green terminal in the coupler.

44 Blow into the white vacuum hose. With the ignition switch On, air should come out the valve air filter and with the ignition switch Off, air should flow through the brown port.

No. 2 three-way solenoid valve (1981 thru 1983)

45 The No.2 three-way solenoid valve is mounted to the radiator support on the right side.

46 Disconnect the green vacuum hose from the No.2 three-way solenoid valve and the yellow vacuum hose from the No.1 water thermo valve.

47 Unplug the green wire and ground it.

48 Turn the ignition switch On and blow through the yellow vacuum hose. Check that air comes out the other port.

49 Turn the ignition switch Off and blow through the yellow vacuum hose. Air should flow through the air filter of the valve (on the back side).

50 If the valve doesn't work properly, replace it with a new one.

No. 3 three-way solenoid valve (1983 A/T only)

51 The No. 3 three-way solenoid valve is mounted near the No. 2 valve.

52 Disconnect the wires from the valve and blow into port A (the end

Fig. 13.18 Location of the No. 2 three-way solenoid valve on the 1981 thru 1983 models (Sec 4)

Fig. 13.19 Location of the vacuum amplifier (Sec 4)

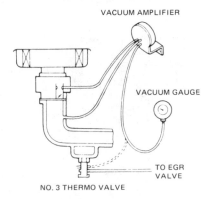

Fig. 13.20 Connect a vacuum gauge as shown to test the vacuum amplifier (Sec 4)

Fig. 13.21 To test the lower intake manifold water thermo valve, disconnect the hose from the EGR valve (right) and on some models, the hose from the purge valve (left) (Sec 4)

port); air should come out of port B (the side port).

53 Using jumper wires, apply battery voltage to the yellow/black wire terminal on the solenoid valve. Ground the brown/red wire terminal.

54 Blow into port A again; air should now come out of port C (the filter end — opposite port A).

Vacuum switch (1983 A/T only)

55 The vacuum switch is attached to the inner fender panel near the battery.

56 Disconnect the wires from the switch.

57 Connect an ohmmeter or self-powered test light to the switch terminals (negative lead to the black wire terminal, positive lead to the brown/red wire terminal).

58 Apply vacuum to the switch port. At a vacuum of 10 to 25 mm Hg, continuity should exist. When a vacuum of 150 mm Hg or greater is applied, no continuity should exist.

Vacuum amplifier

59 The vacuum amplifier is mounted to the right side of the engine compartment, between the battery and shock tower.

60 Disconnect the black vacuum hose from the water thermo valve and connect a vacuum gauge to it.

61 With the engine idling, disconnect the white vacuum hose to the carburetor and read the vacuum gauge. It should be at 2.0 ± 0.2 in-Hg (50 ± 5 mm Hg).

62 Reconnect the white vacuum hose and raise the engine speed to 3500 rpm. The gauge should read 3.54 in Hg (90 mm Hg).

63 Replace the vacuum amplifier with a new one if you don't get the correct vacuum readings.

Lower intake manifold water thermo valve

64 This water thermo valve is mounted to the bottom of the intake manifold.

65 Disconnect the vacuum hose(s) from the EGR valve and the No. 1 purge control valve that connect to the lower water thermo valve.

66 Run the engine at 1500 rpm and feel the hose(s) to be sure no vacuum is evident (1979 and 1980 models need only the hose to the EGR control valve disconnected).

67 Warm the engine to 138°F (59°C) and test the hose(s) for vacuum again. If there is no vacuum at this temperature, replace the water thermo valve with a new one.

68 If you must replace the valve, allow the engine to cool down and drain the coolant into a clean container. Note where the vacuum hoses connect.

69 Wrap the threads of the new valve with thread sealing tape before installing it in the intake manifold. Connect the vacuum hoses and re-fill the radiator.

Front catalytic convertor (1981 thru 1983)

70 The front catalytic convertor is mounted directly to the exhaust manifold. If it clogs or wears out it must be replaced.

71 Disconnect the spark plug leads and the hot air hose to the air cleaner.

72 Disconnect the clamps and hoses to the air injection check valves, then remove the check valves.

73 Unbolt and remove the air injection manifold bracket and the exhaust manifold heat insulator.

74 Raise the vehicle and support it with jackstands.

75 Unbolt the heat insulator and disconnect the hangers from the exhaust pipe.

13

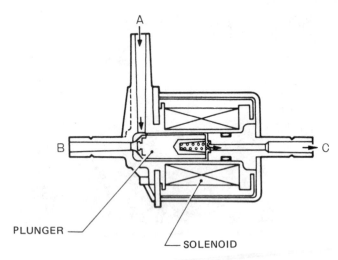

Fig. 13.22 Test the 1980 No. 1 three-way solenoid valve by blowing into port (A) (Sec 4)

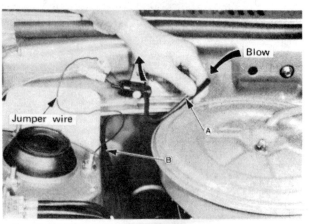

Fig. 13.23 Test the 1981 thru 1983 No.1 three-way solenoid valve by blowing into hose (A) with the ignition switch On, then with the ignition switch Off (Sec 4)

Fig. 13.24 Location of the slow fuel cut solenoid (Sec 4)

Fig. 13.25 Adjusting the accelerator switch (Sec 4)

Fig. 13.26 Location of the altitude compensator (Sec 4)

76 Remove the exhaust pipe-to-catalytic convertor attaching nuts from the front side first, then from the rear side. Detach the exhaust pipe.
77 Lower the vehicle and disconnect the manifold-to-catalytic convertor nuts.
78 Lift out the catalytic convertor.
79 Installation is the reverse of the removal procedure.

No. 1 three-way solenoid valve (1980 only)
80 Disconnect the vacuum hose leading from the intake manifold (A) to this valve.
81 Disconnect the other two vacuum hoses from the No.1 three-way solenoid valve. Unplug the black/white wire from the valve.
82 Blow into the hose from the intake manifold (A) and make sure that air comes out of the port farthest from the one you are blowing into (C).
83 Run a jumper lead from the positive battery terminal to the black/white terminal on the valve.
84 Blow into the vacuum hose (A) again. The air should come out the port nearest the one you are blowing into (B).

No.1 three-way solenoid valve (1981 thru 1983)
85 The No.1 three-way solenoid valve is mounted to the right side of the fire-wall.
86 Disconnect the vacuum hose that connects the servo diaphragm to the No.1 three-way solenoid valve (A) from the servo diaphragm.
87 Disconnect from the solenoid valve the vacuum hose that leads to the intake manifold (B).
88 Unplug the blue wire from the coupler and ground it using a jumper wire.
89 Turn the ignition switch On and blow through the first hose(A) that was disconnected. Air should flow through the air filter of the valve.
90 Turn the ignition switch Off and blow through the same hose. Air

should flow from the port the other vacuum hose (B) connects to.
91 Replace the solenoid valve with a new one if it doesn't operate properly.

Engine speed unit (1979 and 1980)
92 The engine speed unit is attached to the driver's side kick panel, under the dash.
93 To check the operation of the speed unit, attach the leads of a voltmeter to the terminals of the three-way solenoid valve (No. 3 three-way

solenoid valve on 1980 models). Leave the wires attached to the solenoid valve.

94 Start the engine and increase the speed to 2000 rpm. Slowly decrease the engine speed and note when the voltmeter indicates current flow in the circuit. It should be at 1600 to 1800 rpm for 1979 California vehicles with an automatic transmission and all 1980 vehicles and 1400 to 1600 rpm for Canadian vehicles with a manual transmission.

95 Slowly increase the engine speed and note when current flow stops. It should occur when the engine speed is 150 to 250 rpm greater than in Paragraph 94. If not, replace the engine speed unit with a new one.

Engine speed unit (1981 and 1982)

96 The procedure is the same as for 1979 and 1980 models, but the voltmeter leads must be attached to the No. 1 three-way solenoid valve terminals.

97 Current flow should occur when the engine speed decreases to 1430 to 1570 rpm and should cease when it is increased to 2000 to 2200 rpm.

Engine speed unit (1983 only)

98 Have the engine speed unit checked by a dealer service department (the procedure is quite complicated).

Slow fuel cut solenoid valve (1981 thru 1983)

99 The slow fuel cut solenoid valve is connected directly to the carburetor on the right front side.

100 To test it, simply apply power from the battery to the valve terminal and listem for a clicking sound. If there is no clicking, replace the solenoid valve with a new one.

Accelerator switch

101 The accelerator switch is mounted on the back side of the carburetor.

102 Disconnect the electrical coupler to the switch and connect an ohmmeter to the terminals on the switch.

103 With the engine at operating temperature, check for continuity at the switch when the accelerator pedal is depressed (there should be no continuity when the pedal is released).

104 To adjust the accelerator switch, the engine must be at operating temperature and idling at the specified rpm. Completely loosen the accelerator switch adjusting screw.

105 Slowly tighten the adjusting screw until a clicking sound is heard. After it clicks, tighten the adjusting screw another 1-1/4 to 1-1/2 turns.

Altitude compensator

106 The altitude compensator is mounted to the front of the right shock tower.

107 Disconnect the vacuum hose from the air cleaner that leads to the altitude compensator. Blow into the hose. Air should pass at altitudes above 2295 feet (700 m) but not at altitudes below 984 feet (300 m).

No. 1 purge control valve (1980 thru 1983)

108 The No. 1 purge control valve is mounted to the top of the canister located on the right side of the engine compartment.

109 Disconnect the vacuum hose leading to the intake manifold, and the hose leading to the lower intake manifold water thermo valve, from the No.1 purge control valve. Plug the hose connected to the intake manifold.

110 Place a piece of vacuum hose on the purge valve port that normally leads to the lower manifold water thermo valve. With the engine idling, blow lightly into the hose. Air should not pass.

111 Unplug the hose going to the intake manifold and reconnect it to the other purge valve port. Blow into the vacuum hose in the same manner as in the previous step. Air should now pass.

No. 2 purge conrol valve (1981 thru 1983)

112 The No. 2 purge control valve on 1982 and 1983 models is located on top of the canister, on the side nearest the battery.

113 Disconnect the vacuum hose leading to the carburetor from the No. 2 purge control valve.

114 Disconnect the evap hose from the purge control valve and connect a piece of vacuum hose to the port. Blow into the hose to be sure air doesn't pass through the purge valve.

115 Reconnect the hose going to the carburetor and start the engine. With the engine running at 1500 rpm, blow into the vacuum hose. Air should now pass freely.

Air vent solenoid valve

116 The air vent solenoid valve is connected to the air vent hose

coming from the center of the canister.

117 Disconnect the hose from the canister and blow through it. Air should pass through the air vent solenoid valve.

118 Turn the ignition switch On and blow through the hose again. Make sure air does not pass through.

Evaporative shutter valve

119 The evaporative shutter valve is mounted inside the air cleaner snorkel.

120 Start the engine and allow it to idle while you remove the air cleaner element.

121 Use a mirror to see inside of the snorkel or simply feel if the valve is fully open.

122 Unplug the vacuum hose from the diaphragm on top of the snorkel. The evaporative shutter valve should close completely.

---

5 Ignition system

General information

1979 vehicles produced for Canada use a breaker-type distributor. Of these vehicles, the manual transmission equipped models have a dual point distributor.

All cars made between 1980 and 1983 have a transistorized ignition, as do vehicles produced in 1979 for the USA.

---

6 Manual transmission

General information

Manual transmissions used on these vehicles have remained essentially unchanged. The only changes are the addition of a neutral switch to the 1981 thru 1983 models, and the use of spring pins instead of bolts to retain the shift forks and shift rod ends on 1981 thru 1983 models.

Neutral switch — inspection

1 Raise the vehicle and secure it safely on jackstands.

2 Disconnect the electrical connector from the switch.

3 Using an ohmmeter, check for continuity in the switch when the shift lever is in Neutral (there should be no continuity when the shift lever is in any other position).

Transmission — overhaul (1981 thru 1983)

4 The transmission overhaul procedure is basically the same as described in chapter 6A, except that the neutral switch must be unscrewed from the extension housing. Also, to remove the shift rod ends and

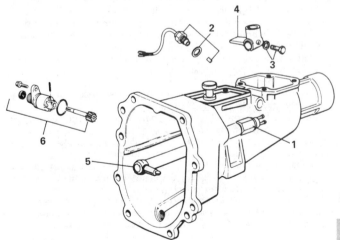

Fig. 13.27 Components of the rear extension housing
(1981 thru 1983 4-speed) (Sec 6)

1 Backup light switch
2 Neutral switch/washer/pin
3 Bolt/washer
4 Control lever end

5 Control lever
6 Speedometer driven gear
   assembly/bolt

13

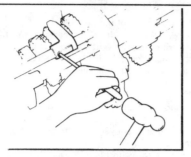

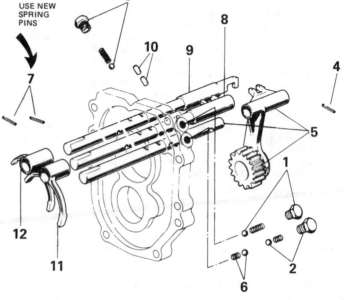

Fig. 3.28 The bearing housing and selector mechanisms (1981 thru 1983); drive the spring pins out with a hammer and punch (insert) (Sec 6)

1	Spring cap bolt/ spring/locking ball	6	Spring/locking ball
2	Spring cap bolt/ spring/locking ball	7	Spring pin
		8	Shift rod (3rd & 4th)
3	Spring cap bolt/ spring/locking ball	9	Shift rod (1st & 2nd)
		10	Interlock pin
4	Spring pin	11	Shift fork (3rd & 4th)
5	Shift fork (Reverse)/rod/ reverse idler gear	12	Shift fork (1st & 2nd)

the shift forks, the spring pins must be driven out with a hammer and an appropriate sized punch. **Note:** *Always replace the spring pins with new units during reassembly.*

7 Propeller shaft

Universal joints — replacement (1982 and 1983)

The universal joints used on 1982 and 1983 models are not replaceable. If the universal joints fail or wear out, the entire propeller shaft must be replaced with a new one.

8 Braking system

Rear brake shoes — removal and refitting (1982 and 1983)

1 The rear brakes on 1982 models are equipped with self adjusters.
2 Procede with the brake shoe removal as described in Chapter 9, Section 14, until you reach paragraph 9.
3 Remove the retaining spring and guide pin from the front brake shoe.
4 Lift out the brake shoe, the self adjuster wheel and the self adjuster operating wheel.
5 Remove the retaining spring and guide pin from the rear brake shoe.
6 Disconnect the parking brake cable and lift off the brake shoe.
7 Remove the retaining clip from the parking brake lever.
8 Unhook the spring to the self adjusting pawl lever. Separate the brake shoe, pawl lever and parking brake lever.
9 Assemble the parking brake lever, pawl lever and spring on the new brake shoe and attach the retaining clip.
10 Connect the parking brake cable to the lever and place the rear brake shoe in position, installing the guide pin and retaining spring.
11 Spread some high-temperature grease on the threads of the self adjuster and in the forks where the adjuster contacts the brake shoe. Screw the self adjuster together completely.
12 Put the self adjuster operating strut and adjusting wheel in place and install the front brake shoe with its guide pin and retaining spring.
13 Continue reassembly by reversing the disassembly sequence.

Rear brake — adjustment (1982 and 1983)

14 The rear brakes on these models are self-adjusting, but after replacing the shoes, or if the length of the adjusitng rod has been changed during another service operation, a manual adjustment will be needed.
15 Block the front wheels. Raise the rear end and support it with jackstands.
16 Remove the rear wheels, then fully release the parking brake.
17 Insert a screwdriver or adjusting tool through the hole in the brake

drum and turn the star wheel until the wheel is locked.
18 Remove the plug in the backing plate and insert a screwdriver into the hole.
19 Push on the self adjuster pawl lever with the screwdriver in the backing plate. Back off the star wheel three or four notches at the same time.
20 Install the wheel and check that it rotates without dragging.
21 Adjust the brake on the other side and install the wheel.
22 Place the backing plate plug back into position and adjust the parking brake as described in Chapter 9.

Vacuum servo unit — removal and refittiing

23 If a vacuum servo unit other than the original is being installed, such as a commercially rebuilt component, check the clearance between the back of the master cylinder piston and the power brake pushrod.
24 Using a depth micrometer, or vernier calipers, measure the distance from the piston to the master cylinder mounting flange.
25 Next, measure the distance from the end of the power brake pushrod to the surface on the vacuum servo unit that the master cylinder mounting flange is in contact with when installed.
26 Subtract the two measurements to get the clearance. If the clearance is more or less than specified, turn the pushrod (after loosening the locknut) until the clearance is correct. Be sure to tighten the locknut on the pushrod when the adjustment is complete.

9 Electrical system

Battery — maintenance

Some later models are equipped with batteries that have a built-in hydrometer. If the indicator on top of the battery is blue, the battery is normal. If not, the electrolyte is low or the battery is undercharged.

Alternator — general information (1982 and 1983)

The alternator used on these models has an IC regulator built into it. Servicing the alternator is basically the same as described in Chapter 10.

Alternator — dismantling, testing, servicing and reassembly (1982 and 1983)

1 Remove the alternator and disassemble it as described in Chapter 10, Section 9. After the rectifier has been removed, use a soldering iron to disconnect the IC regulator and brush holder from it.
2 Continue with disassembly and testing as outlined in Chapter 10, Section 9.
3 Be sure to carefully resolder the brush holder and IC regulator to the rectifier before reinstalling the rectifier assembly.

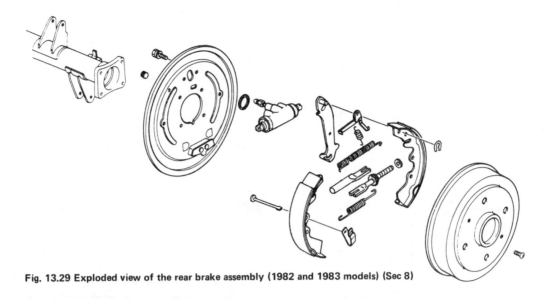

Fig. 13.29 Exploded view of the rear brake assembly (1982 and 1983 models) (Sec 8)

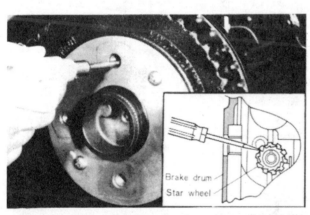

Fig. 13.30 Slip a screwdriver through the hole in the brake drum and turn the star wheel to adjust the brakes (Sec 8)

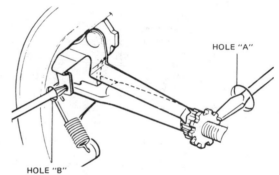

HOLE "A"

HOLE "B"

Fig. 13.31 Push the pawl lever, with a screwdriver, through hole "B" and back off the star wheel through the brake drum, hole "A" (Sec 8)

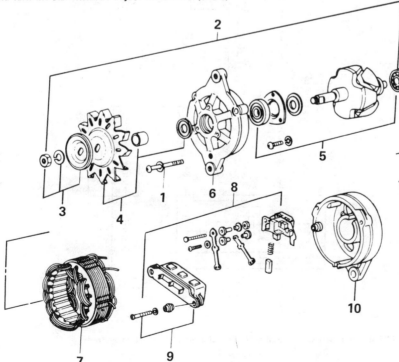

Fig. 13.32 Exploded view of the 1982 and 1983 alternator with IC regulator (Sec 9)

1 Set screw
2 Front bracket assembly
3 Pulley and nut set
4 Fin and spacer
5 Rotor assembly
6 Front bracket
7 Stator
8 Rectifier and regulator
9 Rectifier
10 Rear bracket

13

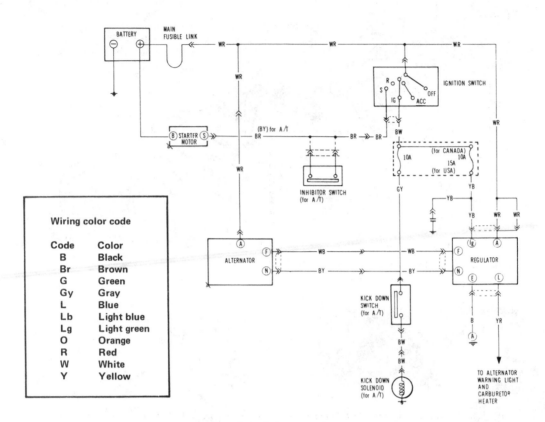

Fig. 13.33 Wiring diagram for 1979,1980 and 1981 charging, starting and kickdown systems

Wiring color code

Code	Color
B	Black
Br	Brown
G	Green
Gy	Gray
L	Blue
Lb	Light blue
Lg	Light green
O	Orange
R	Red
W	White
Y	Yellow

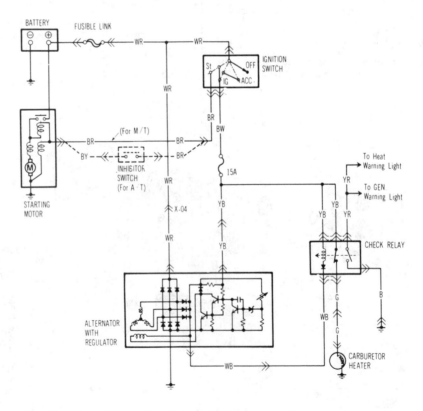

Fig. 13.34 Wiring diagram for 1982 and 1983 charging and starting systems

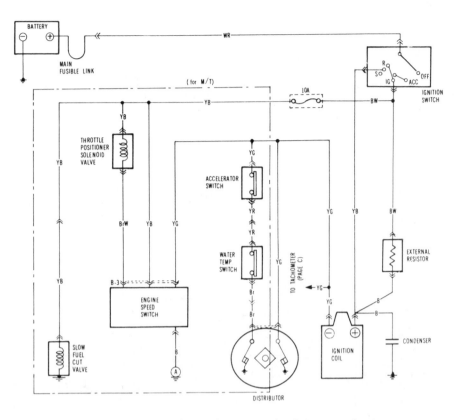

Fig. 13.35 Wiring diagram for 1979 (Canada) ignition and emission control systems

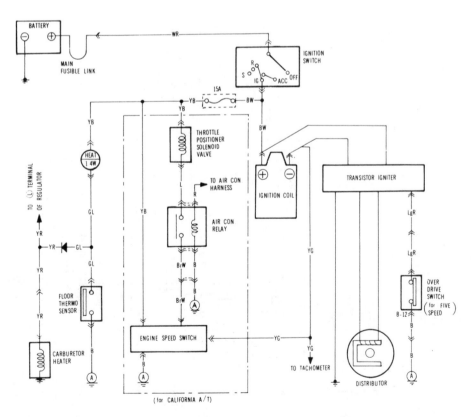

Fig. 13.36 Wiring diagram for 1979 (USA) ignition and emission control systems

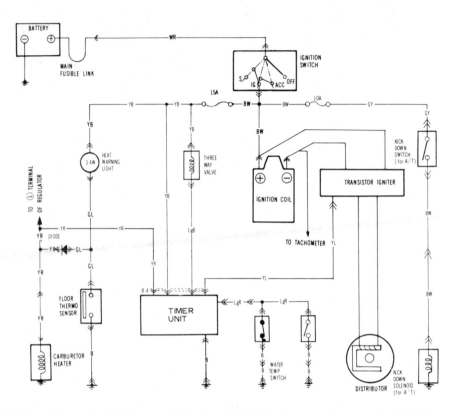

Fig. 13.37 Wiring diagram for 1980 (except California with A/T) ignition, emission control and kickdown systems

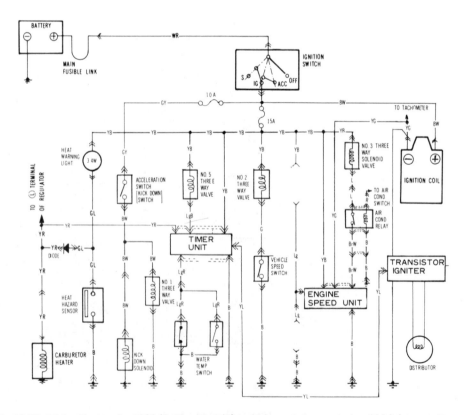

Fig. 13.38 Wiring diagram for 1980 (Calif. with A/T) ignition, emission control and kickdown systems

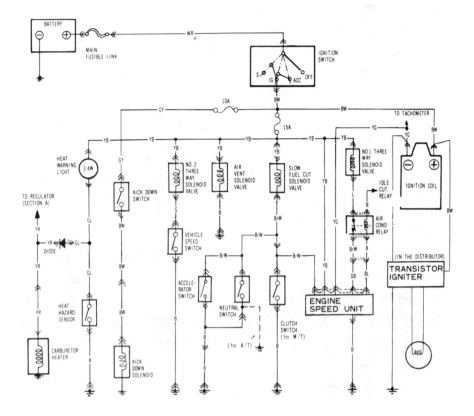

Fig. 13.39 Wiring diagram for 1981 ignition, emission control and kickdown systems

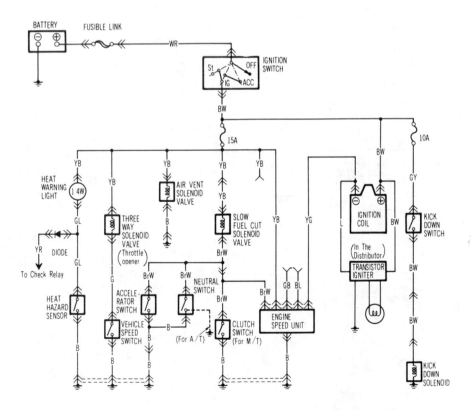

Fig. 13.40 Wiring diagram for 1982 ignition, emission control and kickdown systems

13

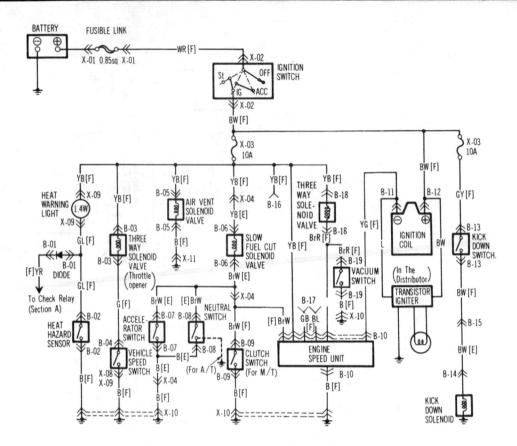

Fig. 13.41 Wiring diagram for 1983 ignition, emission control and kickdown systems

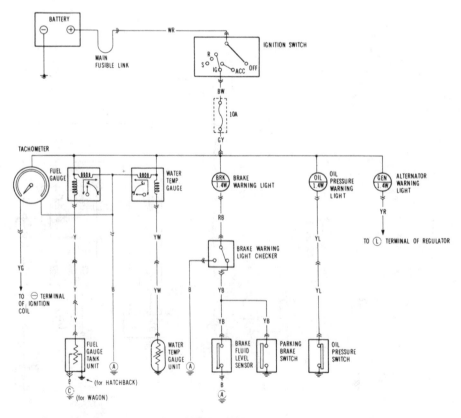

Fig. 13.42 Wiring diagram for 1979 and 1980 meters and warning systems

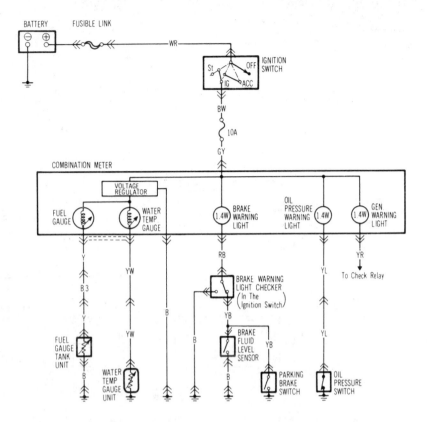

Fig. 13.43 Wiring diagram for 1981 thru 1983 meters and warning systems

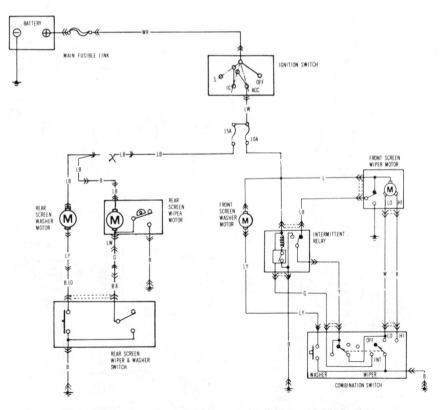

Fig. 13.44 Wiring diagram for 1979, 1980 and 1981 wipers and washer motors

13

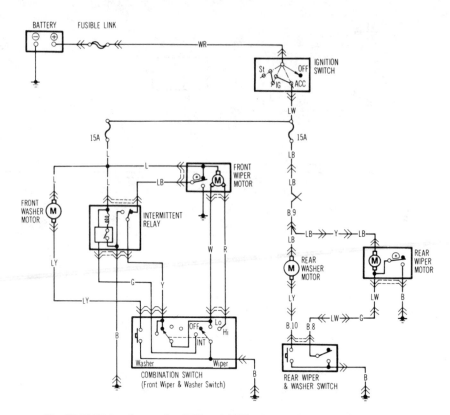

Fig. 13.45 Wiring diagram for 1982 and 1983 wipers and washer motors

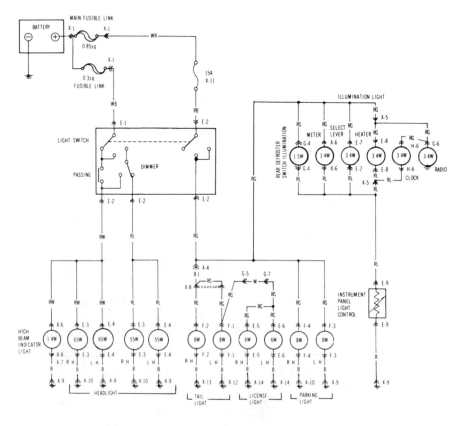

Fig. 13.46 Wiring diagram for 1979, 1980 and 1981 running light systems

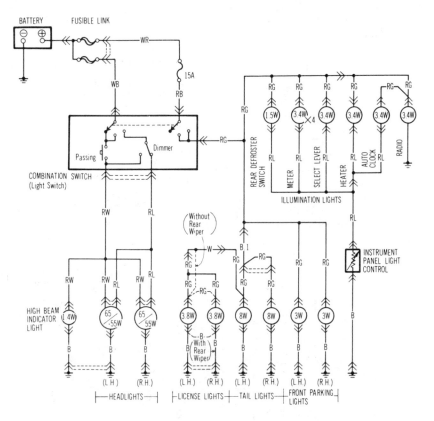

Fig. 13.47 Wiring diagram for 1982 and 1983 running light systems

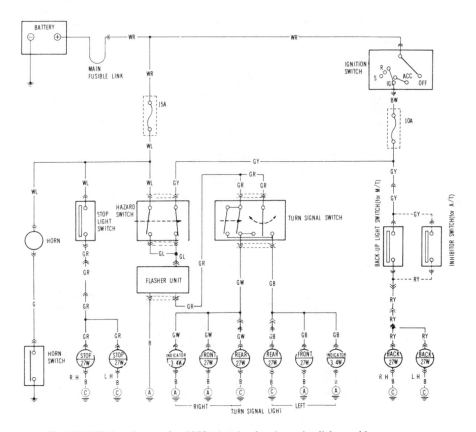

Fig. 13.48 Wiring diagram for 1979 turn signal and warning lights and horn

13

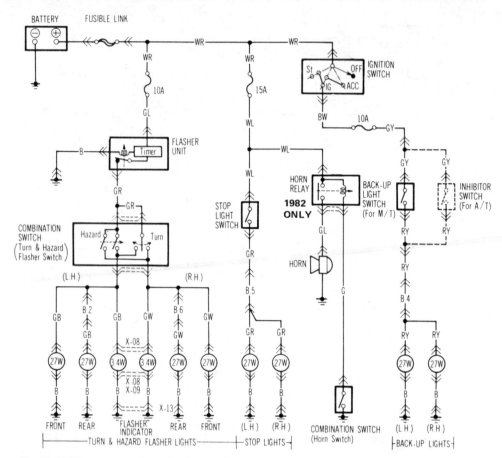

Fig. 13.49 Wiring diagram for 1980 thru 1983 turn signal and warning lights and horn

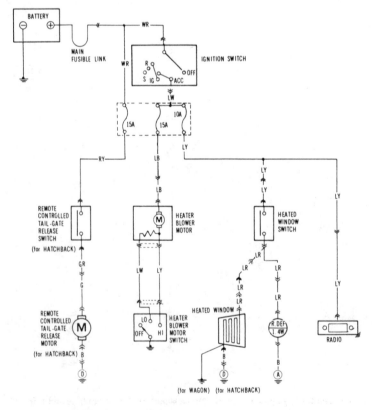

Fig. 13.50 Wiring diagram for 1979 and 1980 rear window defroster, heater, radio and tailgate release

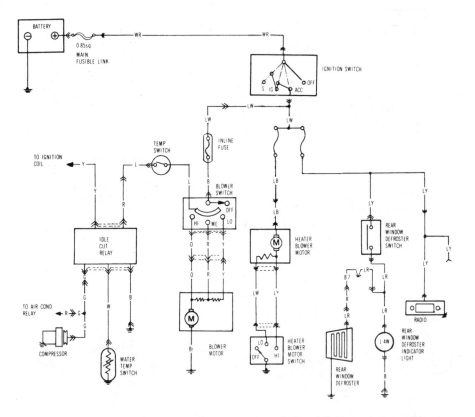

Fig. 13.51 Wiring diagram for 1981 radio, heater, rear window defroster and air conditioner

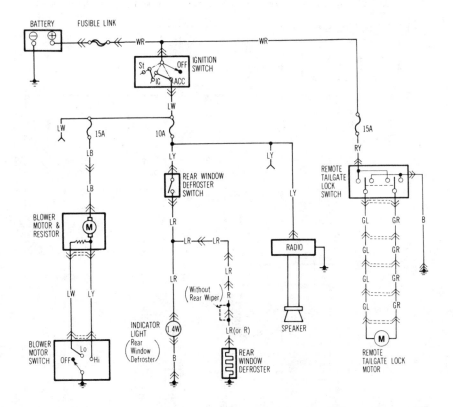

Fig. 13.52 Wiring diagram for 1982 and 1983 radio, rear window defroster, tailgate release and heater

13

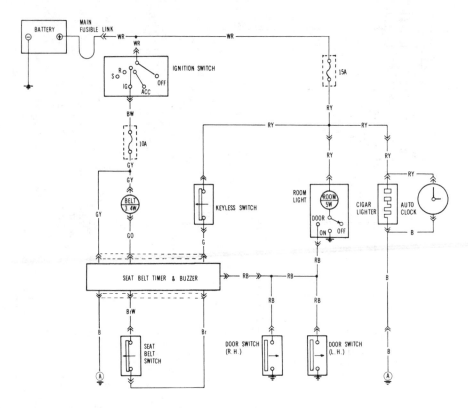

Fig. 13.53 Wiring diagram for 1979 through 1982 clock, dome lamp, seat belt warning system and cigar lighter

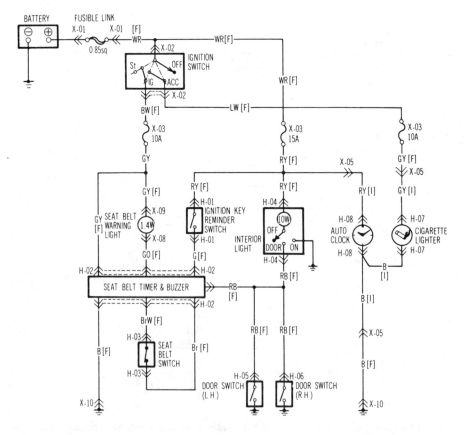

Fig. 13.54 Wiring diagram for 1983 clock, dome lamp, seat belt warning system and cigarette lighter

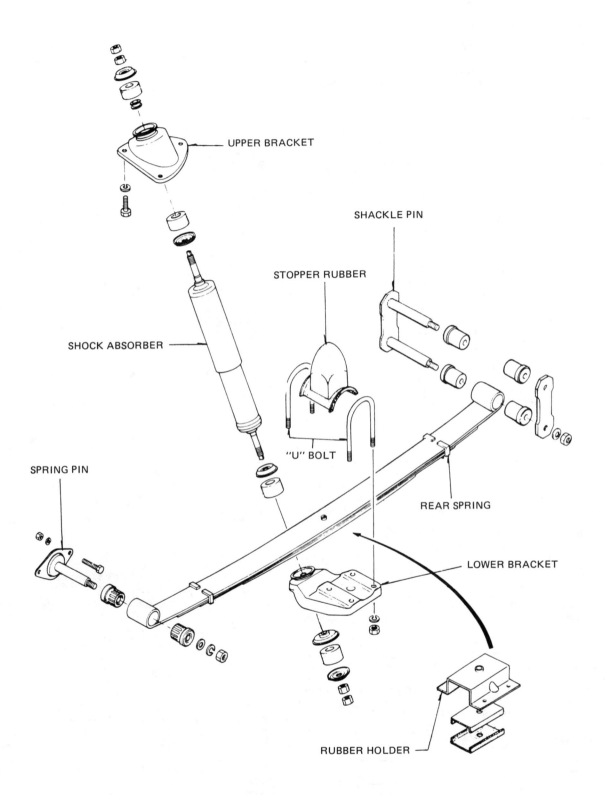

UPPER BRACKET

SHACKLE PIN

STOPPER RUBBER

SHOCK ABSORBER

"U" BOLT

SPRING PIN

REAR SPRING

LOWER BRACKET

RUBBER HOLDER

Fig. 13.55 Exploded view of the rear leaf spring suspension used on station wagons (Sec 10)

13

10 Suspension and steering

Front wheel hub (disc) — reassembly and adjustment

The procedure outlined in Chapter 11, Section 4, is correct, but note that the bearing preload (Paragraph 6) is different for later models. Refer to the Chapter 13 Specifications for the preload value for your particular vehicle.

Rear suspension (wagon) — general information

The rear leaf spring suspension is attached to a solid rear axle assembly. The front of each leaf spring is attached to the frame with a spring pin. At the rear, a shackle is used which allows the spring to flex and change its length slightly while the car is in motion. Dampening is accomplished by telescopic shock absorbers.

Rear shock absorber (wagon) — removal and refitting

1 Raise the rear end of the vehicle and support it under the frame rails with jackstands. Be sure to block the front wheels securely.
2 Remove the bolts that attach the shock absorber upper bracket to the body.
3 Remove the nut and locking nut from the bottom end of the shock absorber.
4 Compress the shock absorber and remove it from the vehicle.
5 Place the shock absorber in a vise and remove the two nuts that attach it to the upper mounting brackets.
6 Installation is the reverse of removal. Tighten the shock absorber nuts until the dimensions shown in the accompanying illustration are obtained.

Rear spring (wagon) — removal and refitting

7 Raise the rear end of the vehicle and support it so that the axle can be lowered. Be sure that the front wheels are blocked securely.
8 Disconnect the lower mounting point of the rear shock absorber.
9 In sequence, remove the U-bolt attaching nuts, the U-bolt, the spring clamp and the plate.
10 Remove the spring pin nut, then remove the two nuts and bolts that secure the spring pin plate to the frame bracket.
11 Remove the spring pin and lower the front of the leaf spring from the vehicle.
12 Remove the shackle pin nuts and shackle plate. **Note:** *It may be easier to remove the shackle plate if the rear muffler is removed first.*
13 When the shackle plate is removed, lower the rear end of the leaf spring from the vehicle.
14 Installation is the reverse of removal. It may help facilitate installation of the shackle pin and spring pin if you apply a soapy water solution to the rubber bushings first.
15 Tighten the U-bolt, the spring pin and the shackle pin to the specified torque.

Steering lock — replacement (1982 and 1983)

16 Remove the steering wheel and the steering column shroud as described in Section 16 of Chapter 11.
17 Disconnect the ignition switch coupler.
18 Use a hacksaw to make grooves in the heads of the bolts that attach the steering lock body to the column shaft. Loosen the bolts with a screwdriver.
19 Install the new steering lock on the column shaft. Tighten the bolts until the heads of the bolts snap off.
20 Reassemble the steering column and steering wheel in the reverse order of disassembly.

11 Bodywork and fittings

Back door (wagon) — removal and refitting

1 The removal of the back door on a wagon is basically the same as removal of the hatchback door described in Chapter 12; note the following differences:
2 The back door interior trim must be removed and the electrical connector and ground wire disconnected.
3 Pull the wiring harness out of the door.
4 If equipped, remove the washer hose and nozzle from the rear door.

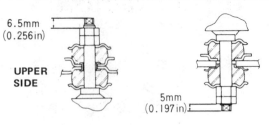

Fig. 13.56 Tighten the rear shock absorbers (wagons only) to the dimensions shown (Sec 10)

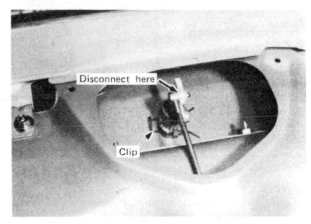

Fig. 13.57 Disconnect the lock rod, then pry off the securing clip to remove the back door lock

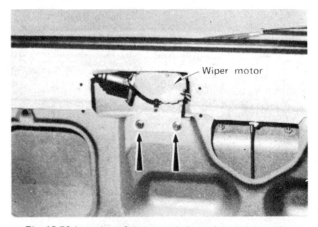

Fig. 13.58 Location of the rear window wiper motor and mounting screws (Sec 11)

5 Remove the mounting bolts and stay dampers as described in Section 11 of Chapter 12.

Back door lock (wagon) — removal and refitting

6 Remove the interior trim from the back door.
7 Disconnect the rod from the lock button.
8 Pry off the clip securing the lock button to the door and remove the button.
9 Remove the bolts securing the door lock to the bottom of the door. Remove the door lock.
10 Installation is the reverse of removal.

Rear window wiper motor — removal and refitting

11 Open the wiper arm cover at its mounting point and remove the attaching nut and rubber bushing.
12 Remove the back door trim, then the wiper hole cover.
13 Disconnect the electrical connector, then remove the wiper motor attaching bolts and lift out the wiper motor.
14 Installation is the reverse of removal.

Conversion factors

Length (distance)

Inches (in)	X	25.4	= Millimetres (mm)	X	0.0394	= Inches (in)
Feet (ft)	X	0.305	= Metres (m)	X	3.281	= Feet (ft)
Miles	X	1.609	= Kilometres (km)	X	0.621	= Miles

Volume (capacity)

Cubic inches (cu in; in³)	X	16.387	= Cubic centimetres (cc; cm³)	X	0.061	= Cubic inches (cu in; in³)
Imperial pints (Imp pt)	X	0.568	= Litres (l)	X	1.76	= Imperial pints (Imp pt)
Imperial quarts (Imp qt)	X	1.137	= Litres (l)	X	0.88	= Imperial quarts (Imp qt)
Imperial quarts (Imp qt)	X	1.201	= US quarts (US qt)	X	0.833	= Imperial quarts (Imp qt)
US quarts (US qt)	X	0.946	= Litres (l)	X	1.057	= US quarts (US qt)
Imperial gallons (Imp gal)	X	4.546	= Litres (l)	X	0.22	= Imperial gallons (Imp gal)
Imperial gallons (Imp gal)	X	1.201	= US gallons (US gal)	X	0.833	= Imperial gallons (Imp gal)
US gallons (US gal)	X	3.785	= Litres (l)	X	0.264	= US gallons (US gal)

Mass (weight)

Ounces (oz)	X	28.35	= Grams (g)	X	0.035	= Ounces (oz)
Pounds (lb)	X	0.454	= Kilograms (kg)	X	2.205	= Pounds (lb)

Force

Ounces-force (ozf; oz)	X	0.278	= Newtons (N)	X	3.6	= Ounces-force (ozf; oz)
Pounds-force (lbf; lb)	X	4.448	= Newtons (N)	X	0.225	= Pounds-force (lbf; lb)
Newtons (N)	X	0.1	= Kilograms-force (kgf; kg)	X	9.81	= Newtons (N)

Pressure

Pounds-force per square inch (psi; lbf/in²; lb/in²)	X	0.070	= Kilograms-force per square centimetre (kgf/cm²; kg/cm²)	X	14.223	= Pounds-force per square inch (psi; lbf/in²; lb/in²)
Pounds-force per square inch (psi; lbf/in²; lb/in²)	X	0.068	= Atmospheres (atm)	X	14.696	= Pounds-force per square inch (psi; lbf/in²; lb/in²)
Pounds-force per square inch (psi; lbf/in²; lb/in²)	X	0.069	= Bars	X	14.5	= Pounds-force per square inch (psi; lbf/in²; lb/in²)
Pounds-force per square inch (psi; lbf/in²; lb/in²)	X	6.895	= Kilopascals (kPa)	X	0.145	= Pounds-force per square inch (psi; lbf/in²; lb/in²)
Kilopascals (kPa)	X	0.01	= Kilograms-force per square centimetre (kgf/cm²; kg/cm²)	X	98.1	= Kilopascals (kPa)
Millibar (mbar)	X	100	= Pascals (Pa)	X	0.01	= Millibar (mbar)
Millibar (mbar)	X	0.0145	= Pounds-force per square inch (psi; lbf/in²; lb/in²)	X	68.947	= Millibar (mbar)
Millibar (mbar)	X	0.75	= Millimetres of mercury (mmHg)	X	1.333	= Millibar (mbar)
Millibar (mbar)	X	0.401	= Inches of water (inH₂O)	X	2.491	= Millibar (mbar)
Millimetres of mercury (mmHg)	X	0.535	= Inches of water (inH₂O)	X	1.868	= Millimetres of mercury (mmHg)
Inches of water (inH₂O)	X	0.036	= Pounds-force per square inch (psi; lbf/in²; lb/in²)	X	27.68	= Inches of water (inH₂O)

Torque (moment of force)

Pounds-force inches (lbf in; lb in)	X	1.152	= Kilograms-force centimetre (kgf cm; kg cm)	X	0.868	= Pounds-force inches (lbf in; lb in)
Pounds-force inches (lbf in; lb in)	X	0.113	= Newton metres (Nm)	X	8.85	= Pounds-force inches (lbf in; lb in)
Pounds-force inches (lbf in; lb in)	X	0.083	= Pounds-force feet (lbf ft; lb ft)	X	12	= Pounds-force inches (lbf in; lb in)
Pounds-force feet (lbf ft; lb ft)	X	0.138	= Kilograms-force metres (kgf m; kg m)	X	7.233	= Pounds-force feet (lbf ft; lb ft)
Pounds-force feet (lbf ft; lb ft)	X	1.356	= Newton metres (Nm)	X	0.738	= Pounds-force feet (lbf ft; lb ft)
Newton metres (Nm)	X	0.102	= Kilograms-force metres (kgf m; kg m)	X	9.804	= Newton metres (Nm)

Power

Horsepower (hp)	X	745.7	= Watts (W)	X	0.0013	= Horsepower (hp)

Velocity (speed)

Miles per hour (miles/hr; mph)	X	1.609	= Kilometres per hour (km/hr; kph)	X	0.621	= Miles per hour (miles/hr; mph)

Fuel consumption*

Miles per gallon, Imperial (mpg)	X	0.354	= Kilometres per litre (km/l)	X	2.825	= Miles per gallon, Imperial (mpg)
Miles per gallon, US (mpg)	X	0.425	= Kilometres per litre (km/l)	X	2.352	= Miles per gallon, US (mpg)

Temperature

Degrees Fahrenheit = (°C x 1.8) + 32   Degrees Celsius (Degrees Centigrade; °C) = (°F - 32) x 0.56

It is common practice to convert from miles per gallon (mpg) to litres/100 kilometres (l/100km), where mpg (Imperial) x l/100 km = 282 and mpg (US) x l/100 km = 235

Safety first!

Regardless of how enthusiastic you may be about getting on with the job at hand, take the time to ensure that your safety is not jeopardized. A moment's lack of attention can result in an accident, as can failure to observe certain simple safety precautions. The possibility of an accident will always exist, and the following points should not be considered a comprehensive list of all dangers. Rather, they are intended to make you aware of the risks and to encourage a safety conscious approach to all work you carry out on your vehicle.

Essential DOs and DON'Ts

DON'T rely on a jack when working under the vehicle. Always use approved jackstands to support the weight of the vehicle and place them under the recommended lift or support points.

DON'T attempt to loosen extremely tight fasteners (i.e. wheel lug nuts) while the vehicle is on a jack — it may fall.

DON'T start the engine without first making sure that the transmission is in Neutral (or Park where applicable) and the parking brake is set.

DON'T remove the radiator cap from a hot cooling system — let it cool or cover it with a cloth and release the pressure gradually.

DON'T attempt to drain the engine oil until you are sure it has cooled to the point that it will not burn you.

DON'T touch any part of the engine or exhaust system until it has cooled sufficiently to avoid burns.

DON'T siphon toxic liquids such as gasoline, antifreeze and brake fluid by mouth, or allow them to remain on your skin.

DON'T inhale brake lining dust — it is potentially hazardous (see *Asbestos* below)

DON'T allow spilled oil or grease to remain on the floor — wipe it up before someone slips on it.

DON'T use loose fitting wrenches or other tools which may slip and cause injury.

DON'T push on wrenches when loosening or tightening nuts or bolts. Always try to pull the wrench toward you. If the situation calls for pushing the wrench away, push with an open hand to avoid scraped knuckles if the wrench should slip.

DON'T attempt to lift a heavy component alone — get someone to help you.

DON'T rush or take unsafe shortcuts to finish a job.

DON'T allow children or animals in or around the vehicle while you are working on it.

DO wear eye protection when using power tools such as a drill, sander, bench grinder, etc. and when working under a vehicle.

DO keep loose clothing and long hair well out of the way of moving parts.

DO make sure that any hoist used has a safe working load rating adequate for the job.

DO get someone to check on you periodically when working alone on a vehicle.

DO carry out work in a logical sequence and make sure that everything is correctly assembled and tightened.

DO keep chemicals and fluids tightly capped and out of the reach of children and pets.

DO remember that your vehicle's safety affects that of yourself and others. If in doubt on any point, get professional advice.

Asbestos

Certain friction, insulating, sealing, and other products — such as brake linings, brake bands, clutch linings, torque converters, gaskets, etc. — contain asbestos. *Extreme care must be taken to avoid inhalation of dust from such products since it is hazardous to health.* If in doubt, assume that they *do* contain asbestos.

Fire

Remember at all times that gasoline is highly flammable. Never smoke or have any kind of open flame around when working on a vehicle. But the risk does not end there. A spark caused by an electrical short circuit, by two metal surfaces contacting each other, or even by static electricity built up in your body under certain conditions, can ignite gasoline vapors, which in a confined space are highly explosive. Do not, under any circumstances, use gasoline for cleaning parts. Use an approved safety solvent.

Always disconnect the battery ground (–) cable *at the battery* before working on any part of the fuel system or electrical system. Never risk spilling fuel on a hot engine or exhaust component.

It is strongly recommended that a fire extinguisher suitable for use on fuel and electrical fires be kept handy in the garage or workshop at all times. Never try to extinguish a fuel or electrical fire with water.

Torch (flashlight in the US)

Any reference to a ''torch'' appearing in this manual should always be taken to mean a hand-held, battery-operated electric light or flashlight. It DOES NOT mean a welding or propane torch or blowtorch.

Fumes

Certain fumes are highly toxic and can quickly cause unconsciousness and even death if inhaled to any extent. Gasoline vapor falls into this category, as do the vapors from some cleaning solvents. Any draining or pouring of such volatile fluids should be done in a well ventilated area.

When using cleaning fluids and solvents, read the instructions on the container carefully. Never use materials from unmarked containers.

Never run the engine in an enclosed space, such as a garage. Exhaust fumes contain carbon monoxide, which is extremely poisonous. If you need to run the engine, always do so in the open air, or at least have the rear of the vehicle outside the work area.

If you are fortunate enough to have the use of an inspection pit, never drain or pour gasoline and never run the engine while the vehicle is over the pit. The fumes, being heavier than air, will concentrate in the pit with possibly lethal results.

The battery

Never create a spark or allow a bare light bulb near a battery. They normally give off a certain amount of hydrogen gas, which is highly explosive.

Always disconnect the battery ground (–) cable *at the battery* before working on the fuel or electrical systems.

If possible, loosen the filler caps or cover when charging the battery from an external source (this does not apply to sealed or maintenance-free batteries). Do not charge at an excessive rate or the battery may burst.

Take care when adding water to a non maintenance-free battery and when carrying a battery. The electrolyte, even when diluted, is very corrosive and should not be allowed to contact clothing or skin.

Always wear eye protection when cleaning the battery to prevent the caustic deposits from entering your eyes.

Mains electricity (household current in the US)

When using an electric power tool, inspection light, etc., which operates on household current, always make sure that the tool is correctly connected to its plug and that, where necessary, it is properly grounded. Do not use such items in damp conditions and, again, do not create a spark or apply excessive heat in the vicinity of fuel or fuel vapor.

Secondary ignition system voltage

A severe electric shock can result from touching certain parts of the ignition system (such as the spark plug wires) when the engine is running or being cranked, particularly if components are damp or the insulation is defective. In the case of an electronic ignition system, the secondary system voltage is much higher and could prove fatal.

Index

HAYNES AUTOMOTIVE MANUALS

ACURA
1776   **Integra & Legend** '86 thru '90

AMC
  Jeep CJ – see JEEP (412)
694   **Mid-size models,** Concord, Hornet, Gremlin & Spirit '70 thru '83
934   **(Renault) Alliance & Encore** all models '83 thru '87

AUDI
615   **4000** all models '80 thru '87
428   **5000** all models '77 thru '83
1117   **5000** all models '84 thru '88

AUSTIN
  Healey Sprite – see MG Midget Roadster (265)

BMW
276   **320i** all 4 cyl models '75 thru '83
632   **528i & 530i** all models '75 thru '80
240   **1500 thru 2002** all models except Turbo '59 thru '77
348   **2500, 2800, 3.0 & Bavaria** '69 thru '76

BUICK
  Century (front wheel drive) – see GENERAL MOTORS A-Cars (829)
*1627   **Buick, Oldsmobile & Pontiac Full-size (Front wheel drive)** all models '85 thru '93
  Buick Electra, LeSabre and Park Avenue; **Oldsmobile** Delta 88 Royale, Ninety Eight and Regency; **Pontiac** Bonneville
*1551   **Buick Oldsmobile & Pontiac Full-size (Rear wheel drive)**
  Buick Electra '70 thru '84, Estate '70 thru '90, LeSabre '70 thru '79
  Oldsmobile Custom Cruiser '70 thru '90, Delta 88 '70 thru '85, Ninety-eight '70 thru '84
  Pontiac Bonneville '70 thru '81, Catalina '70 thru '81, Grandville '70 thru '75, Parisienne '84 thru '86
627   **Mid-size** all rear-drive **Regal & Century** models with V6, V8 and Turbo '74 thru '87
  Regal – see GENERAL MOTORS (1671)
  Skyhawk – see GENERAL MOTORS J-Cars (766)
552   **Skylark** all X-car models '80 thru '85

CADILLAC
*751   **Cadillac Rear Wheel Drive** all gasoline models '70 thru '90
  Cimarron – see GENERAL MOTORS J-Cars (766)

CAPRI
296   **2000 MK I Coupe** all models '71 thru '75
205   **2600 & 2800** V6 Coupe '71 thru '75
375   **2800 Mk II** V6 Coupe '75 thru '78
  Mercury Capri – see FORD Mustang (654)

CHEVROLET
*1477   **Astro & GMC Safari Mini-vans** all models '85 thru '91
554   **Camaro** V8 all models '70 thru '81
*866   **Camaro** all models '82 thru '91
  Cavalier – see GENERAL MOTORS J-Cars (766)
  Celebrity – see GENERAL MOTORS A-Cars (829)
625   **Chevelle, Malibu & El Camino** all V6 & V8 models '69 thru '87
449   **Chevette & Pontiac T1000** all models '76 thru '87
550   **Citation** all models '80 thru '85
*1628   **Corsica/Beretta** all models '87 thru '92
274   **Corvette** all V8 models '68 thru '82
*1336   **Corvette** all models '84 thru '91

704   **Full-size Sedans** Caprice, Impala, Biscayne, Bel Air & Wagons, all V6 & V8 models '69 thru '90
  Lumina – see GENERAL MOTORS (1671)
  Lumina APV – see GENERAL MOTORS (2035)
319   **Luv Pick-up** all 2WD & 4WD models '72 thru '82
626   **Monte Carlo** all V6, V8 & Turbo models '70 thru '88
241   **Nova** all V8 models '69 thru '79
*1642   **Nova and Geo Prizm** all front wheel drive models, '85 thru '90
*420   **Pick-ups '67 thru '87** – Chevrolet & GMC, all full-size models '67 thru '87; Suburban, Blazer & Jimmy '67 thru '91
*1664   **Pick-ups '88 thru '92** – Chevrolet & GMC all full-size (C and K) models, '88 thru '92
*1727   **Sprint & Geo Metro** '85 thru '91
*831   **S-10 & GMC S-15 Pick-ups** all models '82 thru '92
*345   **Vans – Chevrolet & GMC,** V8 & in-line 6 cyl models '68 thru '92

CHRYSLER
*1337   **Chrysler & Plymouth Mid-size** front wheel drive '82 thru '89
  K-Cars – see DODGE Aries (723)
  Laser – see DODGE Daytona (1140)

DATSUN
402   **200SX** all models '77 thru '79
647   **200SX** all models '80 thru '83
228   **B-210** all models '73 thru '78
525   **210** all models '78 thru '82
206   **240Z, 260Z & 280Z** Coupe & 2+2 '70 thru '78
563   **280ZX** Coupe & 2+2 '79 thru '83
  300ZX – see NISSAN (1137)
679   **310** all models '78 thru '82
123   **510 & PL521 Pick-up** '68 thru '73
430   **510** all models '78 thru '81
372   **610** all models '72 thru '76
277   **620 Series Pick-up** all models '73 thru '79
  720 Series Pick-up – see NISSAN Pick-ups (771)
376   **810/Maxima** all gasoline models '77 thru '84
124   **1200** all models '70 thru '73
368   **F10** all models '76 thru '79
  Pulsar – see NISSAN (876)
  Sentra – see NISSAN (982)
  Stanza – see NISSAN (981)

DODGE
*723   **Aries & Plymouth Reliant** all models '81 thru '89
*1231   **Caravan & Plymouth Voyager Mini-Vans** all models '84 thru '91
699   **Challenger & Plymouth Saporro** all models '78 thru '83
236   **Colt** all models '71 thru '77
610   **Colt & Plymouth Champ (front wheel drive)** all models '78 thru '87
*556   **D50/Ram 50/Plymouth Arrow Pick-ups & Raider** '79 thru '91
*1668   **Dakota Pick-up** all models '87 thru '90
234   **Dart & Plymouth Valiant** all 6 cyl models '67 thru '76
*1140   **Daytona & Chrysler Laser** all models '84 thru '89
*545   **Omni & Plymouth Horizon** all models '78 thru '90
*912   **Pick-ups** all full-size models '74 thru '91
*1726   **Shadow & Plymouth Sundance** '87 thru '91
*1779   **Spirit & Plymouth Acclaim** '89 thru '92
*349   **Vans – Dodge & Plymouth** V8 & 6 cyl models '71 thru '91

FIAT
094   **124 Sport Coupe & Spider** '68 thru '78

479   **Strada** all models '79 thru '82
273   **X1/9** all models '74 thru '80

FORD
*1476   **Aerostar Mini-vans** all models '86 thru '92
788   **Bronco and Pick-ups** '73 thru '79
*880   **Bronco and Pick-ups** '80 thru '91
268   **Courier Pick-up** all models '72 thru '82
789   **Escort & Mercury Lynx** all models '81 thru '90
*2046   **Escort & Mercury Tracer** all models '91 thru '93
*2021   **Explorer & Mazda Navajo** '91 thru '92
560   **Fairmont & Mercury Zephyr** all in-line & V8 models '78 thru '83
334   **Fiesta** all models '77 thru '80
754   **Ford & Mercury Full-size,** Ford LTD & Mercury Marquis ('75 thru '82); Ford Custom 500, Country Squire, Crown Victoria & Mercury Colony Park ('75 thru '87); Ford LTD Crown Victoria & Mercury Gran Marquis ('83 thru '87)
359   **Granada & Mercury Monarch** all in-line, 6 cyl & V8 models '75 thru '80
773   **Ford & Mercury Mid-size,** Ford Thunderbird & Mercury Cougar ('75 thru '82); Ford LTD & Mercury Marquis ('83 thru '86); Ford Torino, Gran Torino, Elite, Ranchero pick-up, LTD II, Mercury Montego, Comet, XR-7 & Lincoln Versailles ('75 thru '86)
*654   **Mustang & Mercury Capri** all models including Turbo '79 thru '92
357   **Mustang V8** all models '64-1/2 thru '73
231   **Mustang II** all 4 cyl, V6 & V8 models '74 thru '78
649   **Pinto & Mercury Bobcat** all models '75 thru '80
*1670   **Probe** all models '89 thru '92
*1026   **Ranger & Bronco II** all gasoline models '83 thru '92
*1421   **Taurus & Mercury Sable** '86 thru '92
*1418   **Tempo & Mercury Topaz** all gasoline models '84 thru '91
1338   **Thunderbird & Mercury Cougar/XR7** '83 thru '88
*1725   **Thunderbird & Mercury Cougar** '89 and '90
*344   **Vans** all V8 Econoline models '69 thru '91

GENERAL MOTORS
*829   **A-Cars** – Chevrolet Celebrity, Buick Century, Pontiac 6000 & Oldsmobile Cutlass Ciera all models '82 thru '90
*766   **J-Cars** – Chevrolet Cavalier, Pontiac J-2000, Oldsmobile Firenza, Buick Skyhawk & Cadillac Cimarron all models '82 thru '92
*1420   **N-Cars** – Buick Somerset '85 thru '87; Pontiac Grand Am and Oldsmobile Calais '85 thru '91; Buick Skylark '86 thru '91
*1671   **GM:** Buick Regal, Chevrolet Lumina, Oldsmobile Cutlass Supreme, Pontiac Grand Prix, all front wheel drive models '88 thru '90
*2035   **GM:** Chevrolet Lumina APV, Oldsmobile Silhouette, Pontiac Trans Sport '90 thru '92

GEO
  Metro – see CHEVROLET Sprint (1727)
  Prizm – see CHEVROLET Nova (1642)
  Tracker – see SUZUKI Samurai (1626)

GMC
  Safari – see CHEVROLET ASTRO (1477)
  Vans & Pick-ups – see CHEVROLET (420, 831, 345, 1664)

(continued on next page)

Haynes North America, Inc., 861 Lawrence Drive, Newbury Park, CA 91320 • (805) 498-6703

HAYNES AUTOMOTIVE MANUALS

(continued from previous page)

NOTE: New manuals are added to this list on a periodic basis. If you do not see a listing for your vehicle, consult your local Haynes dealer for the latest product information.

HONDA

- **351 Accord CVCC** all models '76 thru '83
- ***1221 Accord** all models '84 thru '89
- **160 Civic 1200** all models '73 thru '79
- **633 Civic 1300 & 1500 CVCC** all models '80 thru '83
- **297 Civic 1500 CVCC** all models '75 thru '79
- ***1227 Civic** all models '84 thru '91
- ***601 Prelude CVCC** all models '79 thru '89

HYUNDAI

- ***1552 Excel** all models '86 thru '91

ISUZU

- ***1641 Trooper & Pick-up**, all gasoline models '81 thru '91

JAGUAR

- ***242 XJ6** all 6 cyl models '68 thru '86
- ***478 XJ12 & XJS** all 12 cyl models '72 thru '85

JEEP

- ***1553 Cherokee, Comanche & Wagoneer Limited** all models '84 thru '91
- **412 CJ** all models '49 thru '86
- ***1777 Wrangler** all models '87 thru '92

LADA

- ***413 1200, 1300. 1500 & 1600** all models including Riva '74 thru '86

MAZDA

- **648 626** Sedan & Coupe (rear wheel drive) all models '79 thru '82
- **1082 626 & MX-6 (front wheel drive)** all models '83 thru '91
- **370 GLC Hatchback (rear wheel drive)** all models '77 thru '83
- **757 GLC (front wheel drive)** all models '81 thru '86
- ***2047 MPV** '89 thru '93
- **Navajo** – see FORD Explorer (2021)
- ***267 Pick-ups** '72 thru '92
- **460 RX-7** all models '79 thru '85
- ***1419 RX-7** all models '86 thru '91

MERCEDES-BENZ

- ***1643 190 Series** all four-cylinder gasoline models, '84 thru '88
- **346 230, 250 & 280** Sedan, Coupe & Roadster all 6 cyl sohc models '68 thru '72
- **983 280 123 Series** all gasoline models '77 thru '81
- **698 350 & 450** Sedan, Coupe & Roadster all models '71 thru '80
- **697 Diesel 123 Series** 200D, 220D, 240D, 240TD, 300D, 300CD, 300TD, 4- & 5-cyl incl. Turbo '76 thru '85

MERCURY

For all PLYMOUTH titles see FORD Listing

MG

- **111 MGB** Roadster & GT Coupe all models '62 thru '80
- **265 MG Midget & Austin Healey Sprite** Roadster '58 thru '80

MITSUBISHI

- ***1669 Cordia, Tredia, Galant, Precis & Mirage** '83 thru '90
- ***2022 Pick-ups & Montero** '83 thru '91

MORRIS

- **074 (Austin) Marina 1.8** all models '71 thru '80
- **024 Minor 1000** sedan & wagon '56 thru '71

NISSAN

- **1137 300ZX** all Turbo & non-Turbo models '84 thru '89

- ***1341 Maxima** all models '85 thru '91
- ***771 Pick-ups/Pathfinder** gas models '80 thru '91
- ***876 Pulsar** all models '83 thru '86
- ***982 Sentra** all models '82 thru '90
- ***981 Stanza** all models '82 thru '90

OLDSMOBILE

- **Custom Cruiser** – see BUICK Full-size (1551)
- **658 Cutlass** all standard gasoline V6 & V8 models '74 thru '88
- **Cutlass Ciera** – see GENERAL MOTORS A-Cars (829)
- **Cutlass Supreme** – see GENERAL MOTORS (1671)
- **Firenza** – see GENERAL MOTORS J-Cars (766)
- **Ninety-eight** – see BUICK Full-size (1551)
- **Omega** – see PONTIAC Phoenix & Omega (551)
- **Silhouette** – see GENERAL MOTORS (2035)

PEUGEOT

- **663 504** all diesel models '74 thru '83

PLYMOUTH

For all PLYMOUTH titles, see DODGE listing.

PONTIAC

- **T1000** – see CHEVROLET Chevette (449)
- **J-2000** – see GENERAL MOTORS J-Cars (766)
- **6000** – see GENERAL MOTORS A-Cars (829)
- **1232 Fiero** all models '84 thru '88
- **555 Firebird** all V8 models except Turbo '70 thru '81
- ***867 Firebird** all models '82 thru '91
- **Full-size Rear Wheel Drive** – see Buick, Oldsmobile, Pontiac Full-size (1551)
- **Grand Prix** – see GENERAL MOTORS (1671)
- **551 Phoenix & Oldsmobile Omega** all X-car models '80 thru '84
- **Trans Sport** – see GENERAL MOTORS (2035)

PORSCHE

- ***264 911** all Coupe & Targa models except Turbo & Carrera 4 '65 thru '89
- **239 914** all 4 cyl models '69 thru '76
- **397 924** all models including Turbo '76 thru '82
- ***1027 944** all models including Turbo '83 thru '89

RENAULT

- **141 5 Le Car** all models '76 thru '83
- **079 8 & 10** all models with 58.4 cu in engines '62 thru '72
- **097 12 Saloon & Estate** all models 1289 cc engines '70 thru '80
- **768 15 & 17** all models '73 thru '79
- **081 16** all models 89.7 cu in & 95.5 cu in engines '65 thru '72
- **Alliance & Encore** – see AMC (934)

SAAB

- **247 99** all models including Turbo '69 thru '80
- ***980 900** all models including Turbo '79 thru '88

SUBARU

- **237 1100, 1300, 1400 & 1600** all models '71 thru '79
- ***681 1600 & 1800** 2WD & 4WD all models '80 thru '89

SUZUKI

- ***1626 Samurai/Sidekick and Geo Tracker** all models '86 thru '91

TOYOTA

- ***1023 Camry** all models '83 thru '91
- **150 Carina Sedan** all models '71 thru '74
- ***2038 Celica Front Wheel Drive** '86 thru '92
- **935 Celica Rear Wheel Drive** '71 thru '85
- ***1139 Celica Supra** '79 thru '92
- **361 Corolla** all models '75 thru '79
- **961 Corolla** all models (rear wheel drive) '80 thru '87
- ***1025 Corolla** all models (front wheel drive) '84 thru '91
- ***636 Corolla Tercel** all models '80 thru '82
- **230 Corona & MK II** all 4 cyl sohc models '69 thru '74
- **360 Corona** all models '74 thru '82
- ***532 Cressida** all models '78 thru '82
- **313 Land Cruiser** all models '68 thru '82
- **200 MK II** all 6 cyl models '72 thru '76
- ***1339 MR2** all models '85 thru '87
- **304 Pick-up** all models '69 thru '78
- ***656 Pick-up** all models '79 thru '92

TRIUMPH

- **112 GT6 & Vitesse** all models '62 thru '74
- **113 Spitfire** all models '62 thru '81
- **322 TR7** all models '75 thru '81

VW

- **159 Beetle & Karmann Ghia** all models '54 thru '79
- **238 Dasher** all gasoline models '74 thru '81
- ***884 Rabbit, Jetta, Scirocco, & Pick-up** all gasoline models '74 thru '91 & **Convertible** '80 thru '91
- **451 Rabbit, Jetta & Pick-up** all diesel models '77 thru '84
- **082 Transporter 1600** all models '68 thru '79
- **226 Transporter 1700, 1800 & 2000** all models '72 thru '79
- **084 Type 3 1500 & 1600** all models '63 thru '73
- **1029 Vanagon** all air-cooled models '80 thru '83

VOLVO

- **203 120, 130 Series & 1800 Sports** '61 thru '73
- **129 140 Series** all models '66 thru '74
- ***270 240 Series** all models '74 thru '90
- **400 260 Series** all models '75 thru '82
- ***1550 740 & 760 Series** all models '82 thru '88

SPECIAL MANUALS

- **1479 Automotive Body Repair & Painting Manual**
- **1654 Automotive Electrical Manual**
- **1480 Automotive Heating & Air Conditioning Manual**
- **1762 Chevrolet Engine Overhaul Manual**
- **1736 Diesel Engine Repair Manual**
- **1667 Emission Control Manual**
- **1763 Ford Engine Overhaul Manual**
- **482 Fuel Injection Manual**
- **1666 Small Engine Repair Manual**
- **299 SU Carburetors** thru '88
- **393 Weber Carburetors** thru '79
- **300 Zenith/Stromberg CD Carburetors** thru '76

See your dealer for other available titles

---

* Listings shown with an asterisk (*) indicate model coverage as of this printing. These titles will be periodically updated to include later model years – consult your Haynes dealer for more information.

Over 100 Haynes motorcycle manuals also available

1-93

Haynes North America, Inc., 861 Lawrence Drive, Newbury Park, CA 91320 • (805) 498-6703